AF559893

ENGINEERING PHYSICS

ENGINEERING PHYSICS

By

Poonam Yadav

DISCOVERY PUBLISHING HOUSE
NEW DELHI-110002

Reprinted - 2019

First Published - 2006

ISBN: 978-81-8356-074-0

© Author

Engineering Physics

Published by:

DISCOVERY PUBLISHING HOUSE PVT. LTD.
4383/4B, Ansari Road, Darya Ganj
New Delhi-110 002 (India)
Phone: +91-11-23279245, 23253475; 43596065
E-mail: discoverybooksindia@gmail.com
discoverypublishinghouse@gmail.com
web: www.discoverypublishinggroup.com

Printed at:
Infinity Imaging Systems
Delhi

Preface

The present title "Engineering Physics" provides all under-graduate students of Engineering with a broad range of internationally accepted views, facts and theories to prove a useful reference to students, researchers, and professionals of the related fields. The problems of graded difficulties have also been carefully chosen to test their understanding of the basic concepts of Engineering Physics. Many of the problems have been solved step to step to educate the students as to how to tackle these problems systematically. The book is the outcome of author's commitment of offer a comprehensive and effective teaching/learning tool for the benefit of the students of Engineering Physics.

To make the work more comprehensive and informative, the author has consulted many authoritative books, research journals, abstracts, monographs etc. He is grateful to all those great scholars whose work are cited or substantially reproduced.

There can be no claim to originality except in the manner of treatment and much of the information has been obtained from the books and scientific journals available in the different libraries.

The author expresses his thanks to his friends and colleagues whose continue inspirations have initiated him to bring out this book.

The author expresses his gratitude to Mr. Wasan and staff of M/s Discovery Publishing House for their whole hearted co-operation in the publication of this book.

In the mean time, the author will remain sincerely responsible for any shortcomings of the book and be grateful to the readers for their suggestions and constructive criticism for the continues betterment of the book. She takes this opportunity to appeal to the readers to send their suggestions straightway to her publisher.

Author

Contents

1

Special Theory of Relativity

What is relativity

Relativity and measurements of events are comparable; where and when they happen, and by how much any two events are separated in space and in time. In addition, relativity has to do with transforming such measurements and others between reference frames that move relative to each other. (Hence the name *relativity*.) Later on we will discuss such matter.

In 1905 physicist understood about transformation & moving reference frames. Then Albert Einstein published his special theory of relativity. The adjective special means that the theory deals only with inertial reference frames, which are frames that move at constant velocities relative to one another. According to Einstein's general theory of relativity reference frames accelerates, and relativity implies only on inertial reference frames.

Einstein gave two postulates which show that the old ideas about the relativity were wrong and it made scientists unquestionable. This supposed common sense, however, was derived from experience only with things that move rather slowly. Einstein's relativity was applicable on every possible speed and predicted many effects first time.

Einstein demonstrated that space and time are connected, means the duration of time between two events depends upon their distance/ and vice versa. And the entanglement is different for observers who move relative to each other. One result is that time does not pass at a fixed rate, as if it were ticked off with mechanical regularity on some master grandfather clock that controls the universe. Rather, that rate is adjustable: relative motion can change the rate at which time passes. Prior to 1905, no one but a few day-dreamers would have thought that now scientists and engineers are agreed with special relativity because it has reformed their thoughts.

Special relativity seems to be difficult. It is not difficult mathematically, at least not here. But it is difficult in that we must be very careful about who measures what about an event and just how that measurement is made and it can be difficult because it can contradict experience.

The Postulates

Einstein theory based upon two postulates which are as follows:

1. The Relativity Postulate: The laws of physics are the same for observers in all inertial reference frames. No frame is preferred.

According to Galileo in all inertial reference frames the laws of mechanies are same. (Newton first law of motion is one important consequence.) Einstein extended that idea to include *all* the laws of physics, especially electromagnetism and optics. This postulate does not say that the measured values of all physical quantities are the same for admertial observers; most are not the same. The laws of physics relates these equal measurements to each other.

2. The Speed of Light Postulate: The speed of light in vacuum has the same value c in all directions and to all inertial reference frames.

According to second postulate ultimate speed c is equal in all directions in all inertial reference frames. Light happens to travel at this ultimate speed, as do any massless particles (neutrinos might be an example). But any entity that carries energy or information cannot exceed this limit. Thus any particle having mass, never catch speed c inespite of much acceleration.

Two postulates of relativity are applicable on all possibility without exception.

The Michelson-Morley Experiment

In Michelson-Morley Experiment interferometer is arranged (figure 1.1) P is a half glass plate on which a beam of light (from lamp L) falls, place at 45° to the beam. P divides each wave in to two parts. One partial wave, reflected from P, travels off sideways to a mirror M_1, by which it is reflected back to P; part of it is then transmitted through P and enters the telescope T. P transmits another part of the original wave travels ahead and is reflected back by a other mirror M_2 & return to P, after returning it is partially reflected to the telescope on top of the first part of the wave, thus an interference pattern is made.

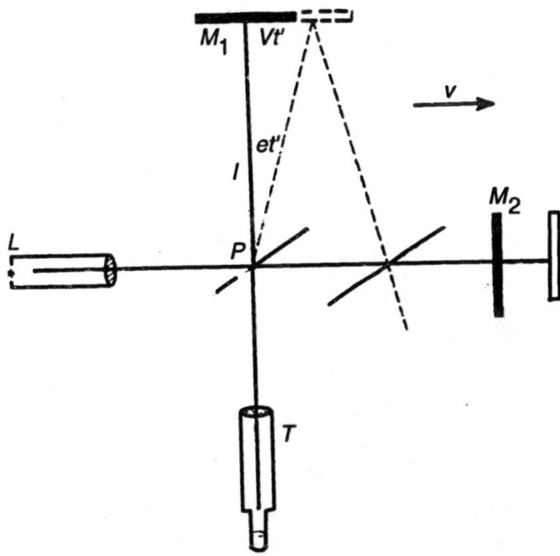

Figure 1.1 : ***The Michelson-Morley experiment***

Plate P is situated on equal distance from both mirrors. If the apparatus is at rest in the ether, the two waves take the same time to return to P and meet in phase both there and in the telescope. Suppose, however, that the apparatus is moving with speed V through the ether in the direction of the initial beam of light. Then, if the initial wave strikes the plate P when it has the position shown in the figure, the paths of the waves and the subsequent positions of reflection from mirrors and plate will be as shown by the dotted lines. The necessary change in the direction of the transverse beam is produced automatically, as an aberration effect, through the operation of Huygens' principle. But now the times taken by the two waves on their journeys are no longer equal. The wave has velocity $c - V$ moving towards M_2 longitudinally on outgoing trip c is the speed of light through the ether and $c + V$ on the return trip. So the time is taken by the wave is t_2 *i.e.*,

$$t_2 = \frac{t}{c-V} + \frac{l}{c+V} = \frac{2lc}{c^2 - V^2} = \frac{2l}{c}\left(1 + \frac{V^2}{c^2} + \cdots\right)$$

l being the distance from the plate to either mirror. The wave moving transversely, on the other hand, travels along the hypotenuse of a right triangle having a side of length l. If it take t time to travel from P to M_1 and cover distance Vt, then in the same time M_1 shifted to vt and $c^2t^2 = V^2t^2 + l^2$ then this wave will return to P in t_1 time.

$$t_1 = 2l' = \frac{2l}{\left(c^2 - V^2\right)^{\frac{1}{2}}} = \frac{2l}{c}\left(1 + \frac{V^2}{2c^2} + \cdots\right)$$

Now the phase difference of two waves, interfered in the telescope is $2\Pi c\left(t_2 - t_1\right)\lambda \approx 2\Pi\, lV^2 / c^2\lambda$ (here d is the wave length of the light) and the fringe pattern is shifted by the motion through $V^2 l / c^2\lambda$ fringes.

In experiment the appratus is arranged to skits on mercury and it rotates again and again through 90°. Since the two light paths are caused to exchange roles by such a rotation, it should cause the fringe pattern to shift twice as much; by reflecting the beam back and forth several times, the effective length l was increased to 11 m. If we insert v with velocity of about 5.9×10^{-5} cm ($V/c = 10^{-4}$), than the shift will be

$$N = 2 \times 10^{-8}\,\frac{11 \times 10^2}{5.9 \times 10^{-5}} = 0.37 \text{ fringe}$$

Michelson and Morley were determined to know a shift of a hundredth of a fringe. Such shifts as were observed amounted only to a small fraction of the theoretical value and were not consistent. Thus the motion did not show expected effects so the results were negative.

If it happens that accidentally earth has no resultant component of velocity parallel to the surface of the earth; for upon its orbital motion there would probably be superposed a motion of the entire solar system through the ether. But for limited time the orbital velocity will be reserved as after 6 months the earth's orbital velocity about the sun would be reserved and through the other should be twice.

Michelson method made observations at various times of the day and at different seasons of the year, in different part of the world but result was always negative. In 1930 Joos in Germany found no evidence of motion through an ether with an interferometer capable of detecting an ether drift one-twentieth of the orbital velocity of the earth. Cedarholm and Townes in 1958 used a new technique of masers by which either drift of one thousand the orbital velocity could be detected but no evidence founded of any drift.

It was difficult to fail the detution of anticipated motion of the earth through the ether inspite of many existing theories of light and matter. The Theoretical calculation rests on a peculiarly simple foundation, for the only properties of light that are made use of are the constancy of its velocity in space and Huygens principle. Negative result can be explain classically three possibilities.

(1) Like baseball carries air with it, as the earth brings other along with it. On this assumption there would never be any motion of the earth through the ether at all, and no difficulties could arise. One objection to this explanation is that the ether next to the earth would then be in motion relative to the other farther away; and this relative motion between different parts of the ether would cause a deflection of the light rays coming from the stars, just as wind is observed to deflect sound waves. This deflection would alter the amount of the stellar aberration. It was difficult to devise a plausible motion for the ether which would give a value of the aberration agreeing with observation and yet preserve the negative result of the Michelson-Morley experiment. Thus a other objection arises if a matter moving with full velocity it does not drag the light waves of a transparent laboratory size object, as it necessarily would do if it completely dragged the ether along with it, and current electro magnetic theory fully observed the partial drag.

(2) If we consider that a moving object projecting light with its natural velocity, as a moving ship with velocity of a projectile fired is equal to the vector sum of its velocity of projection from the gun and the vector of the ship. If this were true, the negative result obtained by Michelson and Morley would at once be explained, for he light from their lamp would have always the same constant velocity relative to the lamp and to the interferometer. Such an assumption, however, is in gross conflict with the wave theory of light. In a medium, the desire of waves is that they have a definite velocity relative to the medium, like sound waves have a definite velocity relative to the air.

(3) fitzGerald and Lorentz gave third explanation according to it motion by ether can cause the material composing the interferometer, parallel to the motion to shorten in a direction. Such a contraction in the ratio $\sqrt{1-V^2/c^2}$ would serve to equalize the light paths and thus to prevent a displacement of the fringes. In 1932 Kennedy and Thorndike conclude inadequated to the explanation according to them no shift work on the two arms of the interferometer at different lengths.

The New Relativity of Einstein

The above explanations and conditions show that nature has a conspiracy to prevent us, to detect motion through the ether. A similar situation can be shown to exist in the field of electromagnetism as well in optics. Many electrical or magnetic experiments try to detect the motion through the ether, but always some other effects disturb there results.

In 1905, Einstein discover some new method to solve this extraordinary situation. He proposed that motion through an ether filling empty space is a *meaningless concept*; only motion relative to material bodies has physical significance. Thus he considered these assumptions to harmonize with the known laws of optics. Possibilities of conflict arise in any argument involving the velocity of light. Consider, for example, a frame of reference S' (say on the earth) moving relative to another frame S (say on the sun), and suppose that S' carries a source of light. Then light from this source must move with the same velocity relative to S as light from a source on S itself, science, as explained above, we cannot suppose that the velocity of light is influenced by motion of its source. But this light must also move *with the same velocity relative to* S'; otherwise the laws of nature would not be the same on S' as on S, and no reason could be assigned for their being different. From the above situations, that if light moving with the same velocity relative to each of two frames which are moving relative to each other looks impossible.

Einstein give the laws of light movement. He based new theory, which is known as the *special* or *restricted theory of relativity,* upon two postulates which may be states as follows:

1. The laws of physical phenomena are the same when stated in terms of either of two inertial frames of reference (and can involve no reference to motion through an ether); no experiment can be performed which can establish that any frame is at absolute rest.
2. The speed of light is independent of the motion of its source and is the same for observers in any inertial frame.

Albert Einstein in year, 1905 showed how motion between an observer and what is being observed affect the measurements of time and space. To say that Einstein's theory of relativity revolutionized science is no exaggeration. Relativity connects space and time, matter and energy, electricity and magnetism—links that are crucial to our understanding of the physical universe. From relativity have come a host of remarkable predictions, all of which have been confirmed by experiment. For deep study all conclusions of relativity can be solved by simple mathematics.

Special Relativity

All motion is relative; the speed of light in free space is the same for all observers.

In elementary physics there is no special points to measure such quantities as length, time interval and mass. Since a standard unit exists for each quantity, who makes a certain determination would not seem to matter—everybody ought to get the same result. For instance, there is no question of principle involved in finding the length of an airplane when were are on board. To measure the length we have to put one end of a tape at the aviplane's anterior end and look at the number on the tape at the airplane's posterior end.

If we are on the ground and we have to measure the length of an airplane is in flight. It is not hard to determine the length of a distant object with a tape measure to establish a baseline, a surveyor's transit to measure angles, and a knowledge of trigonometry. When we measure the moving airplane from the ground, though, we find it to be shorter than it is to somebody in the airplane itself. If motion is involved in measurement we would have to analyse this unexpected difference.

Frames of Reference

The clarification of motion is the first step. When we say that something is moving, what we mean is that its position relative to something else is changing. A passenger moves relative to an airplane; the airplane moves relative to the earth; the earth moves relative to the sun; the sun moves relative to the galaxy of stars (the Milky Way) of which it is a number; and so on. In each case a frame of reference is part of the description of the motion. To measure the movement of anything it is necessary to apply a specific frame of reference.

Newton's first low of motion applies on inertial frame of reference. In such a frame, an object at rest remains at rest and an object in motion continues to move at constant velocity (constant speed and direction) if no force acts on it. If any frame of reference is moving with constant velocity relative to an intertial frame than itself it has intertial frame.

The validity of all inertial frames are equal. Suppose we see something changing its position with respect to us at constant velocity. Is it moving or are we moving? Suppose were are in a closed laboratory in which Newton's first law holds. Is the laboratory moving or is it at rest? These questions are meaningless because all constant-velocity motion is relative. We can not use frame of reference every where because frame of reference is not universal means there is no absolute motion.

The theory of relativity has the suffer with the problems because there is no universal frame of reference. **Special relativity**, which is what Einstein published in 1905, treats problems that involve internal frames of reference. **General relativity**, published by Einstein a decade later, treats problems that involve frames of reference accelerated with respect to one another. An observer in an isolvated laboratory can detect acceleration, as anybody who has been in an elevator or on a merry-go-round knows. In physics we have lot's of impact of special theory, and we should concentrate towards it.

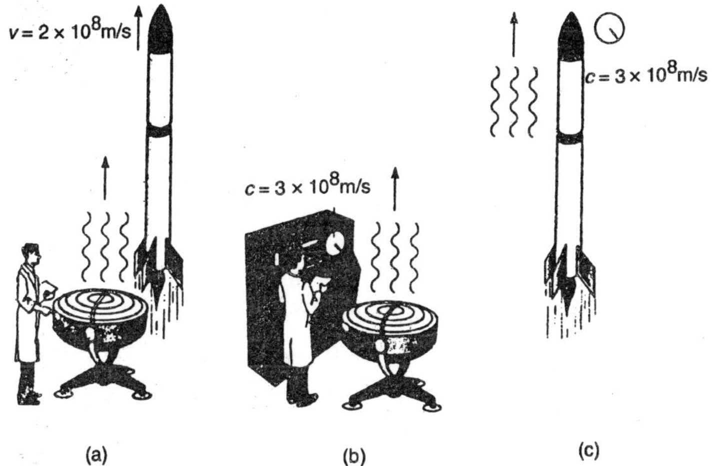

Figure 1.2 : ***The speed of light is the same to all observers.***

The Lorentz Transformation

As we can see in figure 1.3 S' inertial reference frame is moving with velocity v_0 relative to frame S, towards common positive direction of their horizontal axis (x and x'). An observer in S reports space-time coordinates x, y, z, t for an event, and an observer in S' reports x', y', z', t' for the same event. How we can conclude that these sets of number are related.

We suppose that the y and z perpendicular tot he motion and coordinates to each other but not affected by the motion. That is, $y = y'$ and $z = z'$. Our interest then reduces to the relation between x and x' and between t and t'.

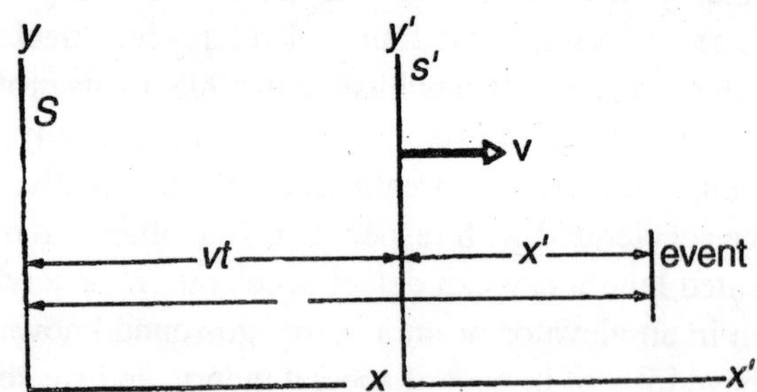

Figure 1.3 : ***Two inertial reference frames; frame S' has velocity v relative to frame S.***

The Galilean Transformation Equations

According to Einstein's special theory of relativity the four co-ordinates of interest can be related by the Galilean transformation equation.

$$x' = x - vt$$
$$t' = t$$

(Galilean transformation equations; approximately Valid at low speeds) (1.1)

In the above equation $t = t' = 0$ because the origin of S and S' coincide. You can verify the first equation with Figure 1.3. The second equation effectively claims that time passes at the same rate for observers in both reference frames. That would have been so obviously true to a scientist prior to Einstein that it would not even have been mentioned. If the value of v is less than C, than equation 1.1 works well.

The Lorentz Transformation Equations

We can derive transformation equation (without any proof) from the postulates of relativity. It is valid for all speeds up to the speed of light. The results, called the **Lorentz transformation equations**, are

$$x' = \gamma (x - vt),\quad y' = y,\quad z' = z,\quad t' = \gamma (t - vx/c^2) \qquad (1.2)$$

(Lorentz transformation equations; valid at all speeds).

We can see that the spatial values x and the temporal values t are bound together in the first and last equations.

If c is infinite than formally relativeness equation should reduce to familiar classical equations. That is, if the speed of light were infinitely great, *all* finite speeds would be "low" and classical equations would never fail. If we let $c \to \infty$ in Eqs. 1.2 $\gamma \to 1$ and these equations reduce—as we expect – to the Galilean equations (Eqs. 1.1). You should check this. Equations 1.2 are written in a form that is useful if we are given x and t and wish to find x' and t'. We may

Table : The Lorentz Transformation Equations for Pairs of Events

1. $\Delta x = \gamma(\Delta x' + v\Delta t')$	1. $\Delta x' = \gamma(\Delta x + v\Delta t)$
2. $\Delta t = \gamma\left(\Delta t' + v\Delta x' / c^2\right)$	2. $\Delta t = \gamma\left(\Delta t' + v\Delta x / c^2\right)$

$$\gamma = \frac{1}{\sqrt{1-(v/c)^2}} = \frac{1}{\sqrt{1-\beta^2}}$$

wish to go the other way, however. On solving the eqs 1.2 to get the values of x and t–

$$x = \gamma (x' + vt'), \qquad (1.3)$$
$$t = \gamma (t' + vx'/c^2).$$

From above equations we can find the other set by interchanging primed and unprimed values and reversing the sign of the relative velocity v from either eqs 1.2 or eqs 1.3.

Equations 1.2 and 1.3 relate the coordinates of a single event as seen by two observers. Sometimes we want to know not the coordinates of a single event but the differences between coordinates for a pair of events.

From earlier discussion we will get

$$\Delta x = x_2 - x_1 \text{ and } \Delta t = t_2 - t_1,$$

as measured by an observer in S, and

$$\Delta x' = x_2' - x_1' \text{ and } \Delta t' = t_2' - t_1',$$

as measured by an observer in S'.

Its we read earlier Lorentz equation can be modified to analyse the part of events. The equations in the table were derived by simply substracting differences (such as Δx and $\Delta x'$) for the four variables in equations 1.2 and 1.3

To substitute the values for differences are should be careful and not mix the values for the first event with second. And Δx is a negative quantity, you must be certain to include the minus sign.

Doppler Effect

Why the universe is believed to be expanding

As we know that the increase in pitch of sound is because of source coming close to us or we are going close to the source and the decrease in the pitch of sound is because of source going far to us or we are going away from the source. These changes in frequency constitute the *doppler effect,* whose origin is straightforward. For instance, successive waves emitted by a source moving toward an observer are closer together than normal because of the advance of the source; because the separation of the waves is the wavelength of the sound, the corresponding frequency is higher. If the frequency of source is v_0 and the frequency of observer is V than the relationship between both is.

Doppler effect in sound

$$v = v_0\left(\frac{1 + v/c}{1 - V/c}\right) \tag{1.4}$$

where c = speed of sound

u = speed of observer (+ for motion toward the source,– for motion away from it)

V = speed of the source (+ for motion toward the observer, – for motion away from him)

If the observer is stationary, $v = 0$, and if the source is stationary, $V = 0$.

If the source or the observer or both are moving than the doppler effect on sound varies with the situation. This appears to violate the principle of relativity: all that should count is the relative motion of source and observer. But sound waves occur only in a material medium such as air or water, and this medium is itself a frame of reference with respect to which motions of source and observer are measurable. Hence there is no contradiction. In the case of light, however, no medium is involved and only relative motion as source and observer is meaningful. Thus doppler effect should be differ in light and sound.

To detect doppler effect in light we should consider a light source as a clock that splits a wave of light with each tick (ticks v_0 times per second). We will examine the three situations shown in Fig 1.4

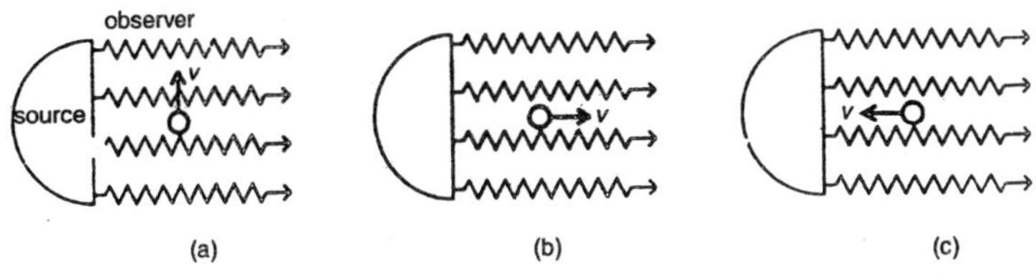

Figure 1.4 : ***The frequency of the light seen by an observer depends on the direction and speed of the observer' motion relative to its source.***

(a) In stitution (a) observer moving perpendicular to a line between him the light source. The proper time between ticks is $t_0 = 1/v_0$, so between one tick and the next the time $t = t_0 / \sqrt{1 - v^2/c^2}$ elapses in the reference frame of the observer. Thus the frequency observer gets is

$$v(\text{transverse}) = \frac{1}{t} = \frac{\sqrt{1 - v^2/c^2}}{t_0}$$

Transverse doppler effect in light

$$v = v_0 \sqrt{1 - v^2/c^2} \qquad (1.5)$$

The frequency of observer v is always lower than the frequency of source v_0.

(b) In situation (b) observer going away from the length source. Now the observer travels the distance *vt* away from the source between ticks, which means that the light wave from a given tick takes *vt/c* longer to reach him than the previous one. Thus the duration of time in which the wave arrives is—

$$T = t + \frac{vt}{c} = t_0 \frac{1 + v/c}{\sqrt{1 - v^2/c^2}} = t_0 \frac{\sqrt{1+v/c}\sqrt{1+v/c}}{\sqrt{1+v/c}\sqrt{1-v/c}} = t_0 \sqrt{\frac{1+v/c}{1-v/c}}$$

and the observed frequency is

$$\nu(\text{receding}) = \frac{1}{T} = \frac{1}{t_0}\sqrt{\frac{1-v/c}{1+v/c}} = \nu_0\sqrt{\frac{1-v/c}{1+v/c}} \qquad (1.6)$$

The frequency of observer ν is lower than the frequency of source *Vo*. Unlike the case of sound waves, which propagate relative to a material medium. It means it does not matter that the observer is receding from the source or the source is receding from the observer.

In situation (c), the observer coming lose to the light source. The observer here travels the distance *vt* toward the source between ticks, so each light wave takes *vt/c* less time to arrive than the previous one. In this case $T = t - vt/c$ and the result is

$$\nu\ (\text{approaching}) = \nu_0\sqrt{\frac{1+v/c}{1-v/c}} \qquad (1.7)$$

The frequency of observer is higher than the frequency of source. Again, the same formula holds for motion of the source toward the observer.

On combining equations (1.6) and (1.7) together—

Longitudinal doppler effect in light

$$\nu = \nu_0\sqrt{\frac{1+v/c}{1-v/c}} \qquad (1.8)$$

by adopting the convention that v is + for source and observer approaching each other and – for source and observer receding from each other.

The eyes are sensitive towards the visible light consists of electro–magnetic waves in a frequency band. Other electromagnetic waves, such as those used in radar and in radio communications, also exhibit the doppler effect in accord with Eq. (1.8). As in radar, waves are used to measure the speed of vehicles, by police men. In radio a set of earth satellites emittswaves from the basis of the highly accurate transit system of marine norvigation

The Expanding Universe

In Astronomy the doppler effect of light is an important tool. Stars emit light of certain characteristic frequencies called spectral lines, and motion of a star toward or away from the earth show up as a doppler shift in these frequencies. The spectral lines of distant galaxies of stars are all shifted toward the low- frequency (red) end of the spectrum and hence are called "red shifts." Such shifts indicate that the galaxies are receding from us and from one another. The speeds of recession are observed to be proportional to distance, which suggests that the entire universe is expanding (Figure 1.8) This proportionality is known as Hubble's law.

The expansion started about 15 billion years ago with the explosion of a hot mass of primeval matter, an event usually called the Big Bang, the matter soon turned into the electrons, protons, and neutrons of which the present universe is composed. Individual aggregates that formed.

Length Contraction

Relative motion affects the measurement of length and duration of time intervals. The length L of an object in motion with respect to an observer always appears to the observer to be shorter than its length L_0 when it is at rest with respect to him. This contraction occurs only in the direction of the relative motion. If the length of an object in rest position is L_0 than its called proper length of object. (We note that in Figure 1.5 the clock is moving perpendicular to v, hence $L = L_0$ there.)

The determination of length contraction can be derived from many ways. Perhaps the simplest is based on time dilation and the principle of relativity. Let us consider what happens to unstable particles called muons that are created at high altitudes by fast cosmic-ray particles (largely protons) from space when they collide with atomic nuclei in the earth's atmosphere. The mass of a muon is 207 times that of the electron and charge is $+e$ or $-e$, it decays into an electron or a positron after an average lifetime of 2.2 μs (2.2×10^{-6}s).

The speed of cosmic ray muons is about 2.994×10^8 m/s, and reach sea level in profusion—one of them passes through each square centimeter of the earth's surface on the average slightly more than once a minute. But in t_0 = 2.2 μs, their average lifetime, muons can travel a distance of only

$$vt_0 = (2.994 \times 10^8 \text{ m/s})(2.2 \times 10^{-6}\text{s}) = 6.6 \times 10^2 \text{ m} = 0.66 \text{ km}$$

Before decaying it created at attitudes of 6km or more.

An observer at rest with respect to a mucon find the lifetime of to = 2.2 μs. Because the muons are hurtling toward us at the considerable speed of $0.998c$, their lifetimes are extended in our frame of reference by time dilation to

$$t = \frac{t_0}{\sqrt{1 - v^2/c^2}} = \frac{2.2 \times 10^{-6} s}{\sqrt{1 - (0.998c)^2/c^2}} = 34.8 \times 10^{-6} s = 34.8 \mu s$$

The lifetime of moving muons is about 10 times longer than those at rest. In a time interval of 34.8 μs, a muon whose speed is $0.998c$ can cover the distance

$$vt = (2.994 \times 10^8 \text{ m/s})\,(34.8 \times 10^{-6}\text{s}) = 1.04 \times 10^4 \text{ m} = 10.4 \text{ km}$$

In its own frame of reference, its lifetime (t_o) is only 2.2 μs, a muon can reach the ground from altitudes of as much as 10.4 km because in the frame in which these altitudes are measured, the muon lifetime is t = 34.8 μs.

For a muon $v = 0.998c$, and somebody were to accompany a muon, so that to him or her the muon is at rest? The observer and the muon are now in the same frame of reference, and in this frame the muon's lifetime is only 2.2 μs. To the observer, the muon can travel only 0.66 km before decaying. The only way to account for the arrival of the muon at ground level is if the distance it travels, from the point of view of an observer in the moving frame, is shortened by virtue of its motion (Figure 1.5). The principle of relativity tells us the extent of the shortening—it must be by the same factor of $\sqrt{1-v^2/c^2}$ that the muon lifetime is extended from the point of view of a stationary observer.

Thus we come to the condition that if an altitude find to be no on the ground than it would look as the lower altitudes in the muon's frame of reference.

$$h = h_0\sqrt{1-v^2/c^2}$$

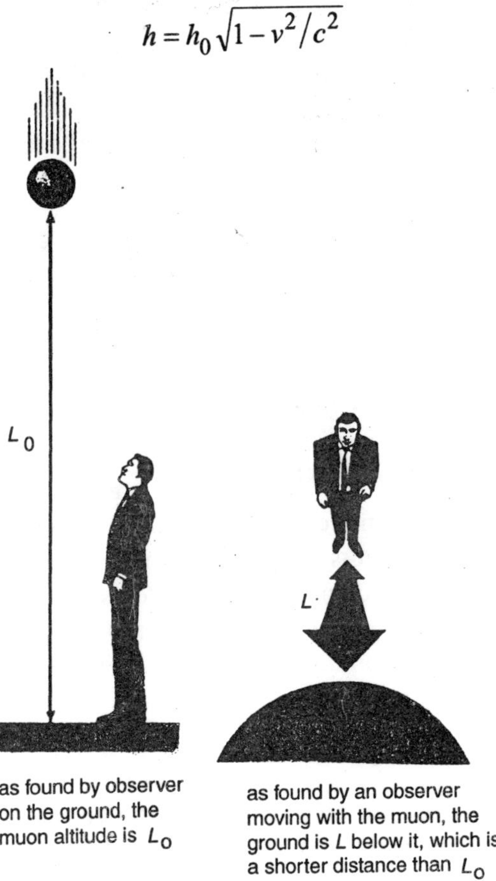

Figure 1.5 : ***Muon decay as seen by different observers. The muon size is greatly exaggerated here; in fact, the muon seems likely to be a point particle with no extension in space.***

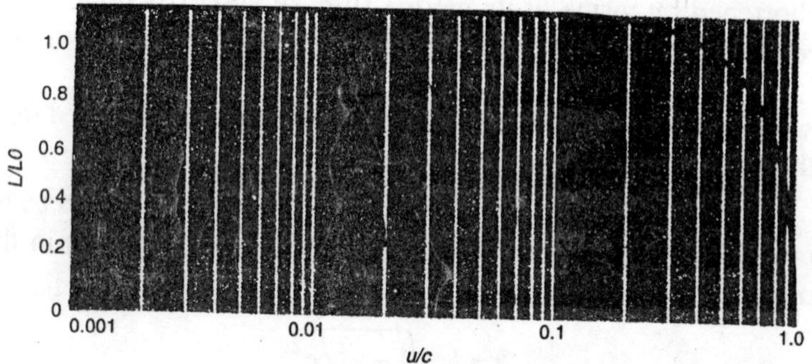

Figure 1.6 : ***Relativistic length contraction. Only lengths in the direction of motion are affected. The horizontal scale is logarithmic.***

In frame reference the mucon travels h_0 = 10.4 km due to time dilation. In the muon's frame of reference, where there is no time dilation, this distance is abbreviated to

$$h = (10.4 \text{ km})\sqrt{1-(0.998c)^2/c^2} = 0.66 \text{ km}$$

Thus a muon travels distance with. 998c in 2.2μs.

The general contraction of lengths in the direction of motion is an example of the relativistic shortening of distance.

Length contraction $L = L_0\sqrt{1-v^2/c^2}$ (1.9)

In Fig (1.6) We can see the graph of L/L_0 versus v/c. It shows that length contraction is most significant at speeds near that of light. A speed of 1000 km/s seems fast to us, but it only results in a shortening in the direction of motion to 99.9994 percent of the proper length of an object moving at this speed. As we can see, if any object travels nine-tenths the speed of light is shortened to 44 percent of its proper length, a significant-change.

The contraction of length is reciprocal to time dilation. To a person in a spacecraft, objects on the earth appear shorter than they did when he or she was on the ground by the same factor of $\sqrt{1-v^2/c^2}$ that the spacecraft appears shorter to some body at rest. The proper length L_0 found in the rest frame is the maximum length any observer will measure. As mentioned earlier, only lengths in the direction of motion undergo contraction. If a observer, observe a spacecraft– from outside he find it shorter but not narrower.

Twin Paradox

A longer life, but it will not seem longer

Now we would understand the relativistic effect which is called the twin paradox. $(3 - 2) \times 10^8$ m/s, or only 1×10^8 m/s, for the same light waves. If you are moving parallel to the waves at 2×10^8 m/s than you find their speed 3×10^8 m/s to me.

These results can be counted only without violating the principle of relativity. It must be true that measurements of space and time are not absolute but depend on the relative motion between an observer and what is being observed. If I were to measure from the ground the rate at which your clock ticks and the length of your meter stick, I would find that the clock ticks more slowly than it did at rest on the ground and that the meter stick is shorter in the direction of motion of the spacecraft. To you, your clock and meter stick are the same as they were on the ground before you took off. To me they are different because of the relative motion, different in such a way that the speed of light you measure is the same 3×10^8 m/s I measure. In free space the speed of light is equal for all observers but the time intervals and lengths are relative quantities.

Maxwell developed a unified theory before Einsteins's work. It was a conflict between the principles of mechanics (based on Newton's law of motion) and those of electricity and magnetism. Newtonian mechanics had worked well for over tow centuries. Maxwell's theory not only covered all that was then known about electric and magnetic phenomena but had also predicted that electromagnetic waves exist and identified light as an example of them. Thus the equation of electromagnetism and those of newtonion mechanics are differ to each others on the basis of measurements. As one work in a different inertial frame respectively.

According to Einstein the Newtonian mechanics is not consistent with special relativity but Maxwell's theory do. The branches of physics come in accord due to the modifications of newtonian mechanics. As we will find, relativistic and newtonian mechanics agree for relative speeds much lower than the speed of light, which is why newtonian mechanics seemed correct for so long. Newtonian mechanics does not work at high speeds but convert in to relativistic from replaced by the relativistic version.

Time Dilation

A moving clock ticks more slowly than a clock at rest

Relative motion affects on measurement of time intervals for observer and what he observe. As a result, a clock that moves with respect to an observer ticks more slowly than it does without such motion and in a different inertial frame the all activities compliest slowly to an observer. The duration of time interval is different in space and ground as if the time interval between two events t duration longer in ground. The quantity t_0, which is determined

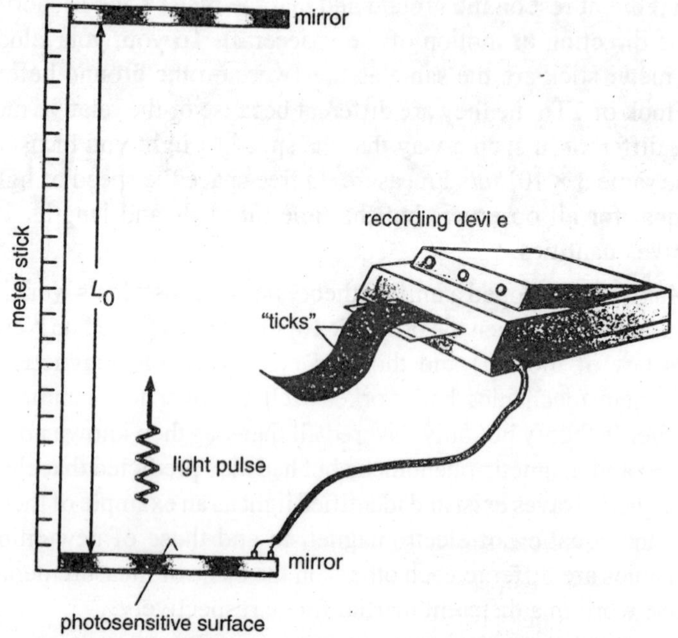

Figure 1.7 : ***A simple clock Each "tick" corresponds to a round trip of the light pulse from the lower mirror to the upper one and back.***

by events that occur *at the same place* in an observer's frame of reference, is called the proper time of the interval between the events. When witnessed from the ground, the events that mark the beginning and end of the time interval occur at different places, and in consequence the duration of the interval appears longer than the proper time. This effect is known as times dilation (dilation is become larger).

In fig 1.7 we can see the time dilation considering simple type of two clocks. In each clock a pulse of light is reflected back and forth between two mirrors L_0 apart. Whenever the light strikes the lower mirror, an electric signal is produced that marks the recording tape. We have each mark corresponds to the tick of an ordinary clock.

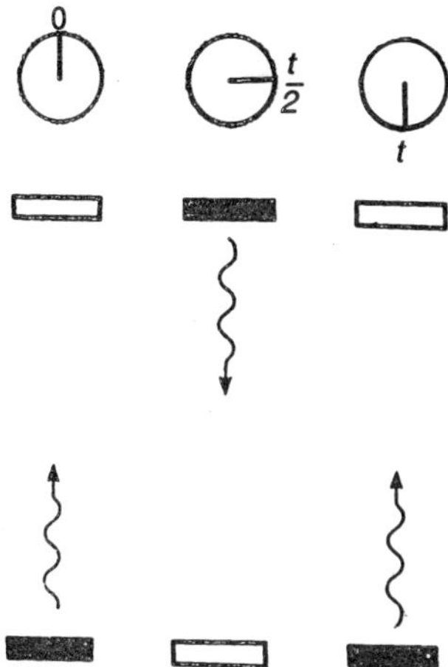

Figure 1.8 : ***A light-pulse colck at rest on the ground as seen by an observer on the ground. The dial represents a conventional clock on the ground.***

One clock moving in a space craft. A clock in space craft moves with speed v relative to the clock is at rest on the ground if some one observe both clocks can he find the same rate of their tick.

In fig 1.8 we can see the arrangement of a clock.

In fig 1.9 we can see the clock which is moving. The time interval between ticks is the proper time t_0 and the time needed for the light pulse to travel between the mirrors at the speed of light c is $t_0/2$. Hence $t_0/2 = L_0/c$ and

$$t_0 = \frac{2\,L_0}{c} \tag{1.10}$$

We can see in fig 1.8 moving clock which have mirrors perpendicular to the direction of motion relative to the ground. The time interval between ticks is t. Because the clock is moving, the light pulse, as seen from the ground, follows a zigzag path. On its way from the lower mirror to the upper one in the time $t/2$, the pulse travels a horizontal distance of $v(t/2)$ and a total distance of $c(t/2)$. The vertical, distance between the mirrors is L_0.

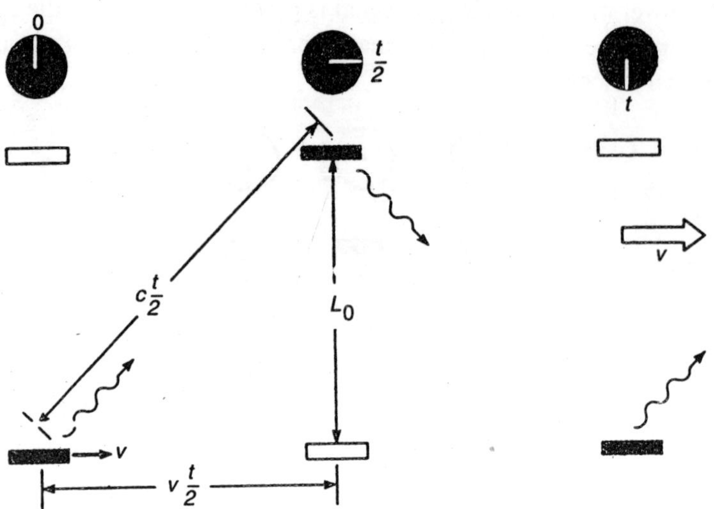

Figure 1.9 : ***A light-pulse clock in a spacecraft as seen by an observer on the ground. The mirrors are parallel to the direction of motion of the spacecraft. The dial represents a conventional clock on the ground.***

$$\left(\frac{ct}{2}\right)^2 = L_2^0 + \left(\frac{vt}{2}\right)^2$$

$$\frac{t^2}{4}\left(c^2 - v^2\right) = L_0^2$$

$$t^2 = \frac{4L_0^2}{c^2 - v^2} = \frac{\left(2L_0\right)^2}{c^2\left(1 - v^2/c^2\right)}$$

$$t = \frac{2L_0/c}{\sqrt{1 - v^2/c^2}} \qquad (1.11)$$

But $2L_0/c$ is the time interval t_0 between ticks on the clock on the ground, as in Eq. (1.10) and so

Time dilation
$$t = \frac{t_0}{\sqrt{1 - v^2/c^2}} \qquad (1.12)$$

Symbols in above equation represents

$t_0 =$ time interval on clock at rest relative to an observer = proper time

$t =$ time interval on clock in motion relative to an observer

$v =$ speed of relative motion

$c =$ speed of light

For a moving object t is always greater than t_0 and quantity smaller than 1. The moving clock in the spacecraft appears to tick at a slower rate than the stationary one on the ground, as seen by an observer on the ground.

Similarly the same observed by the pilot of the spacecraft for measurement of the clock on the ground. To him, the light pulse of the ground clock follows a zigzag path that requires a total time t per round trip. In the spacecraft the clock of pilot at rest, ticks at intervals of t_0. He also finds so.

$$t = \frac{t_0}{\sqrt{1 - \frac{v^2}{c^2}}}$$

The Relativity of Velocities

If two observer in different inertial reference frames s and s' measuring for the same moving particle, than to compare their velocities we will use the Lorentz Transformation equation. Suppose relative to s, s' moves with velocity v.

We assume that a particle moving parallel to the axes x, x' with constant velocity. Sends out two signals as it moves. Each observer measures the space interval and the time interval between these two events. On converting these measurements

$$\Delta x = \gamma(\Delta x' + v\,\Delta t')$$

and
$$\Delta t = \gamma\left(\Delta t' + \frac{v\,\Delta x'}{c^2}\right).$$

On dividing first equation by second we find

$$\frac{\Delta x}{\Delta t} = \frac{\Delta x' + v\,\Delta t'}{\Delta t' + v\Delta x'/c^2}$$

Dividing the numerator and denominator of the right side by $\Delta t'$, we find

$$\frac{\Delta x}{\Delta t} = \frac{\Delta x'/\Delta t' + v}{1 + v(\Delta x'/\Delta t')/c^2}$$

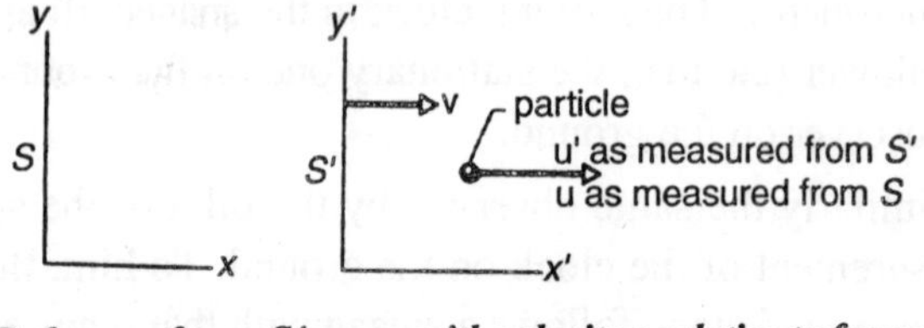

Figure 1.10 : ***Reference frame S′ moves with velocity v relative to frame S. A particle has velocity u′ relative to reference frame S′ and velocity u relative to reference frame S.***

The velocity of the particles as measured in *s*, in differential limit $\Delta x/\Delta t = u$ and the velocity of the particle as measured in s' is $\Delta x'/\Delta t' = u'$. There putting the values in above equation.

$$u = \frac{u' + v}{1 + u'\,v/c^2} \quad \text{(velocity transformation)} \qquad (1.12)$$

Now we finally get the relativistic velocity transformation equation. On considering Galilean velocity transformation equation.

$$u = u' + v \quad \text{(classical velocity transformation)}, \qquad (1.13)$$

when we apply the formal test of letting $c \to \infty$.

Relativity of Mass

Rest mass is least

If a object moving freely and a force applied on the object than it increases its kinetic energy. The object goes faster and faster as a result. Because the speed of light is the speed limit of the universe, however, the object's speed cannot keep increasing in proportion as more work is done on it. But conservation of energy is still valid in the world of relativity. On increasing the speed of object mass increases and work done converts in to kinetic energy but V never exceed than C. As the object's speed in creases to know about it's mass imagine an elastic collision of two particles A and B, observed by an observer in the reference frames S and S' (these are in uniform relative motion), and the kinetic energy is conserved in the collision, The properties of A and B are identical when determined in reference frames in which they are at rest. As shown in fig 1.11 S and S' are oriented with S' moving in the +y direction with respect to S at the velocity V.

The particles A and B had been at rest position in frame sand S′ respectively before collision. Then, at the same instent, A was thrown in the +y direction at the speed V_A while B was thrown in the $-y'$ direction at the speed V'_B where

$$V_A = V'_B \tag{1.14 a}$$

The behaviour of A and B exactly same frame S and S' respectively.

The collision of two particles show following results; Rebound of A in $-y$ direction with speed V_A and resound of B in $+y'$ direction with speed $V'B$. If the particles are thrown from positions Y apart, an observer in S finds that the collision occurs at $y = ½\, Y$ and one in S' finds that it occurs at $y' = y = ½\, Y$. For particle A the round trip time to which measured in frame S is

$$T_0 = \frac{Y}{V_A} \tag{1.14b}$$

and it is the samc for B in S'.

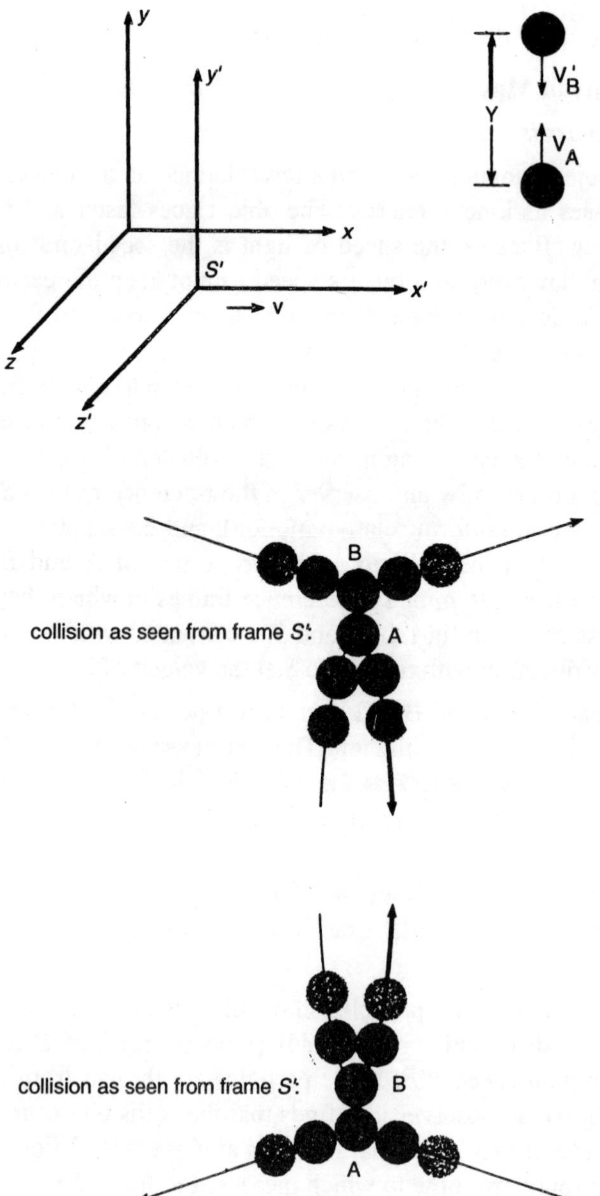

Figure 1.11 : ***An elastic collision as observed in two different frames of reference. The balls are initially* Y *apart, which is the same distance in both frames since S' moves only in the* x *direction.***

$$T_0 = \frac{Y}{V_B}$$

If linear momentum is conserved in the S frame it must be true that

$$m_A V_A = m_B V_B \tag{1.15}$$

In above equation A has mass m_A and B has mass m_B and the velocities are V_A and V_B respectively according to s frame. Now the speed V_B according to S *is*

$$V_B = \frac{Y}{T} \tag{1.16}$$

In frame S B make its round trip and time T. In S' however, B's trip requires the time T_0, where

$$T = \frac{T_0}{\sqrt{1 - v^2/c^2}} \tag{1.17}$$

As we discussed earlier. Although observers in both frames see the same event, they disagree about the length of time the particle thrown from the other frame requires to make the collision and return.

As we discuss earlier. Although observer in both frames see the same event, they disagree about the length of time the particle thrown from the other frame requires to make the cellision and return.

Replacing T in Eq. (1.16) with its equivalent in terms of T_0. We have

$$V_B = \frac{Y\sqrt{1 - v^2/c^2}}{T_0}$$

From Equation (1.14b) $\quad V_A = \frac{Y}{T_0}$

Putting the values of V_A and V_B in Equation we get the momentum as

$$m_A = m_B \sqrt{1 - v^2/c^2} \tag{1.18}$$

An observer observes that A and B are quite same, the difference between m_A and m_B means that measurements of mass, like those of space and time, it dependent on the relative speed of the observer.

If we consider the S than A and B both are moves in above example. In order to obtain a formula that gives the mass m of a body measured while in motion in terms of its mass m_0 when measured at rest, we need only consider a similar example in which V_A and V'_B are very small compared with v. On considering S we get that B coming towards A with velocity V, and than keep continue after failing a collision.

$$m_A = m_0 \text{ and } m_B = m$$

and so

Relativistic mass $$m = \frac{m_0}{\sqrt{1 - v^2/c^2}} \qquad (1.19)$$

If something is moving with velocity V considering on observer than it is larger in mass comparison to the rest position relative to an observer with factor $/\sqrt{1-v^2/c^2}$. This mass increase is reciprocal, to an observer in s', $m_A = m$ and $m_B = m_0$. Measured from the earth, a spaceraft in flight is shorter than its twin still on the ground and its mass is greater. To somebody on the spacecraft in flight the ship on the ground also appears to be shorter and to have a greater mass. This effect is very less to observe for rocket speed.

The increase in relativistic mass considered only up to speeds of light. At a speed one-tenth that of light the mass increase amounts to only 0.5 percent, but this increase is over 100 percent at a speed nine-tenths that of light. Only atomic particles such as electrons, protons, mesons, and so on have sufficietnly high speed for relativistic effects to be measurable, and in dealing with these particles, the "ordinary" laws of physics cannot be used. Historically, the first confirmation of Equation (1.19) was the discovery by Bucherer in 1908 that the ratio e/m of the electron's charge to its mass is smaller for fast electrons than for slow ones.

The equation about which we are discussing hare is now has converted in to a formula because it has shown its validity up to many extent after many experiments.

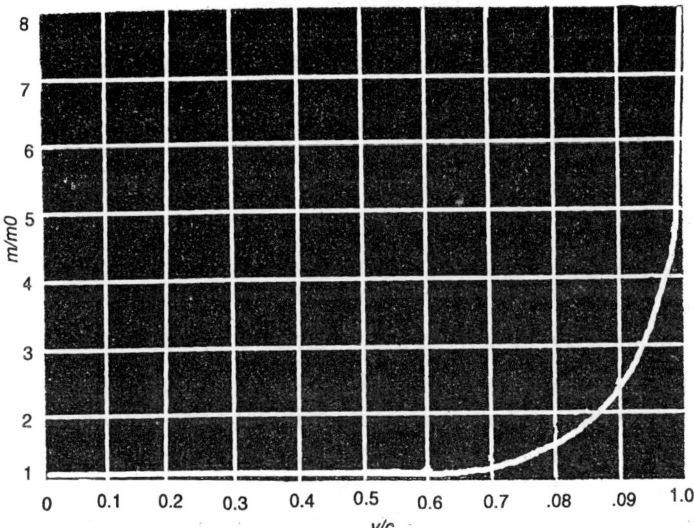

Figure 1.12 : ***The relativity of mass. Since*** **m = ∞** ***when*** **v = c.** ***No material object can equal the speed of light in free space.***

Relativistic Momentum

Provided that linear momentum **p** is defined as

Relativistic momentum

$$\mathbf{p} = m\mathbf{v} = \frac{m_0 \mathbf{v}}{\sqrt{1 - v^2/c^2}} \tag{1.20}$$

In classical physics the conservation of momentum works just in case of special relativity. However, Newton's second law of motion is correct only in the from

Relativistic second law

$$F = \frac{d}{dt}(mv) = \frac{d}{dt}\left(\frac{m_0 v}{\sqrt{1 - v^2/c^2}}\right) \tag{1.21}$$

or

$$F = ma = m\frac{dv}{dt}$$

from equation

$$\frac{d}{dt}(mv) = m\frac{dv}{dt} + v\frac{dm}{dt}$$

and *dm/dt* does not vanish if the speed of the bo iy varies with time. The force working on any body is equavalent to its time rate of change in the momentum.

Mass and energy

E = mc²

The Einstein's equation about mass and energy has been obtained from the postulates of special relativity. Now we see that this equation can be derived from which we already know.

From elementary physics if we consider that the work done on an object with constant force of magnitudes f and it work from a distance S, where f is in the same direction as S, is given by $W = FS$. If no other forces act on the object and the object starts from rest, all the work done on it becomes kinetic energy KE, so KE = FS. For a simple situation F should not be constant neccessarily.

Kinetic energy $\quad \text{KE} = \int_0^s F\,ds$

For an object the kinetic energy will be. KE = $\frac{1}{2} m_0 V^2$ if mass = m_0 and speed = V considering nonrelativistic physics. To find the correct relativistic formula for KE we start from the relativistic form of the second law of motion. Considering eq (1.21) we get

$$\text{KE} = \int_0^s \frac{d(mv)}{dt}\,ds = \int_0^{mv} v\;d(mv) = \int_0^v v\,d\left(\frac{m_0 v}{\sqrt{1 - v^2/c^2}}\right)$$

Integrating by parts $\left(\int x\,dy = xy - \int y\,dx\right)$,

$$\text{KE} = \frac{m_0 v^2}{\sqrt{1 - v^2/c^2}} - m_0 \int_0^v \frac{v\,dv}{\sqrt{1 - v^2/c^2}}$$

$$= \frac{m_0 v^2}{\sqrt{1 - v^2/c^2}} + \left[m_0 c^2 \sqrt{1 - v^2/c^2} \right]_0^v$$

$$= \frac{m_0 c^2}{\sqrt{1 - v^2/c^2}} - m_0 c^2$$

$$= mc^2 - m_0 c^2 \qquad (1.22)$$

from the above derivations we can say that for an object the kinetic energy is equal to the increase in its mass because the relative motion multiplied by the square of the speed of light. So from above equation—

Total energy $$mc^2 = m_0 c^2 + \text{KE} \qquad (1.23)$$

If we consider the mc^2 as the total energy E of the object and see that at rest KE = 0, it nevertheless possesses the energy m_0c^2. According m_0c^2 is called the rest energy E_0 of something whose mass at rest is m_0. So now we get—

$$E = E_0 + \text{KE}$$

where

Rest energy $$E_0 = m_0 c^2 \qquad (1.24)$$

If the object is moving, its total energy is

Total energy $$E = mc^2 = \frac{m_0 c^2}{\sqrt{1 - v^2/c^2}} \qquad (1.25)$$

As we know that the energy and mass can not be considered as independent quantities, their separate conservation principles are properly a single one—the principle of conservation of mass energy. Mass can be created or destroyed, but when this happens, an equivalent amount of energy simultaneously vabnishes or comes into being. and vice versa. Thus we get that the mass and energy are independent to each other for any object.

The unit of mass is kilogram (kg) and unit of energy is joule (J) and their factor of conversion is denoted as c^2, so 1 kg of matter—the mass of this book is about that—has an energy content of $m_0c^2 = (1 \text{ kg}) (3 \times 10^8 \text{ m/s})^2 = 9 \times 10^{16}$J. This is enough to send a payload of a million tons to the moon. It was difficult to know that how much energy comes from a modut amount of matter, before Eintein's work.

The process of liberation of rest energy is known to us. It is simply that we do not usually think of them in such terms. In every chemical reaction that evolves energy, a certain amount of matter disappears, but the lost mass is so small a fraction of the total mass of the reacting substances that it is imperceptible. Hence the "law" of conservation of mass in chemistry. 1 kg of dynamite vanishes 6×10^{-11} kg of matter on explosion which can not be measure but more than 5 million joule of energy can be measure.

General Relativity

Gravity is a warping of spacetime

Special relativity is related to inertial frames of reference without acceleration. Einstein's 1916 **general theory of relativity** goes further by including the effects of acceleration on what we observe. Its essential conclusion is that the force of gravity arises from a wrapping of spacetime around a body of matter (Figure 1.17). If a object moving in space, follows a curved path and can be trapped there.

The principle of equivalence is central to general relativity:

It is difficult for a observer to differentiate between the effect of gravitational field and acceleration of the laboratory.

This experiment come true from the many observation and experiments that the inertial mass of an object, which governs the object's acceleration when a force acts on it, is always equal to its gravitational mass, which governs the gravitational force another object, exerts on it. (The two masses are actually proportional, the constant of proportionality is set equal to 1 by an appropriate choice of the constant of gravitation G).

Gravity and Light

The light is the subject to gravity this statements comes from the principle of equivalence. If a light beam is directed across an açcelerated laboratory, its path relative to the laboratory will be curved. It shows that the acceleration of laboratory is equal to the light beam, subject to the gravitational field, the beam follows the same curved path.

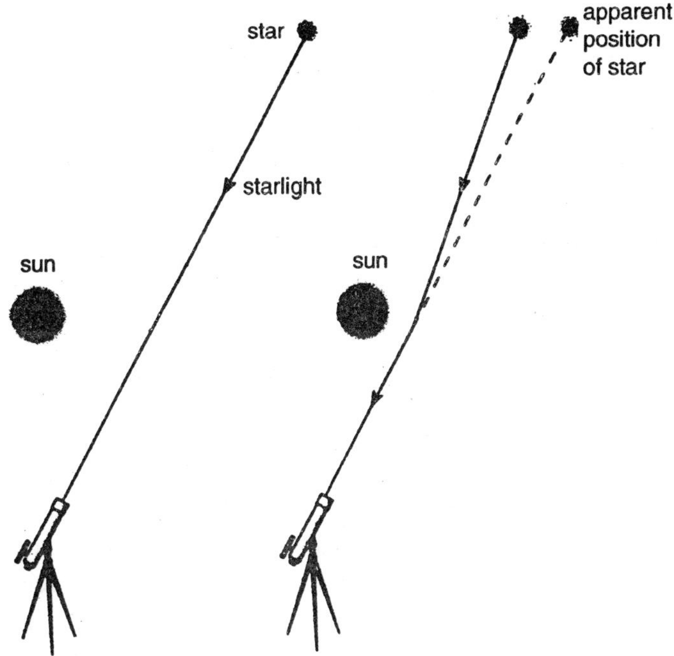

Figure 1.13 : ***Starlight passing near the sun is deflected by its strong gravitational field. The deflection can be measured during a solar eclipse when the sun's disk is obscured by the moon.***

The general relativity shows that the light rays should charge their path toward it by 0.005° to graze sun, the diameter of a dime seen from a mile away. This prediction was first confirmed in 1919 by photographs of stars that appeared in the sky near the sun during an eclipse, when they could be seen because the sun's disk was covered by the moon. The photograph were then compared with other photographs of the same part of the sky taken when the sun was in a distant part of the sky (Figure 1.13). The conclusions made Einstein a star.

In a gravitational field where light deflects, a dense concentration of mass assuming the galaxy of stars. It can be useful to make multiple images of these distant light source.

A **quasar**, the nucleus of a young galaxy, is brighter than 100 billion stars but is no larger than the solar system. The first observation of gravitational lensing was the discovery in 1979 of what seemed to be a pair of nearby quasars but was actually a single one whose light was deviated by an intervening massive object. As now we have known the number of other gravitational lenses, it shows effect on radio and light waves from distant sources.

Other Findings of General Relativity

The further work on general relativity was very difficult it was like a solving a difficult puzzle. The perihelion of a planetary orbit is the point in the orbit nearest the sun Mercury's orbit has the peculiarity that its perihelion shifts (precesses) about 1.6° per century (Figure 1.15). All but 43" (1" = 1 arc second = $\frac{1}{3600}$ of a degree) of this shifts due to the attractions of other planets, and for a while the discrepancy was used as evidence for an undiscovered planet called Vulean whose orbit was supposed to be inside that of Mercury. When gravity is weak, general relativity gives very nearly the same results as Newton's formula $F = Gm_1m_2/r^2$. Because Mercury is near to the sun so that it moves under heavy gravitational field, but Einstein shown a precession of 43" per century expectation for its orbit, through general relativity.

General relativity predicted the presence of gravitational waves having speed equal to light and confirmed after long wait. To visualise gravitational waves, we can think in terms of the model in which two dimensional space is represented by a rubber sheet distroted by masses embedded in. If one of the masses vibrates, waves will be sent out in the sheet that set other masses vibrations. Electromagnetic waves comes from a vibrating electric charge and start to vibrate other charges.

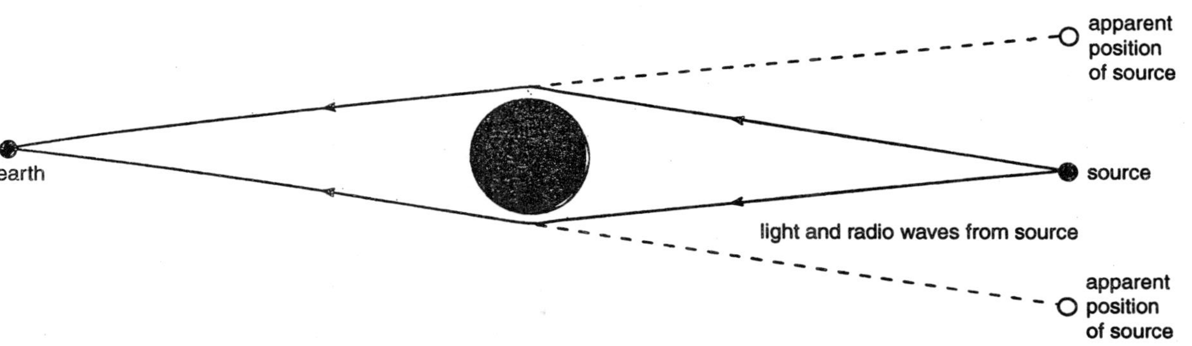

Figure 1.14 : ***A gravitational lens. Light and radio waves from a source such as a quasar are deviated by a massive object much as a galaxy so that they seem to come from two or more identical sources. A number of such gravitational lenses have been identified.***

Because the gravitational waves are quite weak, so they show difference between two kinds of waves. But inspite of much effort it can not be detected directly. However, in 1974 strong evidence for gravitational waves was found in the behaviour of a system of two nearby stars, one a pulsar, that revolve around each other. A pulsar is a very small, dense star, composed mainly of neutron, that spins rapidly and sends out flashes of light and radio waves at a regular rate, much as the rotating beam of a lighthouse does. The pulsar in this particular binary system emits pulses every 59 milliseconds (ms), and it and its companion (probably another neutron star) have an orbital period of about 8 h. According to general relativity, such a system should give off gravitational waves and lose energy as a result, which would reduce the orbital period as the stars spiral in toward each other. A change in orbital period means a change in the he arrival times of the pulsar's flashes, and in the case of the observed binary system the orbital period was found to be decreasing at 75 ms per year. This is so close to the figure that general relativity predicts for the system

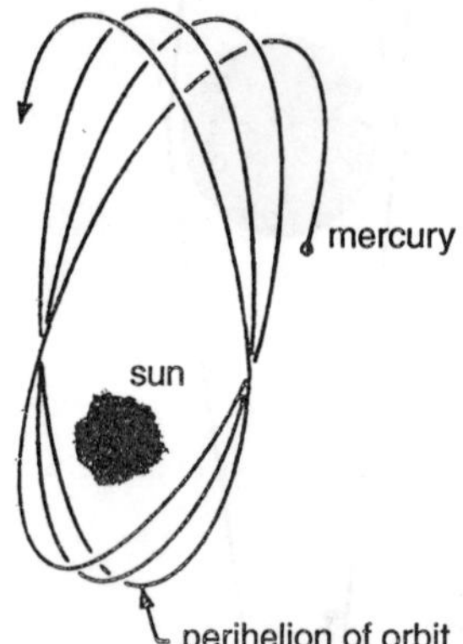

Figure 1.15 : ***The precession of the perihelion of Mercury's orbit.***

that there seems to be no doubt that gravitational radiation is responsible. This excellent work in physics done by Joseph Taylor and Russell Hulse and they got Nobel prize for that.

2

Optics

Interference of Light Waves

In this section we will discuss about light waves which produce interference pattern, and for light waves it is difficult to observe stationary interference pattern. For example, if we use two conventional light sources (like two sodium lamps) illuminating two pinholes (see Figure2.1 we will not observe any interference pattern on the screen. This can be understood from the following reasoning: In a conventional light source, light comes from a large number of independent atoms; each atom emitting light for about 10^{-9} sec, i.e., light emitted by an atom is essentially a pulse lasting for only 10^{-9} sec. We can see that in similar conditions if atom emits waves that would be different for different atoms in their initial phases.

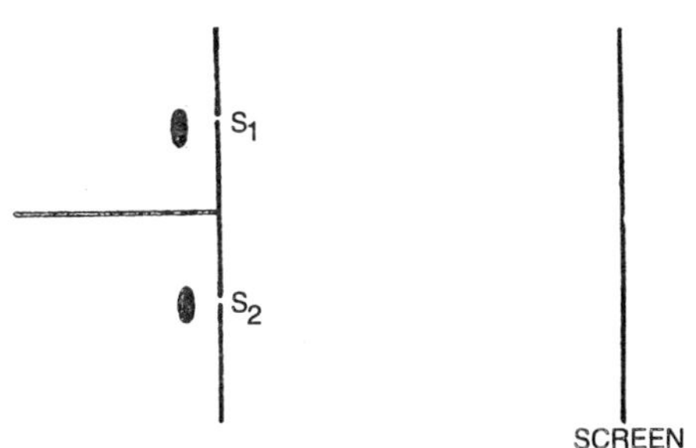

Figure 2.1 : ***If two sodium lamps illuminate two pinholes S_1 and S_2, no interference pattern will be observed on the screen.***

As we can see in the fig that holes S_1 and S_2 have a fixed phase relationship for a period of about 10^{-9} sec. from where the light is coming out and the interference pattern changes every billionth of a second. The eye can notice intensity changes which last at least for a tenth of a second and hence we will observe a uniform intensity over the screen. However, if we have a camera whose time of shutter opening can be made less than 10^{-9} sec then the film will record an interference pattern. We can confirm the above discussion by noting that light beam of two independent sources do not have any stationary interference pattern.

In 1802 Thomas Young suggested a simple method to connect the phase relationship between two sources. The trick lies in the division of a single wavefront into two; these two split wavefronts act as if they emanated from two sources having a fixed phase relationship and, therefore, when these two waves were allowed to interfere, a stationary interference pattern was obtained. In the fig we can see that the pinhole S is illuminated by a light source.

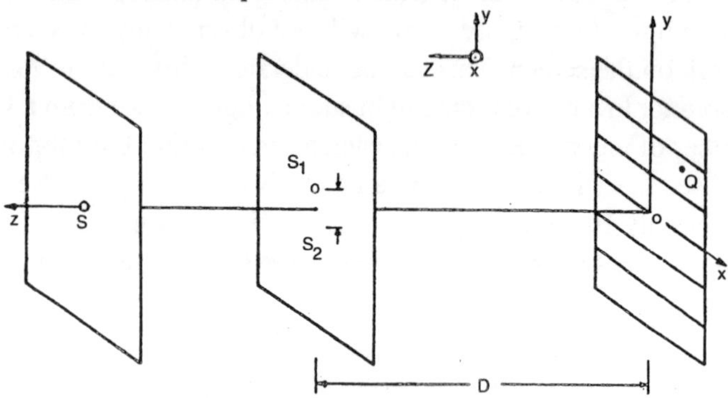

Figure 2.2 : ***Young's arrangement to produce interference pattern.***

Light diverging from this pinhole fell on a barrier which contained two pingoles S_l and S_2 which were very close to one another and were located equidistant from S. Spherical waves emanating from S_1 and S_2 (see Figure 2.3) were coherent and on the screen beautiful interference fringes were obtained. In order to show that this was indeed an interference effect. Young showed that the fringes on the screen disappear when S_1 (or S_2) is covered up. Young explained the interference pattern by considering the principle of superposition, and by measuring the distance between the fringes he calculated the wavelength.

In the above figure we can see the section of the wavefron which have S, S_1 and S_2

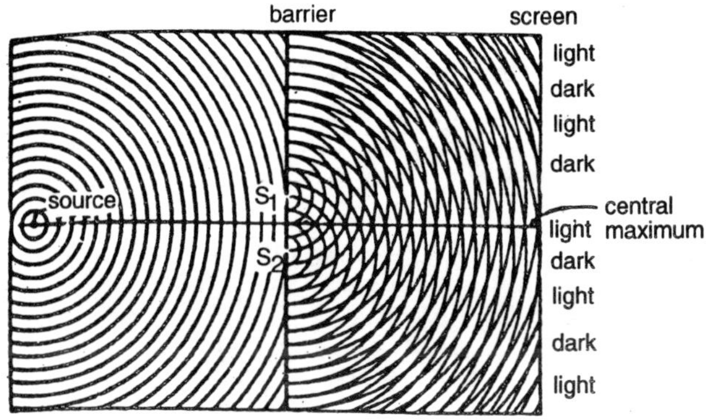

Figure 2.3 : ***Sections of the spherical wavefronts emanating from* S, S*1* *and* S*2*.**

The Interference Pattern

Assuming that two pinholes S_1 and S_2 represent young's interference experiment. We would determine the positions of maxima and of minima on the line LL' which is parallel to the y-axis and lies in the plane containing the points S, S_1 and S_2 (see figure 2.4) We would see that interference pattern around O is a series of dark and bright lines which and perpendicular to the plane (see fig2.4).

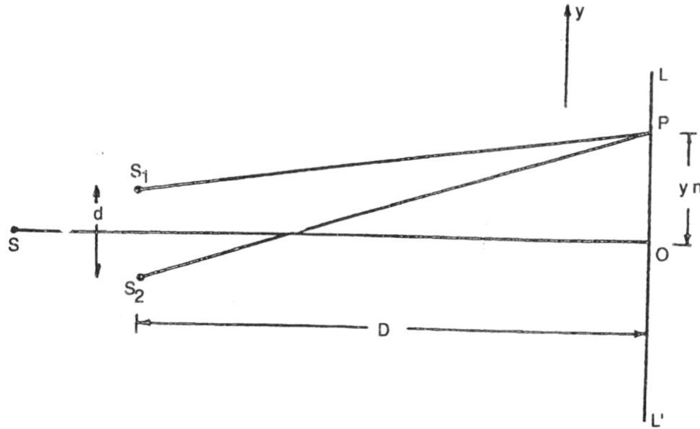

Figure 2.4 : ***Arrangement for producing young's interference pattern.***

Considering the point P on line LL' we get

$$S_2P - S_1P = n\lambda; n = 0, 1, 2, \qquad (2.1)$$

Now,

$$(S_2P)^2 - (S_1P)^2 = \left[D^2 + \left(y_n + \frac{d}{2}\right)^2\right] - \left[D^2 + \left(y_n - \frac{d}{2}\right)^2\right]$$

$$= 2y_n d$$

where

$$S_1S_2 = d \text{ and } OP = y_n$$

Thus

$$S_2P - S_1P = \frac{2y_n d}{S_2P + S_1P} \qquad (2.2)$$

If $y_n d << D$ then negligible error will be introduced if $S_2P + S_1P$ is replaced by $2D$. For example, for $d = 0.02$ cm, $D = 50$ cm, $OP = 0.5$ cm (which corresponds to typical values for a light interference experiment)

$$S_2P + S_1P = [(50)^2 + (0.51)^2]^{1/2} + [(50)^2 + (0.49)^2]^{1/2}$$

$$\approx 100.005 \text{ cm}$$

from above we can put $2D$ in plane of $S_2P + S_1P$ and we get the error about 0.005%. Thus we get

$$S_2P - S_1P \approx \frac{y_n d}{D} \qquad (2.3)$$

Using Equation (2.1) we obtain

$$y_n = \frac{n\lambda D}{d} \qquad (2.4)$$

Thus the dark and bright fringes are equally spaced and the distance between two consecutive dark (or bright) fringes is given by

$$\beta = y_{n+1} - y_n = \frac{(n+1)\lambda D}{d} - \frac{n\lambda D}{d}$$

or

$$\beta = \frac{\lambda D}{d} \qquad (2.5)$$

β denotes the fringe width.

In order to determine the shape of the interference pattern we first note that the locus of the point P such that

$$S_2P - S_1P = \Delta \tag{2.6}$$

Here we see hyperbola for any plane having points S_1 and S_2 Consequently, the locus is a hyperbola of revolution obtained by rotating the hyperbola about the axis S_1 S_2. *In* order to find the shape of the fringe on the screen we assume the origin to be at the point O and the z-axis to be perpendicular to the plane of the screen as shown in figure (2.2). The y-axis is assumed to be parallel to S_2S_1. *If* we take a point Q on the plane of screen. (figure 2.2). Let its coordinates be $(x, y, 0)$. The coordinates of the points S_1 and S_2 are $\left(0, \frac{d}{2}, D\right)$ and $\left(0, -\frac{d}{2}, D\right)$ respectively. Thus

$$S_2P - S_1P = \left[x^2 + \left(y + \frac{d}{2}\right)^2 + D^2\right]^{1/2} - \left[x^2 + \left(y - \frac{d}{2}\right)^2 + D^2\right]^{1/2}$$

$$= \Delta \text{ (says)}$$

or $$\left[x^2 + \left(y + \frac{d}{2}\right)^2 + D^2\right] = \left\{\Delta + \left[x^2 + \left(y - \frac{d}{2}\right)^2 + D^2\right]^{1/2}\right\}^2$$

or $$\left[2yd - \Delta^2\right]^2 = (2\Delta)^2\left[x^2 + \left(y - \frac{d}{2}\right)^2 + D^2\right]$$

Hence, $$\left(d^2 - \Delta^2\right)y^2 - \Delta^2 x^2 = \Delta^2\left[D^2 + \frac{1}{4}\left(d^2 - \Delta^2\right)\right]$$

which is the equation of a hyperbola. Now we get that the fringes is hyperbola in shape, thus we can rearranging the equation as

$$y = \pm\left(\frac{\Delta^2}{d^2 - \Delta^2}\right)^{1/2}\left[x^2 + D^2 + \frac{1}{4}\left(d^2 - \Delta^2\right)\right]^{1/2} \tag{2.7}$$

For values of x such that

$$x^2 << D^2 \tag{2.8}$$

the loci straight lines parallel to the x-axis. Thus we obtain approximately straight line fringes on the screen. It should be emphasized that the fringes are straight lines although the sources S_1 and S_2 are point sources. If we consider the slits sources in place of point sources than we get the straight line fringes with increased intensities. The fringes so produced are said to be non-localixed; they can be photographed by just placing a film on the screen; and we can see these fringes with the help of eyepieces.

Fresnel Biprism

To snow the interference pattern, frensnel suggested another arrangement. He used a biprism, which was actually a simple prism, the base angles of which are extremely small $\left(\sim\frac{1}{3}^{\circ}\right)$.The base of the prism is shown in figure 2.5 and the prism is assumed to stand perpendicular to the plane of the paper. *S* represents the slit which is also placed perpendicular to the plane of the paper. Light from the slit *S* gets refracted by the prism and produces two virtual images S_1 and S_2. These images act as coherent sources and produce interference fringes on the right of the biprism. We can see the fringes with the help of eyepiece. If *n* represents the refractive index of the material of the biprism and α the base angle, then $(n - 1)\ \alpha$ is approximately the nagular deviation produced by the prism and, therefore, the distance S_1S_2 is $2\ a\ (n - 1)\ \alpha$, where *a* represents the distance from *S* to the base of the prism. Now we get the values of $n = 1.5$, $d = \frac{10}{3} \approx 5.8 \times 10^{-3}$ radians, $\alpha = 2$ cm, and one gets $d = 0.012$ cm.

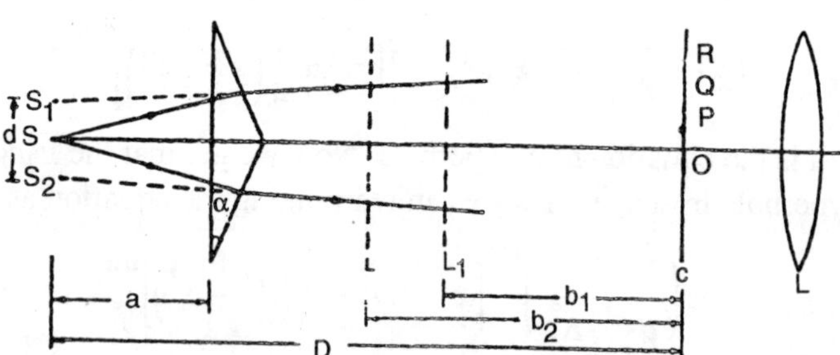

Figure 2.5 : ***Fresnel's biprism arrangement.*** **C** ***and*** **L** ***represent the positions of the crosswires and the eyepiece respectively. In order to determine*** **d** ***one introduces a lens between the biprism and the crosswires;*** $\mathbf{L_1}$ ***and*** $\mathbf{L_2}$ ***represent the two positions of the lens where the slits are clearly seen.***

Biprism arrangement can also be used to detect the wavelength of an almost monochromatic light of sodium lamp. Light from the sodium lamp illuminates the slit S and interference fringes can be easily viewed through the eyepiece. The fringe width (β) can be determined by means of a micrometer attached to the eyepiece. After getting the value of β we can get the value of λ through following equation.

$$\lambda = \frac{d\beta}{D} \tag{2.9}$$

To determine the value of d it is not necessary to get the value of α. In fact the distances d and D can easily be determined by placing a convex lens between the biprism and the eyepiece. For a fixed position of the eyepiece there will be two positions of the lens (shown as L_1 and L_2 in Figure (2.5) where the images of S_1 and S_2 can be seen at the eyepiece. Let d_1 be the distance between the two images when the lens is at the position L_1 (at a distance b_1 from the eyepiece). Let d_2 and b_2 be the corresponding distances when the lens is at L_2. From above discussion now we get that

$$d = \sqrt{d_1 d_2}$$

and
$$D = b_1 + b_2$$

Typically for $d \approx 0.01$ cm, $\lambda \approx 6\times10^{-5}$ cm, $D \approx 50$ cm $\beta \approx 0.3$ cm.

For obove if we have a slit in place of a point source. Since each pair of points S_1 and S_2 produce straight line fringes, the slit will also produce straight line fringes of increased intensity.

Displacement of fringes

As we can see in the figure 2.6 that we will get a change in the interference pattern by placing a thin transparent plate in the path of any one of the two interference beams. Let t be the thickness of the plate and let n be its refractive index. It is easily seen from the figure that light reaching the point P from S_1 has to traverse a distance t in the plate and a distance S_1P-t in air. Now the time taken by the light to travel from point S_1 to P is as

$$\frac{S_1P-t}{c}+\frac{t}{v}=\frac{1}{c}\left[S_1P-t+nt\right]$$

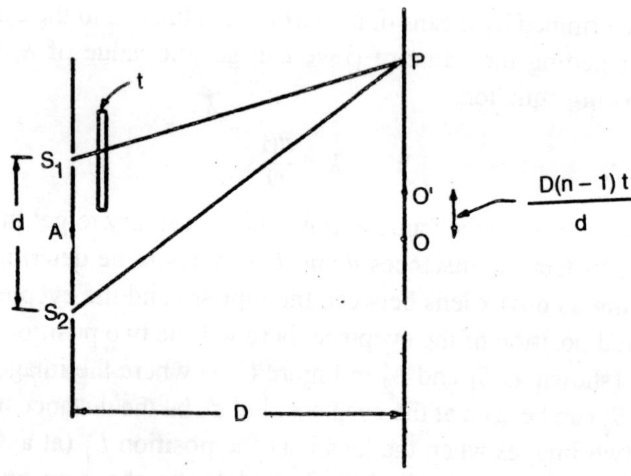

Figure 2.6 : *If a thin transparent sheet (of thickness t) is introduced in one of the beams, the fringe pattern gets shifted by a distance (n–1)tD/d.*

$$=\frac{1}{c}\left[S_1P+(n+1)\,t\right] \tag{2.10}$$

As we know that V represent the speed of light in plate i.e. $\frac{c}{n}$. Equation (2.10) shows that by introducing the thin plate the effective optical path increases by $(n-1)t$. Thus, when the thin plate is introduced the central fringe (which corresponds to equal optical path from S_1 and S_2) is formed at the point O' where

$$S_1O'+(n-1)\,t=S_2O'$$

Since [see Eq. (19)]

$$S_2O'-S_2O'\approx\frac{d}{D}OO'$$

therefore

$$(n-1)t=\frac{d}{D}OO' \tag{2.11}$$

Now we get that the shifting of fringe pattern to distance Δ is given by as follows

$$\Delta = \frac{D(n-1)\,t}{d} \tag{2.12}$$

The above principle enables us to determine the thickness of extremely thin transparent sheets (like that of mica) by measuring the displacement of the central fringe. So the displacement of central fringe is easy to detect in case of white light used as a source.

The LLoyd's Mirror Arrangement

In fig 2.7 we can see the LLoyd's mirror arrangement in which a light from a slit falls on a plane mirror at grazing incidence. The light directly coming from the slit S_1 interferes with the light reflected from the mirror forming an interference pattern in the region BC of the screen. One may thus consider the slit S_1 and its virtual image S_2 to form two coherent sources which produce the interference pattern. It should be noted that at graxing incidence one really need not have. We should know that it is more necessary to have a dielectric surface than a mirror for a grazing incidence to get high reflectivity.

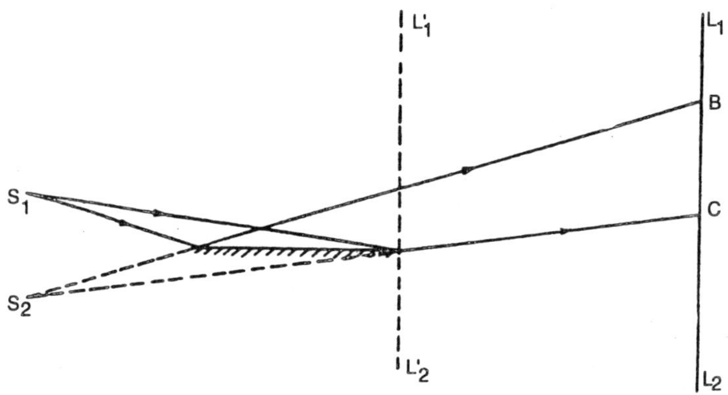

Figure 2.7 :. ***The Lloyd's mirror arrangement.***

If is shown in the above fig that the screen can not observe the central fringe while it has shifted in position L_1 L_2, where it touches the end of the reflector. Alternatively, one may introduce a thin mica sheet in the path of the direct beam so that the central fringe appears in the region BC. Indeed, if the central fringe is observed with white light, it is found to be dark. This implies that the reflected beam undergoes a sudden phase change of π on reflection. Considering that the screen has a point P on its plane.

$$S_2P - S_1P = n\lambda, \quad n = 0, 1, 2, 3, \ldots$$

we will get minima (i.e. destructive interference). On the other hand, if

$$S_2P - S_1P = \left(n + \tfrac{1}{2}\right)\lambda$$

we will get maxima.

Now considering the principle of optical reversibility we will get the abrupt phase change of Π, by the reflection of light in denser medium, then no such abrupt phase change occurs when reflection takes place at a rarer medium.

Phase change on Reflection

Now we will consider the principle of optical reversibility to get the reflection of light at an interface between two medium. According to this principle, in the absence of any absorption, a light ray that is reflected or refracted will retrace its original path if its direction is reversed.

Consider a light ray incident on an interface of two media of refractive indices n_1 and n_2 as shown in figure 2.8 (a). Let the amplitude reflection and transmission coefficients be r_1 and t_1 respectively. Now we get that if the amplitude of the incident ray is a, then the amplitudes of the reflected and refracted rays would be ar_1, and at_1 respectively.

If we consider that two rays of amplitude at_1 and ar_1 incident on medium 1 and 2 respectively as shown in figure 2.8. The ray of amplitude at_1 will give rise to a reflected ray of amplitude at_1 r_2 and a transmitted ray of amplitude at_1 t_2 where r_2 and t_2 are the amplitude reflection and transmission coefficients when a ray is incident from medium 2 on medium 1. Similarly, the ray of amplitude ar_1 will give rise to a ray of amplitude $ar_1{}^2$ and a refracted ray of amplitude ar_1t_1. If two rays of amplitudes $ar_1{}^2$ and at_1t_2 combine to give the incident ray than it is because of the principle of optical reversibility.

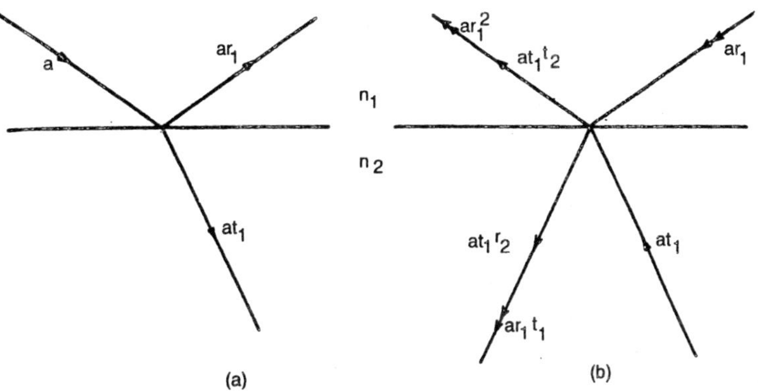

Figure 2.8 : ***(a) A ray travelling in a medium of refractive index* n_1 *incident on a medium of refractive index* n_2. *(b) Rays of amplitudees* ar_1 *and* at_1 *incident on a medium of refractive index* n_1.**

$$ar_1^2 + at_1t_2 = a$$

or
$$t_1t_2 = 1 - r_1^2 \tag{2.13}$$

Further, the two rays of amplitudes $at_1\,r_2$ and $ar_1\,t_1$ must cancel each other, i.e.

$$at_1r_2 + ar_1t_1 = 0$$

or
$$r_2 = -\,r_1 \tag{2.14}$$

The Lloyd's arrangement shows that an abrupt phase change of π happens in denser medium, from equation we can confirm that such abrupt phase change does not happen in rare medium, that no such abrupt phase change occurs when light gets reflected by a rarer medium. This is indeed borne out by experiments. Equations (2.13)(2.14) are known as stokes' relations.

Further we will calculate the amplitude reflection and transmission coefficients for plane waves incident on a dielectric and also on a conductor. We would discuss the further coefficients satisfy stokes relations, the phase change on reflection will also be discussed there.

Newton's Rings

In fig we can see that on a plane glass surface a plane convex lens is placed than a thin film of air is formed between the curved surface of the lens (AOB) and the plane glass plate (*POQ*). At the point of contact O the thickness of the air film is zero and increases with the movement towards contact point. If we allow monochromatic light (such as from a sodium lamp) to fall on the surface of the lens, then the light reflected from the surface *AOB* interferes with the light reflected from the surface *POQ*.

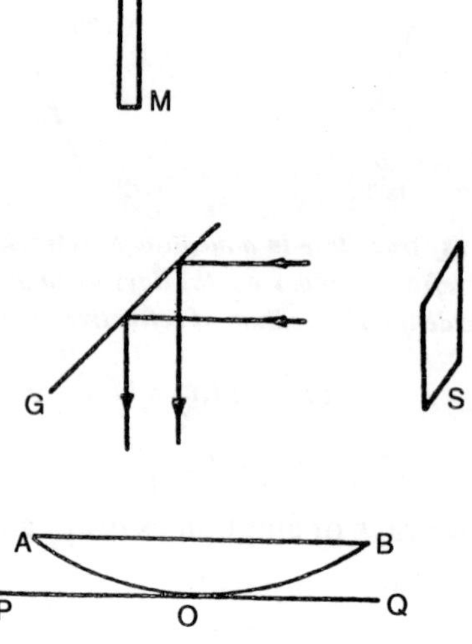

Figure 2.9: ***An arrangement for observing Newton's rings. Light from an extended source* S *is allowed to fall on a thin film formed between the planoconvex lens* AOB *and the plane glass plate* POQ. M *represents a travelling microscope.***

For near normal incidence (and considering points very close to the point of contact) the optical path difference between the two waves is very nearly equal to $2nt$, where n is the refractive index of the film and t the thickness of the film. In this way we can say that the thickness of the air film satisfies the condition.

$$2nt = \left(m + \frac{1}{2}\right)\lambda;\ m = 0, 1, 2, \ldots$$

we will have maxima. Similarly the condition

$$2nt = m\lambda \tag{2.15}$$

From above equation we can say that it is correspond to minima. Since the convex side of the lens is a spherical surface, the thickness of the air film will be constant over a circle (whose centre will be at *O*) and we will obtain concentric dark and bright rings. These rings are known as Newton's rings. If we will consider the upper surface of the film we will get the observation of fringes.

The radio of various rings can easily be calculated. As mentioned earlier, the thickness of the air film will be constant over a circle whose centre is at the point of contact. If we have radius of *m*th dark ring *r*m and thickness of air film is *t* than *m*th dark ring appears to be formed then

$$r_m^2 = t(2R - t) \tag{2.16}$$

where *R* represents the radius of curvature of the convex surface of the lens (see Figure2.10). If $R \approx 100$ cm and $t \leq 10^{-3}$ cm than *t* is neglegible in comparison to 2*R*

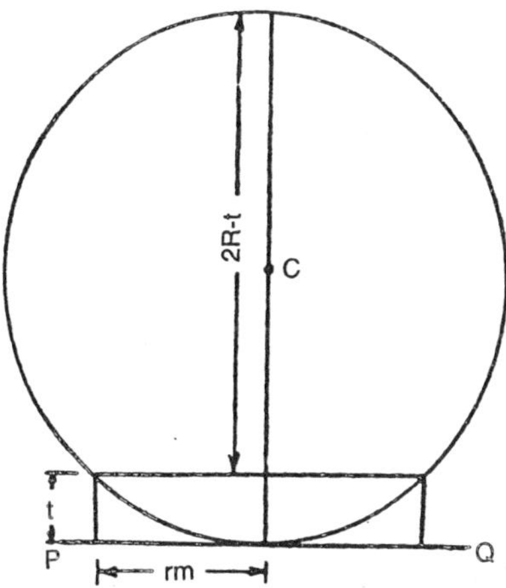

Figure 2.10 : r_m ***represents the radius of the* mth *dark ring; the thickness of the air film (where the* mth *dark ring is formed) is* t.**

$$r_m^2 \approx 2Rt$$

or

$$2t \approx \frac{r_m^2}{R} \tag{2.17}$$

Substituting this in Eq. (63), we get

$$r_m^2 \approx m\lambda R;\ m = 0, 1, 2, \ldots \tag{2.18}$$

from above we come to the conclussion that the variation in the radii of the ringe depend according to the square root of natural number. Thus the rings will become close to each other as the radius increases (see figure2.11). Between the two dark rings there will be a bright ring whose radius will be $\sqrt{m+\frac{1}{2}}\,\lambda R$.

We can see in fig 2.9 newton's ring apparatus which we use in the laboratory. Light from an extended source (emitting almost monochromatic light, like a sodium lamp) is allowed to fall on a glass plate which partially reflects the beam. This reflected beam falls on the plano-convex lens-glass plate attsmhr, rmy smf Newton's rings can easily be observed by viewing directly or through a travelling microscope *M*. Actually, one really need not have plano-conved lens; the rings can also be visible with a biconvex lens.

Typically for $\lambda = 6 \times 10^{-5}$ cm, and R = 100 cm

$$r_m = 0.0774\sqrt{m}\ \text{cm} \tag{2.19}$$

Accordingly the radii of first ring would be approximately 0.0774 cm and second 0.110 cm and radii of third would be 0.134 cm. Notice that the spacing between the second and third dark rings is smaller than the spacing between the second and third dark rings is smaller than the spacing between the first and second dark rings.

From equation 2.15 we concludes that the central spot should be dark. Normally, with the presence of minute dust particles the point of contact is really not perfect and the central spot may not be perfectly dark. Thus while carrying out the experiment one should measure the radii of the *m*th and th $(m + p)$th ring $(p \approx 10)$ and take the difference in the squares of the radii $\left(r_{m+p}^2 - r_m^2 = p\lambda R\right)$, which is indeed independent of. Usually, the diameter can be more accurately measured and in terms of the diameters the wavelength is given by the following expression:

$$\lambda = \frac{D_{m+p}^2 - D_m^2}{4pR} \tag{2.20}$$

We can measured the radius of any curvature by a spherometer and diameters of dark ring, one can experimentally determine the wavelength.

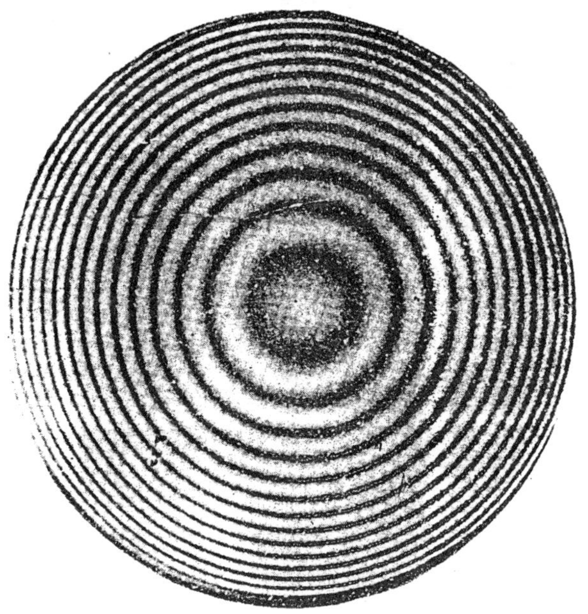

Figure 2.11: ***Newton's rings as observed in reflection. The rings observed with transmitted light are of much poorer contrast.***

If we take liquid of refractive index n and put in between the lens and the glass plate, than the radii of the dark rings will be as

$$r_m = (m\lambda R/n)^{1/2} \tag{2.21}$$

On comparing above equation with equation (2.18). Further, if the refractive indices of the material of the lens and of the glass plate are different and if the refractive index of the liquid lies in between the two values, the central spot will be bright and Eq. (2.21) would give the radii of the bright rings.

An important practical application of the principle involved in the Newton's rings experiment lies in the determination of the optical flatness of a glass plate. Suppose we have a glass surface which is placed on a another surface known flatness. If a monochromatic light beam is allowed to fall on this combination and the reflected light is viewed by a microscope, then, in general, dark and bright patches will be seen. The space between the two glass surfaces forms an air film of varying thickness and whenever this thickness becomes $m\lambda/2$, we see a dark spot and when this thickness becomes $\left(m+\frac{1}{2}\right)\lambda/2$ we see a bright spot. Two consecutive dark fringes will be separated by the air film whose thickness will differ by $\lambda/2$. To calculate the optical flatness of a glass plate we should measure the distance between consecutive dark and bright fringes.

3

Diffraction

Assuming that in figure 3.1 a plane wave incident on a slit *b* of narrow width. According to geometrical optics one expects the region *AB* of the screen *SS'* to be illuminated and the remaining portion (known as the geometrical shadow) to be absolutely dark. However, if the observations are made carefully then one finds that if the width of the slit is not very large compared to the wavelength, then the light intensity in the region *AB* is not uniform and there is also some intensity inside the geometrical shadow. We get the larger amounts of energy on making smaller the width of the slit. This spreading-out of a wave when it passes through a narrow opening is known as the phenomenon of diffraction and the intensity distribution on the screen is known as the diffraction pattern. We will discuss the phenomenon of diffraction in this chapter and wills that the spreading out decreases. With decrease in wavelength. Due to the shortness of light wave lengths $\left(\lambda \sim 5 \times 10^{-5} \text{ cm}\right)$ it is difficult to observe the effect of diffraction.

Figure 3.1 : ***If a plane wave in incident on an aperture then according to geometrical optics a sharp shadow will be cast in the region AB of the screen.***

We should know that there is not much difference between interference and diffraction, indeed interference corresponds to the situation when we consider the super position of waves coming out from a number of point (or line) sources and diffraction corresponds to the situation when we, consider waves coming out from an area source like a circular or rectangular aperture or even a large number of rectangular apertures (like the diffraction grating).

We can divide diffraction phenomena in to two parts :

(i) Fresnel diffraction and (ii) Fraunhofer diffraction.

In case of fresnet diffraction (we can see in figure 3.2) from the diffraction aperture, the source of light and the screen are at finite distance. In the Fraunhofer class of diffraction, the source and the screen are at infinite distances from the aperture; this is easily achieved by placing the source on the focal plane of a convex lens and lacing he screen on the focal plane of another convex lens [see Figure 3.2 (b)]. The two lenses effectively move

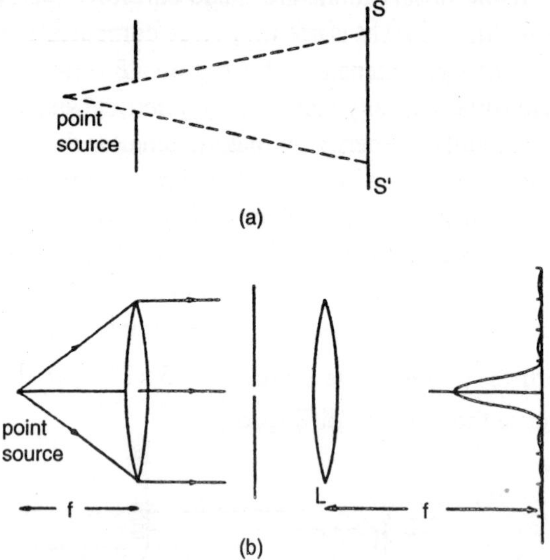

Figure 3.2 : ***(a) When either the source or the screen (or both) are at finite distances from the aperture, the diffraction pattern corresponds to the Fresnel class. (b) In the Fraunhofer class both the source and the screen are at infinity.***

the source and the screen to infinity because the first lens makes the light beam parallel and the second lens effectively makes the screen receive a parallel beam of light. Thus we get to know that it is very easy to get the intensity distribution of a fraunhofer deffraction pattern which we plan to do in this chapter further, the Fraunhofer diffraction pattern is not difficult to observe; all that one needs is an ordinary laboratory spectrometer; the collimator renders a Parallel beam of light and the telescope receives parallel beams of light on its focal plane. As we can see in the figure that the aperture of diffraction is situated on the prism table.

Single Slit Diffraction Pattern

We will see that an infinitely long slit of width to produces diffraction pattern. A plane wave is assumed to fall normally on the slit and we wish to calculate the intensity distribution on the focal plane of the lens L [see Figure 3.3 (a)]. We assume that the slit consists of a large number of equally spaced point sources and that each point on the slit is a source of Huygens' secondary wavelets which interfere with the Wavelets emanating from other points. Let the point sources be at A1, A2, A_3,... and let the distance between two consecutive points be Δ [see Figure 3.3] if we assume that the number of point sources are n than.

$$b = (n-1)\,\Delta \tag{3.1}$$

Now we would found the resultant field of point P which produced by these n source. P being an arbitrary point (on the focal plane of the lens) receiving parallel rays making an angle θ with the normal to the slit [see Figure 3.3 (b)]. Since the slit actually consists of a continuous distribution of sources, we will, in the final expression, let n go to infinity and Δ go to zero such that $n\Delta$ tends to b.

Now, at the point P, the amplitudes of the disturbances reaching from A_1, A_2, A_3, ... will be very nearly the same use the point P is at a distance which is very large in comparison to b. Because for point P the difference in the length of path, the field produced by A_1 and A_2 will be different.

For an incident plane wave, the points A_1, A_2,... are in phase and, therefore, the additional path traversed by the disturbance emanating from the point A_2 will bc A_2A_2' where A_2' is the foot

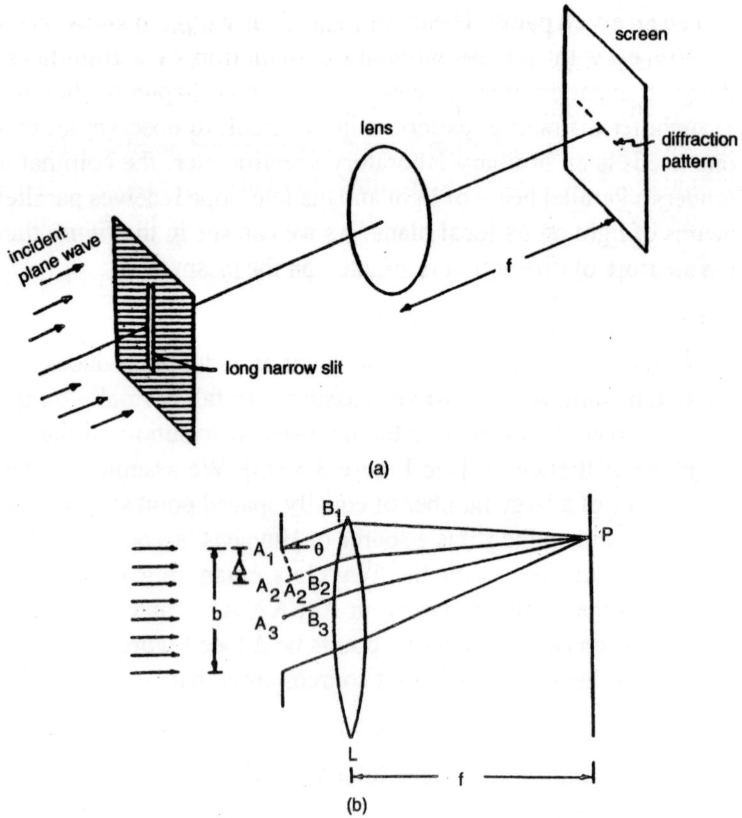

Figure 3.3 : ***(a) Diffraction of a plane wave incident normally on along narrow slit of width b. Notice that the spreading occurs along the width of the slit. (b) In order to calculate the diffraction pattern, the slit is assumed to consist of a large number of equally spaced points.***

of the perpendicular drawn from A_1 on A_2B_2. This follows from the fact that the optical paths A_1B_1P and $A_1'B_2P$ are the same. If the angle of diffracted rays is to the slet then we have the path difference as —

$$A_2A_2' = \Delta \sin \theta$$

And the difference of corresponding phase would be—

$$\phi = \frac{2\pi}{\lambda} \Delta \sin \theta \tag{3.2}$$

In this way because of the disturbance emanating from the point A_1 at the point P is $\alpha \cos wt$ then the field due to the disturbance emanating from A_2 would be $\cos(\omega t - \phi)$. Now the difference in the phases of the disturbance reaching from the points A_2 and A_3 will also be ϕ and thus the resultant field at the point P would be given by

$$E = a\left[\cos \omega t + \cos(\omega t - \phi) + \ldots + \cos(\omega t - (n-1)\,\phi)\right] \quad (3.3)$$

where $$\phi = \frac{2\pi}{\lambda}\,\Delta \sin\theta$$

As we discussed earlier.

$$\cos \omega t + \cos(\omega t - \phi) + \ldots + \cos\left[\omega t - (n-1)\,\phi\right]$$

$$= \frac{\sin n\ \phi/2}{\sin\phi/2}\cos\left[\omega t - \frac{1}{2}(n-1)\phi\right] \quad (3.4)$$

Thus

$$E = E_\theta \cos\left[\omega t - \frac{1}{2}(n-1)\phi\right] \quad (3.5)$$

Now the amplitude of E_0 of the resultant field will be—

$$E_\theta = a\frac{\sin(n\phi / 2)}{\sin\phi / 2} \quad (3.6)$$

In the limit of $n \to \infty$ and $\Delta \to 0$ in such a way that $n\Delta \to b$, we have

$$\frac{n\phi}{2} = \frac{\pi}{\lambda} n\,\Delta\sin\theta \to \frac{\pi}{\lambda} b\sin\theta$$

Further

$$\phi = \frac{2\pi}{\lambda}\Delta\sin\theta = \frac{2\pi}{\lambda}\frac{b\sin\theta}{n}$$

would tend to zero and we may, therefore, write

$$E_\theta \approx \frac{a\sin\left(\frac{n\phi}{2}\right)}{\frac{\phi}{2}} = na\frac{\sin\frac{\pi b\sin\theta}{\lambda}}{\frac{\pi b\sin\theta}{\lambda}}$$

$$= A\frac{\sin\beta}{\beta} \tag{3.7}$$

where

$$A = na$$

and

$$\beta = \frac{\pi b \sin\theta}{\lambda} \tag{3.8}$$

Thus

$$E = A\frac{\sin\beta}{\beta}\cos(\omega t - \beta) \tag{3.9}$$

Thus the intensity distribution would be—

$$I = I_0 \frac{\sin^2\beta}{\beta^2} \tag{3.10}$$

where I_0 represents the intensity at $\theta = 0$.

Positions of Maxima and Minima

We can see in figure 3.4 (a) the variation in intensity of β. It is obvious from Eq. 3.10 that the intensity is zero when

$$\beta = m\pi,\ m \neq 0 \tag{3.11}$$

[When $\beta = 0, \left(\frac{\sin\beta}{\beta}\right) = 1$ and $I = I_0$ which corresponds to the maximum of the intensity.] Substituting the value of β one obtains

$$b\sin\theta = m\lambda;\ m = \pm 1, \pm 2, \pm 3, \ldots \text{(minima)} \tag{3.12}$$

as the conditions for minima. The first minimum occurs at $\theta = \pm\sin^{-1}\left(\frac{\lambda}{b}\right)$; the second minimum at $\theta = \pm\sin^{-1}\left(\frac{2\lambda}{b}\right)$, etc. Since $\sin\theta$ cannot exceed unity, the maximum value of m is the integer which is less than (and closest to) $\frac{b}{\lambda}$.

We can get the position of minima directly from simple qualitative arguments. Let us consider the case $m = 1$. The angle θ satisfies the equation

$$b\sin\theta = \lambda \tag{3.13}$$

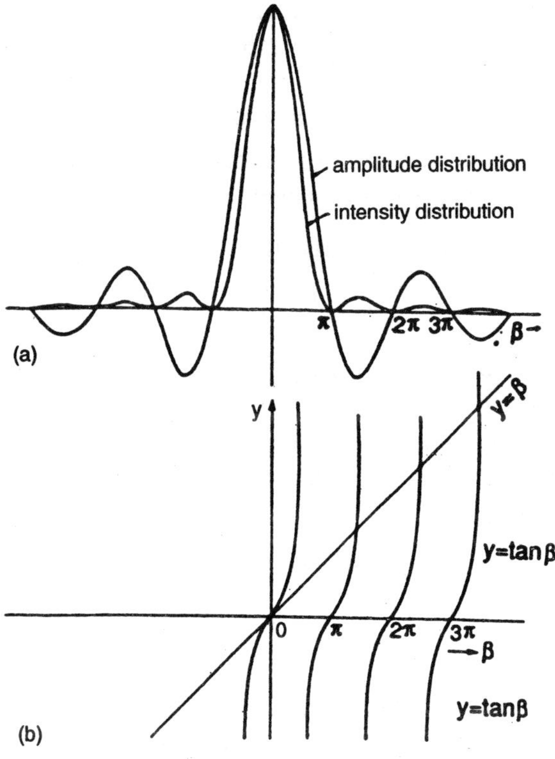

Figure 3.4 : ***(a) The intensity distribution corresponding to the single slit Fraunhofer diffraction pattern. (b) Graphical method for determining the roots of the equation tan $\beta - \beta$.***

Figure 3.5 shows that the slit is divided into two halves. Consider two points *A* and *A′* separated by a distance *b*/2. Clearly the path difference between the disturbances (reaching the point *P*) emanating from *A* and *A′* is $\frac{b}{2}\sin\theta$ which in this case is $\frac{\lambda}{2}$. The corresponding phase difference will be π and the resultant disturbance will be zero. Similarly, the disturbance from the point *B* will be cancelled by the disturbance reaching from the point *B′*. In this way the resultant intensity will be zero because the resultant disturbance of the upper half of the slit will cancel by the disturbance in resultant intensity of slit from the lower half.

$$b \sin 0 = 2\lambda \tag{3.14}$$

In a similar manner when we divide the slit into four parts; the first and second quarters cancelling each other and the third and fourth quarters cancelling each other. Such as if $m = 3$ then the slit will be divided into six parts and so on .

To detect the position of maxima we would differentiate the equation 3.10 to β and set it equal to zero, accordingly—

$$\frac{dI}{d\beta} = I_0\left[\frac{\sin\beta\cos\beta}{\beta^2} - \frac{2\sin^2\beta}{\beta^3}\right] = 0$$

or

$$\sin\beta[\beta - \tan\beta] = 0 \tag{3.15}$$

The condition $\sin\beta = 0$, or $\beta = m\pi(m \neq 0)$ correspond to minima. The conditions for maxima are roots of the following transcendental equation

$$\tan\beta = \beta \text{ (maxima)} \tag{3.16}$$

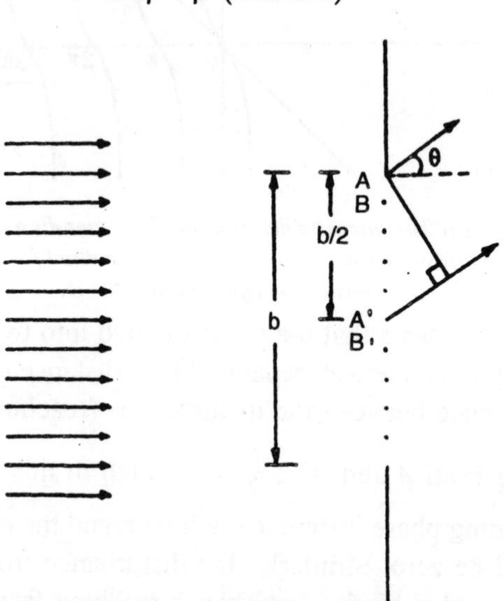

Figure 3.5 : *The slit is divided into two halves for deriving the condition for the first minimum.*

Now taking root $\beta = 0$ considering central maximum. The other roots can be found by determining the points of intersections of the curves $y = \beta$ and $y = \tan\beta$ [see Figure 3.4 (b)]. The intersections occur at $\beta = 1.43\pi$, $\beta = 2.46\pi$, etc. and are known as the first maximum, the second maximum, etc. Since is about 0.0496, the

$$\left[\frac{\sin(1.43\,\pi)}{1.43\pi}\right]^2$$

intensity of the first maximum is about 4.96% of the central maximum. Thus the intensitie of second maxima is 1.68% and of third maxima is 0.83 of the central maximum.

Diffraction by a Circular Aperture

As we have read earlier that if a plane wave incident on a long narrow slit of width b (see in figure 3.6) then the wave spreads

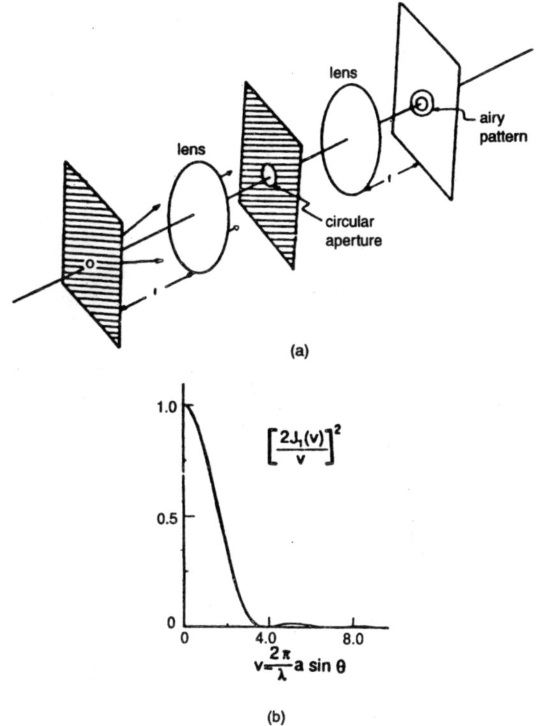

Figure 3.6 : ***(a). Experimental arrangement for observing the Fraunhofer diffraction pattern by a circular aperture.***
(b) The corresponding intensity distribution.

along the width of the slit with angular divergence $\sim \lambda / b$. In a similar manner one can discuss the diffraction os a plane wave by a circular aperture. Figure 3.6 shows the arrangement for observing the diffraction pattern; a plane wave is incident normally on the circular aperture and a lens whose diameter is much lamer than that of the aperture is placed close to the aperture and the Fraunhofer diffraction pattern is observed on the focal plane of the lens. Because of the rotational symmetry of the system, the diffraction pattern will consist of concentric dark and bright rings. Figure 3.7 shows the diffraction pattern which is known as the Airy pattern. fig 3.7 shows the diffraction pattern for a circular aperture, it's proper derivation is quite complicated so we give hare the final result of intensity distribution as —

$$I = I_0 \left[\frac{2J_1(v)}{v} \right]^2 \tag{3.17}$$

where

$$v = \frac{2\pi}{\lambda} a \sin\theta \tag{3.18}$$

Figure 3.7 : ***The Airy pattern.***

In above equations α = radius of the aperture, λ = wave length of light and θ = angle of diffraction and I_o = intensity at $\theta = 0$ it represents the central maximum, $\lambda_1(v)$ = Bessel function of the first order For those not familiar with Bessel functions, we may mention that the variation of $J_1(u)$ is somewhat like a damped sine curve (see Figure 3.8) and although $J_1(0) = 0$ we have

$$\underset{v\to 0}{\text{Lt}} \frac{2J_1(v)}{v} = \tag{3.19}$$

similar to the relation

$$\underset{v\to 0}{\text{Lt}} \frac{\sin x}{x} = 1$$

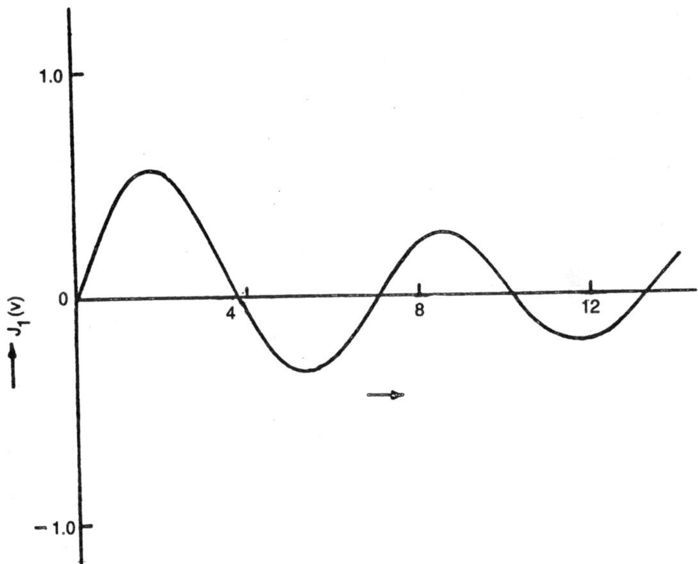

Figure 3.8 : ***The variation of*** **J_1 *(v) with* v.**

Other zeros of $J_1(v)$ occur at

$$v = 3.832, 5.136, 7.016, \ldots \tag{3.20}$$

which correspond to the successive dark circles in the Airy pattern (see Figure 3.7). Thus the first dark ring appears when

$$\sin\theta = \frac{3.832\,\lambda}{2\pi a} \simeq \frac{1.22\lambda}{D} \tag{3.21}$$

where D (= $2a$) is the diameter of the aperture. Considering the first dark ring we will get that mathematical analyses shows that about 84% of the energy is contained thus we may say that the angular spread of the beam is approximately given by

$$\Delta\theta \simeq \frac{\lambda}{D} \tag{3.22}$$

As we discussed earlier we can associate the angular devergence with diffraction pattern as—

$$\Delta\theta \sim \frac{\lambda}{\text{linear dimension of the aperture}} \tag{3.23}$$

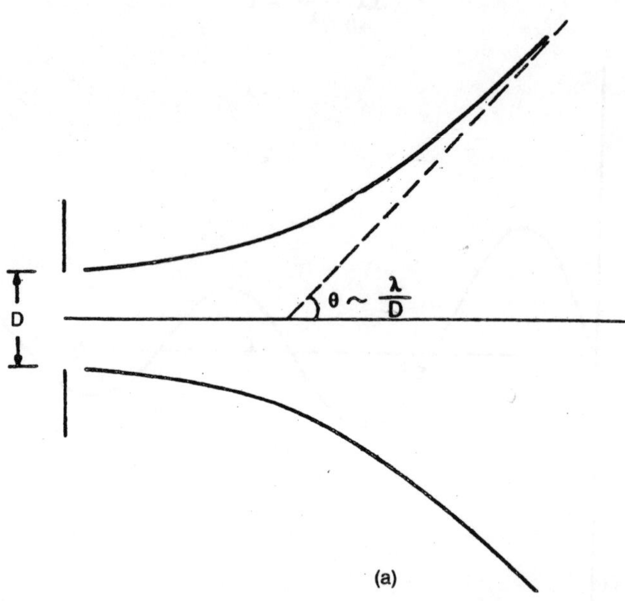

Figure 3.9 : *(a): The diffraction divergence of a plane wave incident normally on a circular aperture of diameter* D.

Figure 3.9 shows an interesting application of the above phenomenon. A layman would expect that in order to obtain more directionality of sound one should use a louds as shown in Figure 3.9 (a); however, this will result in a greater diffraction divergence and only a small fraction of energy will reach the observe. Accordingly on using a loudspeaker of large diameter, greater directionality is achieved.

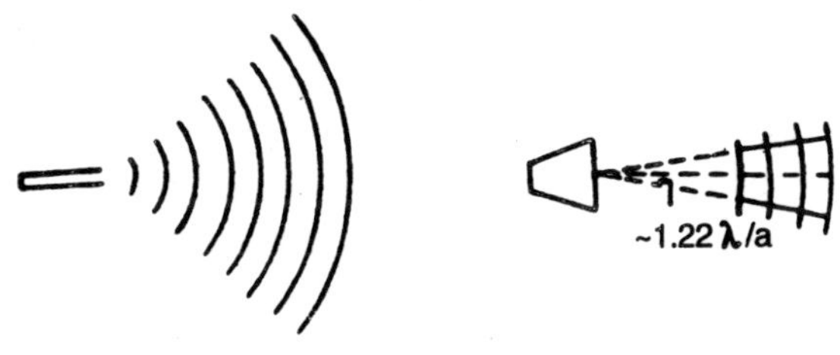

(b)

Figure 3.10 : *(b): The directionality of sound waves increases with increase in the diameter of the speaker.*

Directionality of Laser Beams

If we take a sodium lamp as a source of light it will radiate in all direction. On the other hand, the output from a laser is usually such that its divergence is primarily due to diffraction effects. Assuming that, the laser output is a circular beam of diameter D (with a plane phase front), the diffraction divergence will be approximately given by (see Figure 3.9).

$$\theta \simeq \frac{\lambda}{D} \tag{3.24}$$

and the beam width at a large distance z will be approximately given by

$$w(z) \simeq \frac{\lambda z}{D} \tag{3.25}$$

As an example, we consider a laser beam with $\lambda \simeq 6 \times 10^{-5}$ cm and D $\simeq$ 2 cm so that the divergence semiangle is given by

$$\theta \simeq 3 \times 10^{-5} \text{ radians} \simeq 6 \text{ seconds of arc}$$

Accordingly the width of a beam will be 3 cm which propagate for 1 km. Greater the width of the beam at its waist (where the phase front is plane), the greater will be the directionality of the beam.

Because the beam has high direction so it focus to very small dimension. Now, according to geometrical optics, a plane wave would focus to a point by an aberrationless converging lens; however, because of diffraction one obtains the Airy pattern on the focal plane of the convex lens-the convex lens itself acting as a circular aperture provided the beam dimension is greater than the lens aperture. The radius of the first dark ring (which is a measure of the dimension of the focussed spot) will be about $\lambda f/a$ (see Figure 3.11) where λ is the wavelength of light, f is the focal length of the lens and a is the radius of the circular aperture or the radius of the beam—whichever is smaller. If there is a ordinary lense and we take $f \simeq$ 5cm, $\lambda \simeq 6 \times 10^{-5}$ cm and $\alpha \simeq$ 1 cm then the area of the focussed spot would be—

$$\pi\left(\frac{\lambda f}{a}\right)^2 \simeq 3\times10^{-7}\text{cm}^2 = 3\times10^{-11}\text{m}^2 \tag{3.26}$$

Thus a 3 MW laser beam will produce an intensity of about

$$I \simeq \frac{P}{\pi(glf/a)^2} \simeq \frac{3\times10^6}{3\times10^{-11}} = 10^{17}\text{W}/\text{m}^2 \tag{3.27}$$

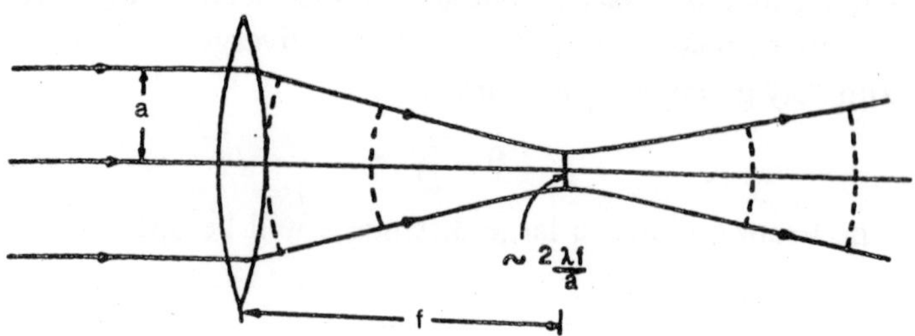

Figure3.11: ***If a truncated plane wave (of radius a) is incident on an aberrationless lens (of focal length f) then the emerging wave is a spherical wave which gets focussed to a spot whose radius is about *.***

at the focal plane of the lens. The spark created in air *at* the focus of a 3 MW peak power pulse from a ruby laser. The electric field strengths produced at the focus are of the order of 10^9V/m. As we know that the intensity is $\frac{1}{2}\varepsilon_0 c E_0^2$ then we can qualitatively verified it in this way—

$$\frac{1}{2}\times 8.85\times10^{-12}\times 3\times10^{8} E_0^2 \simeq 10^{17} \Rightarrow E_0 \simeq 8.7\times10^{9}\ \text{V}/\text{m}$$

An extremely small region with radius ~2μm, produces large intensity. Such high intensities lead to numerous industrial applications of the laser like welding, hole drilling, cutting materials, etc. for further details.

As we discuss above the greater the radius of the beam, the smaller will be the size of the focussed spot and hence greater will be the intensity, the focussed spot. Indeed one may use a beam expander (see Figure to produce a beam of greater size and hence a smaller focussed spot size. However, after the focussed spot, the beam would have a greater divergence and would therefore expand within a very short distance. One usually defines *a depth of focus* as the distance over which the intensity of the beam (on the axis) decreases by a certain factor of the value at the focal point. In this way a small focused spot leads to a small depth of focus.

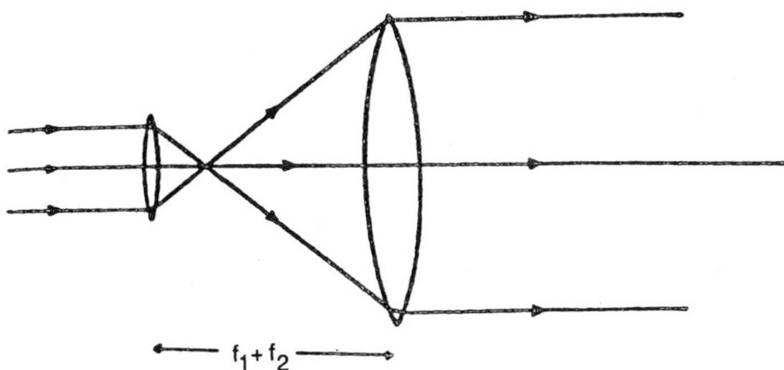

Figure 3.12: ***Two convex lenses separated by a distance equal to the sum of their focal lengths act like a beam expander.***

Limit of Resolution

To focus the stars by a telescope objective of diameter D we assume them as a point sources which entering in the aperture as plane waves (figure 3.13). As discussed in the previous section, the system can be thought of as being equivalent to a circular aperture of diameter *D*, followed by a converging lens of focal length *f*. As such, each point source will produce its Airy pattern as schematically shown in Figure 3.13. To detect the diameter of Airy rings, the diameter of the objective, its focal length and the wavelength of light is used.

As we can see in figure 3.13 the Airy pattern is shown and both are situated away from each other consequently the two objects are said to be well resolved. Since the radius of the first dark ring is $122\lambda f / D$ the Airy patterns will overlap more for smaller values of D and he t for better resolution one requires a larger diameter of the objective. Therefore the diameter of the objective is the characteristic of the telescope. As the objective of 4" diameter belongs to 40 inch telescope.

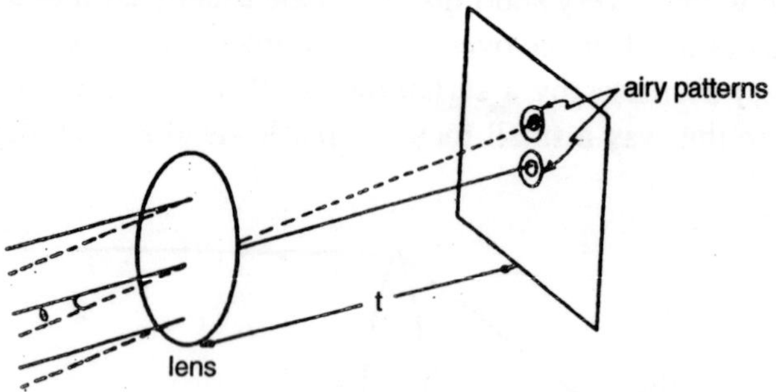

Figure 3.13: ***The image of two distant objects on the focal plane of a convex lens. If the diffraction patterns are well separated, they are said to be resolved as the objective of 40" diameter belongs to 40 inch telescope.***

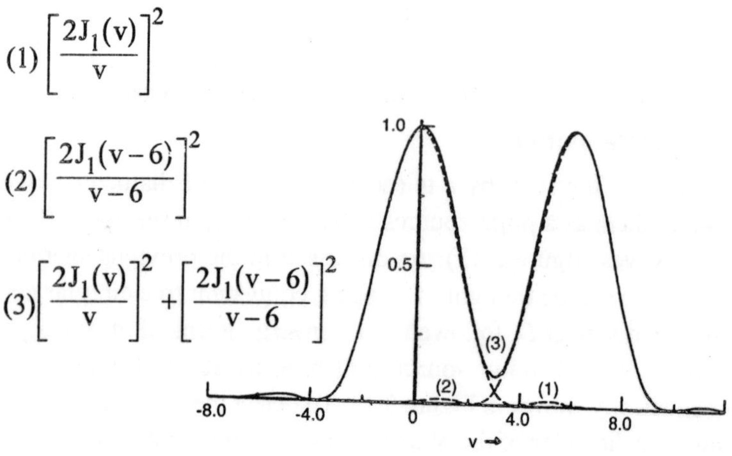

(a)

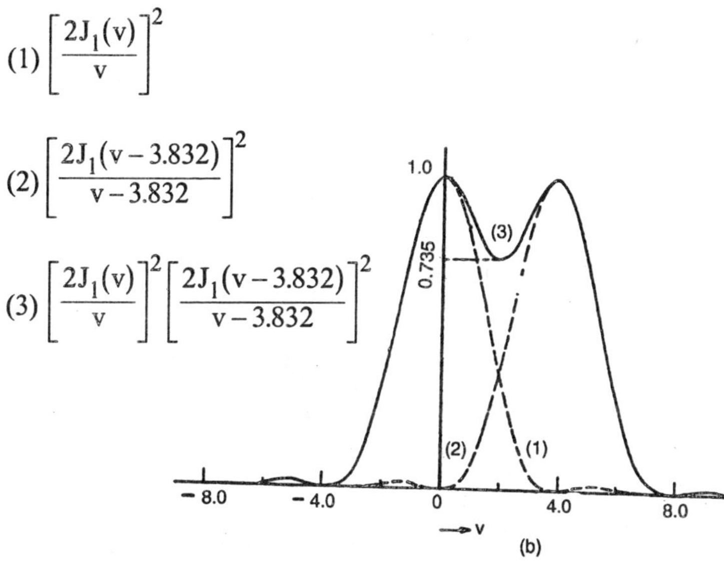

Figure 3.14: ***The dashed curves correspond to the intensity distribution produced by two point sources (producing the same intensity at the central spot) independently; the solid curves represent the resultant. (a) and (b) correspond to the angular separation of the two objects equal to 6X* D and 1.22* respectively. In the first case the objects are well resolved and in the second case (according to the Rayleigh criterion) they are just resolved.***

In figure 3.14 (a) and 3.14 (b) shows the independent intensity distribution and their resultant which produced by two distant point objects of different angular separations; in each case we have assumed that the two sources produce the same intensity at their respective central spots. Obviously, the resultant intensity distributions are quite complicated (see Figure 3.16); what we have plotted in Figures 3.14 and 3.15 are the intensity distributions on the line joining the two centres of the Airy patterns; we should mention here that since the point sources are independent sources, their intensity distributions (Airy patterns) will add. If we consider this line as x axis then according to the figures 3.14 and 3.15, v will be

$$v = \frac{2\pi a}{\lambda f} x \tag{3.28}$$

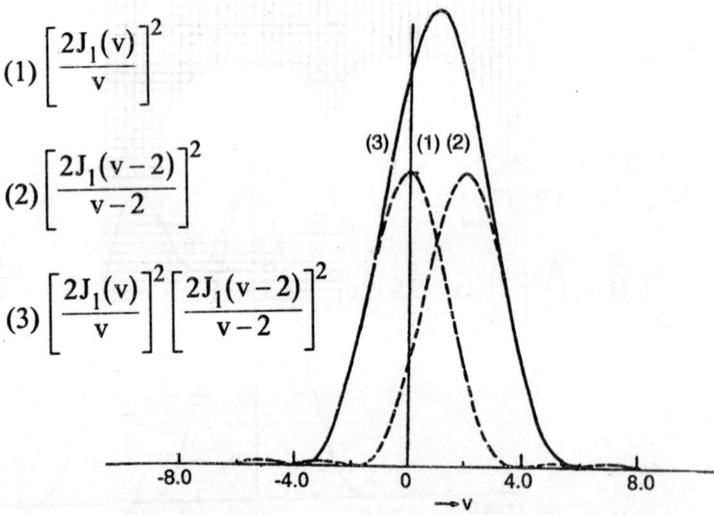

Figure 3.15: ***The dashed curves correspond to the intensity distribution produced independently by two distant point objects having an angular separation of 2λ/πD. The resultant intensity distribution (shown as a solid curve) has only one peak and hence the objects are unresolved.***

In Figure 3.14 (*a*) the distribution of intensity is shown. The two distant point objects, which have an angular separation of $6\lambda / \pi\omega$ and as can be seen the two images are clearly resolved

$$\Delta\theta \simeq \frac{2}{\pi}\frac{\lambda}{D}$$

corresponds to fig. 3.15 and as can be seen, the resultant intensity distribution has only one peak and therefore the two points cannot be resolved at all. Finally, if the angular separation of the two objects is $1.22\lambda/D$ then the central spot of the one pattern falls on the first minimum of the second and the objects are said to be just resolved. This criterion of limit of resolution is called the Rayleigh criterion of resolution and the intensity distribution corresponding to this is plotted in Figure (3.14 *b*). We can see the real pattern of diffraition in figure 3.16.

If we consider a telescope objective of diameter 5 cm and focal length 30 cm we get a numerical appreiation of the above discussion.

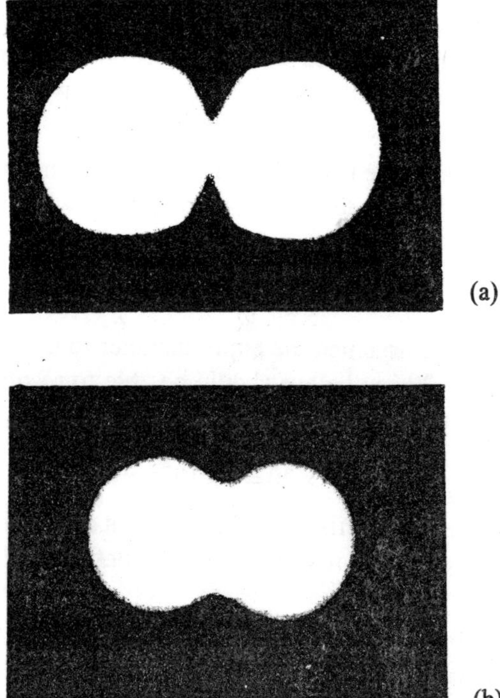

(a)

(b)

Figure 3.16 : *The acutal diffraction patterns corresponding to two point sources when they are : (a) well resolved, and (b) just resolved.*

Assuming the light wavelength to be 6×10^{-5} cm, now we get that the two distant objects with minimum angular separation will be resolved as

$$\frac{1.22\lambda}{D} = \frac{1.22 \times 6 \times 10^{-5}}{5} \simeq 1.5 \times 10^{-5}$$

Further, the radius of the first dark ring (of the Airy pattern) will be

$$\frac{1.22\lambda}{D} \times \text{focal length} = \frac{1.22 \times 6 \times 10^{-5}}{5} \times 30$$

Now we know that if the diameter of the objective is larger then its resolving power will also be better. For example, the diameter of the Largest telescope objective is about 80" and the corresponding angular separation of the objects that it can resolve is

$\simeq$0.07 sec. of arc. This very low limit of resolution is never achieved in ground based telescopes due to the turbulence of the atmosphere. However a aperture still have a power to gather larger light and ability to see deeper in space.

We can calculate the angular resolution of the human eye because primarliy it is due to diffraction effects when it would be—

$$\Delta\theta \sim \frac{\lambda}{D} \approx \frac{6\times10^{-5}}{2\times10^{-1}} = 3\times10^{-4} \text{ rad.} \tag{3.29}$$

where we have assumed the pupil diameter to be 2 mm. Thus, at a distance of 20 m, the eye should be able to resolve two points which are separated by a distance

$$3\times10^{-4}\times20 = 6\times10^{-3}\,\text{m} = 6\text{ mm}$$

One can indeed verify that this result is qualitatively valid by finding the distance at which the millimetre scale will become blurred.

As we discuss above, a identical but displaced pattern is produced by two object points. If that is not the case then the two central maxima will have different intensities. In this way for the limit of resoultion *i.e.*, two magma stand out, one should modified criterion.

Resolving Power of a Microscope

As figure 3.17 shows that we will take a microscope objective of diameter D in regarding the resolving power. Considering two closely spaced self luminous point object as P and Q and observe through microscope. Assuming the absence of any geometrical aberrations, rays emanating from the points P and Q will produce spherical wavefronts (after refraction through the lens) which will form Airy patterns around their paraxial image points P' and Q. For the points P and Q to be just resolved, the point Q' should lie on the first dark ring surrounding the point P' and therefore we must have

$$\sin\alpha' \approx \frac{1.22\lambda}{D} = \frac{1.22\lambda_0}{n'D} \tag{3.30}$$

where n and n' represent the refractive indices of the object and image spaces, λ_0 and λ $(= \lambda_0/n')$ represent the wavelength of light in free space and in the medium of refractive index n' respectively. In figure 3.17 the angle α is shown

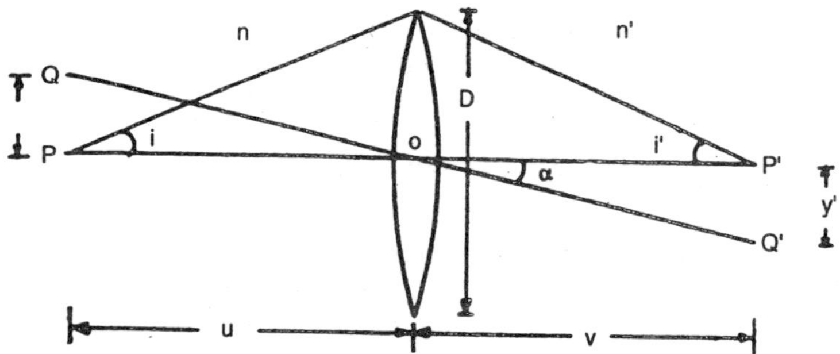

Figure 3.17 : *The resolving power of a microscope objective.*

$$\sin\alpha' \approx \frac{y'}{OP'} = \frac{y' \tan i'}{D/2} \approx \frac{y' \sin i'}{D/2} \tag{3.31}$$

where we have assumed $\sin i' \approx \tan i'$, this is justified since the image distance (OP') is large compared to D. Using Equations (3.30) and (3.31), we get

$$y' \approx \frac{0.61\lambda_0}{n' \sin i'}$$

If we now use the sine law $n'y' \sin i' = ny \sin i$, we get

$$y \approx \frac{0.61\lambda_0}{n \sin i} \tag{3.32}$$

It shows the smallest distance which microscope can resolve. The quantity $n \sin i$ is the numerical aperture of the optical system and the resolving power increases with increase in the numerical aperture. It is for this reason that in some microscopes the space between the object and the objective is filled with an oil—and they are referred to as "oil immersion objectives". Equation (3.32) also tells us that the resolving power increases with decrease in λ. Similarly for the illumination of the object we can use blue light or even ultraviolet light.

As we discussed earlier, to increase the intensities of two object points they are self luminous. However, in actual practice, the objects are illuminated by the same source -and, therefore, in general, there is some phase relationship between the waves emanating from the two object points; for such a case the intensities will not be strictly additive nevertheless Equation (3.32) shows the proper limit of resolution.

Two-slit Fraunhofer Diffraction Pattern

As we discussed earlier about the fraunhofer diffraition pattern, it produced by a slit and the intensity distribution consisted of maxima and minima. In this section we will study the Fraunhofer diffraction pattern produced by two parallel slits (each of width *b*) separated by a distance. We would find that the resultant intensity distribution is a product of the single slit diffraction pattern and the interference pattern produced by two point sources separated by a distance *d*.

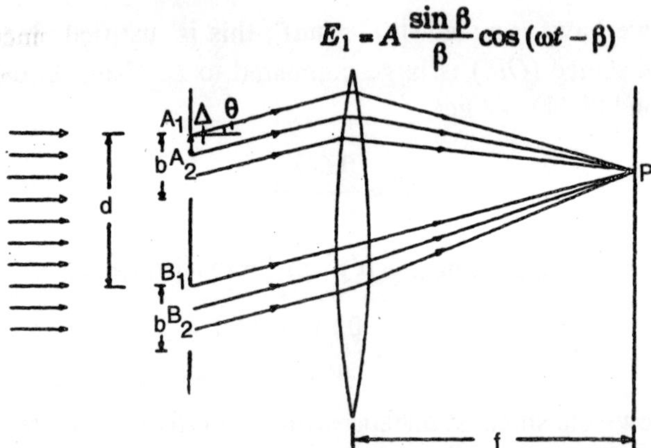

Figure 3.18 : ***Fraunhofer diffraction of a plane wave incident normally on a double slit.***

To calculate the diffraction pattern of a large number of equally spaced point sources, we use a method similar to that used for a single slit. And must remember that each point (on the slit) is a source of Huygens' secondary wavelets. Let the point sources A_1, A_2, A_3, (in the first slit) and at B_1, B_2, B_3,... (in the second slit) [see Figure 3.18] As before, we assume that the distance between two consecutive points in either of the slits is Δ. Considering θ as *a* an angle with diffraction ray normally to the plane of slit, then the path difference between the disturbances reaching the point *P* from two consecutive points in a slit will be Δ sin θ. As we can see in the figure the field produced by the first slit at point *P* would be as—

$$E_1 = A\frac{\sin\beta}{\beta}\cos(\omega t - \beta) \tag{3.33}$$

Similarly, the second slit will produce a field

$$E_2 = A\frac{\sin\beta}{\beta}\cos(\omega t - \beta - \Phi_1) \tag{3.34}$$

at the point *P*, where

$$\Phi_1 = \frac{2\pi}{\lambda} d \sin\theta \tag{3.35}$$

represents the phase difference between the disturbances (reaching the point P) from two corresponding points on the slits; by corresponding points, we imply pairs of points like (A_1, B_1), (A_2, B_2),... which are separated by a distance d. Hence the resultant field will be

$$E = E_1 + E_2$$

$$= A\frac{\sin\beta}{\beta}\left[\cos(\omega t - \beta) + \cos(\omega t - \beta - \Phi_1)\right] \tag{3.36}$$

which represents the interference of two waves, each of amplitude $A\frac{\sin\beta}{\beta}$ and differing in phase by Φ_1. Equation (3.36) can be rewritten in the form

$$E = A\frac{\sin\beta}{\beta}\cos\gamma\cos\left(\omega t - \tfrac{1}{2}\beta - \tfrac{1}{2}\Phi_1\right) \tag{3.37}$$

where

$$\gamma = \frac{\Phi_1}{2} = \frac{\pi}{\lambda} d \sin\theta \tag{3.38}$$

The intensity distribution *will* be of the form

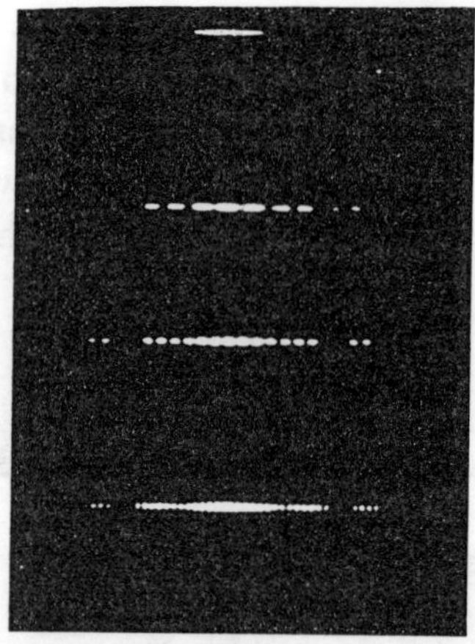

Figure3.19. ***The double-slit Fraunhofer diffraction pattern corresponding to*** b = ***0.0088 cm and*** λ = ***6.328*** $\times 10^{-5}$ ***cm. The values of d are 0, 0.0176, 0.035 and 0.070 cm respectively.***

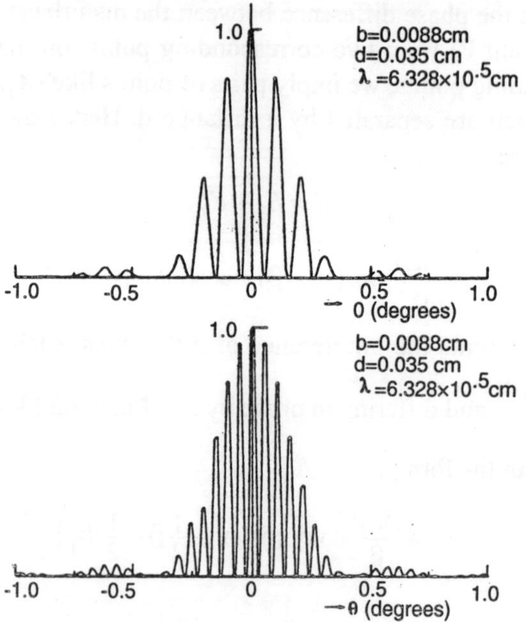

Figure 3.20 : ***The double-slit intensity distribution as predicted by Eq. (41) corresponding to d = 0.035 cm and 0.070 cm respectively (b = 0.0088 cm,*** λ ***= 6.328 ×*** 10^{-5} ***cm).***

$$I = 4I_0 \frac{\sin^2 \beta}{\beta^2} \cos^2 \gamma \tag{3.39}$$

In above equation $I_0 \dfrac{\sin^2 \beta}{\beta^2}$ is the intensity distribution produced by one slit. As can be seen, the intensity distribution is a product of two terms; the first term $\left(\sin^2 \beta / \beta^2\right)$ represents the diffraction pattern produced by a single slit of width b and the second term $\left(\cos^2 \gamma\right)$ represents the interference pattern produced by two point sources separated by a distance d. Indeed, if the slit widths are very small (so that there is almost no variation of the $\sin^2 \beta / \beta^2$ term with θ) then one simply obtains the Young's interference pattern we have shown the two slit diffraction patterns corresponding to d = 0, 0.0176, 0.035 and 0.070 cm with b = 0.0088 cm and $\lambda = 6.328 \times 10^{-5}$cm. As the intensity distribution is concluded in equation 3.39 we can see in figure 3.20.

Positions of Maxima and Minima

From equation (3.39) we can say that the intensity is zero than—

$$\beta = \pi, 2\pi, 3\pi, \ldots \tag{3.40}$$

or when

$$\gamma = \frac{\pi}{2}, \frac{3\pi}{2}, \frac{5\pi}{2}, \ldots \tag{3.41}$$

The corresponding angles of diffraction will be given by the following equations:

$$b \sin \theta = m\lambda; \ (m = 1, 2, 3, \ldots) \tag{3.42}$$

and

$$d \sin \theta = \left(n + \tfrac{1}{2}\right)\lambda; \ (n = 0, 1, 2, 3, \ldots) \tag{3.43}$$

The interference maxima occur when

$$\gamma = 0, \pi, 2\pi, \ldots \tag{3.44}$$

or when,

$$d \sin \theta = 0, \lambda, 2\lambda, 3\lambda, \ldots \tag{3.45}$$

As we can see the angles in above equation, at positions of the maxima provides the variation of the diffraction which is not too rapid. Further, a maximum may not occur at all if θ corresponds to a diffraction minimum, i.e if $b \sin \theta = \lambda, 2\lambda, 3\lambda, \ldots$ These are usually referred to as missing order e.g., as we can se in figure 3.20, for $b = 0.0088$ cm the interference maxima are extremely weak around $\theta = 0.41°$ if is because of the following—

$$\theta = \sin^{-1}\left(\frac{\lambda}{b}\right) = \sin^{-1}\left[\frac{6.328 \times 10^{-5}}{8.8 \times 10^{-3}}\right] = \sin^{-1}\left[7.19 \times 10^{-3}\right]$$

$$\simeq 0.00719 \text{ radians}$$

$$\simeq 0.412°$$

the first minimum of the diffraction term occurs.

N- Slit fraunhofer Diffraction Pattern

A parallel slit N produces a diffraction pattern, each of width b, and the distance between two consecutive slit is assumed to be d.

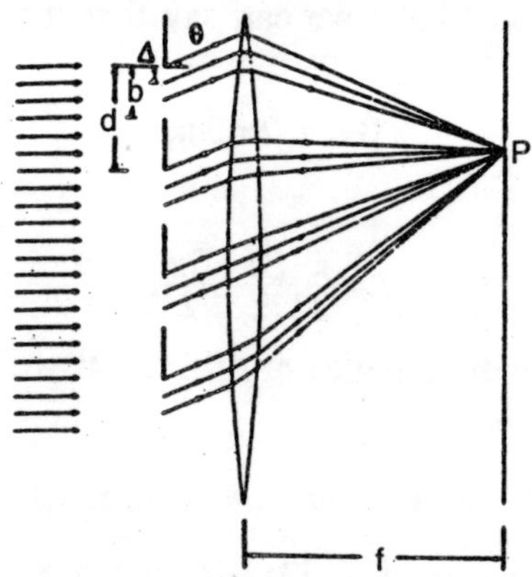

Figure 3.21 : ***Fraunhofer diffraction of a plane wave incident normally on a multiple slit.***

As before, we assume that each slit consists of n equally spaced point sources with spacing Δ (see Figure3.21). So we can say that (at an arbitrary point) is the sum of N term.

$$E = A\frac{\sin\beta}{\beta}\cos(\omega t - \beta) + A\frac{\sin\beta}{\beta}\cos(\omega t - \beta - \Phi_1)$$

$$+\ldots + A\frac{\sin\beta}{\beta}\cos(\omega t - \beta - (N-1)\Phi_1) \qquad (3.46)$$

where the first term represents the amplitude produced by the first slit, the second term by the second slit, etc. and the various symbols have the same meaning Rewriting Equation (3.46) we get

$$E = \frac{A\sin\beta}{\beta}\Big[\cos(\omega t - \beta) + \cos(\omega t - \beta + \Phi_1)$$

$$+\ldots + \cos\left(\omega t - \beta - (N-1)\Phi_1\right)\Big]$$

$$= \frac{A\sin\beta}{\beta}\frac{\sin N\gamma}{\sin\gamma}\cos\left[\omega t - \beta - \tfrac{1}{2}(N-1)\Phi_1\right] \qquad (3.47)$$

where

$$\gamma = \frac{\Phi_1}{2} = \frac{\pi}{\lambda} d\sin\theta$$

Thus the corresponding intensity distribution is as—

$$I = I_0\frac{\sin^2\beta}{\beta^2}\frac{\sin^2 N\gamma}{\sin^2\gamma} \qquad (3.48)$$

Here, $I_0 \sin^2\beta/\beta^2$ is the intensity distribution production of a single slit. As can be seen, the intensity distribution is a product of two terms; the first term $\left(\dfrac{\sin^2\beta}{\beta^2}\right)$ represents the diffraction pattern produced by a single slit and the second term $\left(\dfrac{\sin^2 N\gamma}{\sin^2\gamma}\right)$ represents the interference pattern produced by N equally spaced point sources. For $N = 1$, Equation (3.48) reduces to the single slit diffraction pattern and for $N = 2$, to the double slit diffraction pattern [see Equation (3.39)]

Positions of Maxima and Minima

To get intense maxima, value of N should be very large and $\gamma \simeq m\pi$ that is—

$$d\sin\theta = m\lambda \ \ (m = 0, 1, 2, \ldots) \tag{3.49}$$

This can be easily seen by noting that

$$\operatorname*{Lt}_{\gamma \to m\pi} \frac{\sin N\gamma}{\sin \gamma} = \operatorname*{Lt}_{\gamma \to m\pi} \frac{N \cos N\gamma}{\cos \gamma} = \pm N;$$

thus, the resultant amplitude and the corresponding intensity distributions are given by

$$E = N \frac{A \sin\beta}{\beta} \tag{3.50}$$

and

$$I = N^2 I_0 \frac{\sin^2\beta}{\beta^2} \tag{3.51}$$

where

$$\beta = \frac{\pi b \sin\theta}{\lambda} = \frac{\pi b}{\lambda}\frac{m\lambda}{d} = \frac{\pi b m}{d} \tag{3.52}$$

In above equation we can see the principal maxima. Physically, at these maxima the fields produced by each of the slits are in phase and, therefore, they add up and the resultant field is N times the field produced by each of the slits; Thus the value of intensity is large unless $\frac{\sin^2\beta}{\beta}$ itself is very small. Since $|\sin\theta| \le 1$, m cannot be greater than d/λ [see Eq. (3.49)]; thus, there will only be a finite number of principal maxima.

From Eq. (3.48) it can easily be seen that the intensity is zero when either

$$b\sin\theta = n\lambda, \ n = 1, 2, 3, \ldots \tag{3.53}$$

or

$$N\gamma = p\pi, \ p \ne N, 2N, \ldots \tag{3.54}$$

From the equation we can get the minima corresponding to the single slit diffraction pattern. The angles of diffraction corresponding to Equation (3.54) are

$$d \sin\theta = \frac{\lambda}{N}, \frac{2\lambda}{N}, \ldots, \frac{(N-1)\lambda}{N}, \frac{(N+1)\lambda}{N}, \frac{(N+2)\lambda}{N},$$

$$, \frac{(2N-1)\lambda}{N}, \frac{(2N+1)\lambda}{N}, \frac{(2N+2)\lambda}{N}, \ldots \tag{3.55}$$

Consequently we get $(N-I)$ minima between two principles. Between two such consecutive minima the intensity has to have a maximum; these maxima are known as secondary maxima. Typical diffraction patterns for $N = 1, 2, 3$ and 4 are shown in Figure 3.22 and the intensity distribution as predicted by Equation (50) for $N = 4$ is shown in Figure 3.23. When N is very large the principal maxima will be much more intense in comparison to the secondary maxima. Now we discuss here two points—

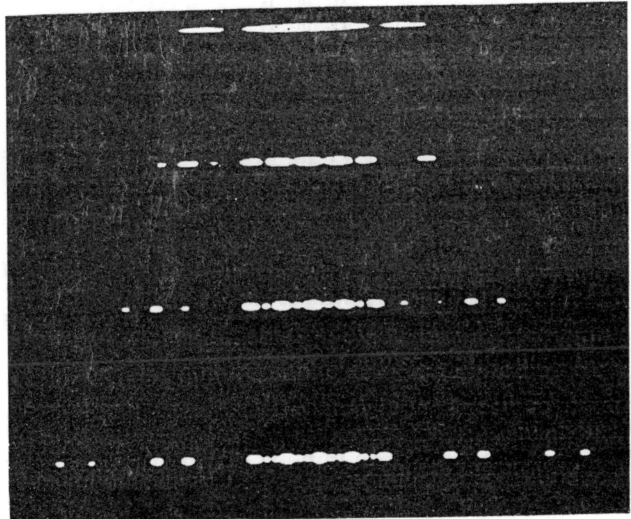

Figure 3.22 : ***The multiple-slit Fraunhofer diffraction patterns corresponding to*** **b** ***= 0.0044 cm, d = 0.0132 cm and $\lambda = 6.328 \times 10^{-5}$ cm. The number of slits are 1, 2, 3 and 4 respectively.***

(i) A particular principal maximum may be absent if it corresponds to the angle which also determines the minimum of the single slit diffraction pattern. This will happen when

$$d\sin\theta = m\lambda \tag{3.56}$$

and

$$b\sin\theta = \lambda, 2\lambda, 3\lambda, \ldots \tag{3.57}$$

are satisfied simultaneously and is usually referred to as a missing order. If in equation b sin θ is close to an integral multiple of λ than the intensity of the corresponding principal maximum will be very weak.

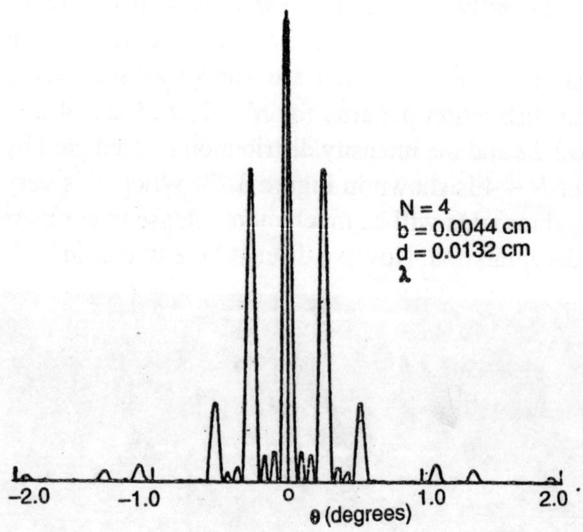

Figure 3.23: ***The intensity distribution corresponding to the four-slit Fraunhofer diffraction pattern as predicted by Equation (5) corresponding to b = 0.0044 cm, d = 0.0132 cm and λ = 6.328 × 10^{-5} cm. The principal maxima occur at * 0.275°, 0.55°, 0.82°, 1.1 °,..,. Notice the (almost) absent third order.***

(ii) In addition to the minima predicted by Equation (3.54), we will also have the diffraction minima (see Equation 3.53) If N is very large, than the number of such minima would be very small.

Width of the Principal Maxima

As we discused earlier the principal maxima of diffraction patter of N th produced slit by m th order is as

$$d\sin\theta_m = m\lambda,\ m = 0, 1, 2, \ldots \tag{3.58}$$

Further, the minima occur at the angles given by Equation (3.55). If $\theta_m + \Delta\theta_{1m}$ represent the angles of diffraction corresponding to the first minimum on either side of the principal maximum, then $\frac{1}{2}\left(\Delta\theta_{1m} + \Delta\theta_{2m}\right)$ is known as the angular half width of the m th order principal maximum. If the value of *N* is large, than $\Delta\theta_{1m} \simeq \Delta\theta_{2m}$, and we can write it $\Delta\theta_m$ too so—

$$d \sin\left(\theta_m \pm \Delta\theta_m\right) = m\lambda \pm \frac{\lambda}{N} \qquad (3.59)$$

But

$$\sin\left(\theta_m \pm \Delta\theta_m\right) = \sin\theta_m \cos\Delta\theta_m \pm \cos\theta_m \sin\Delta\theta_m$$

$$\simeq \sin\theta_m \pm \Delta\theta_m \cos\theta_m \qquad (3.60)$$

Thus Equation (3.59) gives us.

$$\Delta\theta_m \simeq \frac{\lambda}{Nd\cos\theta_m} \qquad (3.61)$$

The above equation shows that as *N* increases the principal maximum became sharper.

The Diffraction Grating

We discussed above about the diffraction pattern, which a system of paralleled equidistant slits produces. An arrangement which a large number of equidistant slits is known as a diffraction grating; the corresponding diffraction pattern is known as the exact positions of the principal maxima in he diffraction pate Since depend on the wavelength, the principal maxima corresponding to different spectral lines (associated with a source) will correspond to different angles of diffraction. Accordingly, to detect the wavelength goating spectrum provides an easy experimental set up. From Equation (3.6) we see that for narrow principal maxima (i.e. sharper spectral lines) a large value of *N* is required. A good quality grating, therefore, requires a large number of slits (typically about 15000 per inch). This is achieved by ruling grooves with a diamond point on an optically groove is ruled, the machine lifts the diamond point and on an optically transparent sheet of material; the grooves act as opaque spaces. After each groove is rules, the machine lifts the diamond point and moves the sheet forward for the ruling of the next groove. We will get the movement of the sheet with the rotation of a screw driving the carriage (carrying it) if the distance between two consecutive groves is very small, the movement of the sheet is obtained with the help of the rotation of a screw which drives the carriage carrying it. Further, one of the important requirements of a good quality is that the lines should be as equally s spaced as possible; consequently, the pitch of the screw must be constant, and it was not until the manufacture of a nearly perfect screw (which was achieved by Rowland in 1882) that the problem of construction of grating was successfully solved. Rowlands arrangements gave 14438 lines per inch, corresponding to $d = 2.54/14438 = 1.759 \times 10^{-4}$ cm. For such a grating, for $\lambda = 6 \times 10^{-5}$ cm, the maximum value of m would be 2, and, therefore, only the first two orders of the spectrum will be observed. The third order spectrum would be visible to if we have $d = 5 \times 10^{-5}$ cm.

If we take a actual grating cast on a transparent film like that of cellulose acetate, than we can produce a commercial grating. An appropriate strength solution of cellulose acetate is poured on the ruled surface and allowed to dry to form a strong thin film, detachable form the parent grating. These impressions of agrating are preserved by mounting the film between two glass sheets. Nowadays gratings are also produced holographically, where one records the interference pattern between two plane or spherical waves. If we compare ruled and holographic gratings then we get that holographic gratings, have much number of lines/cm.

The Grating Spectrum

As we show earlier the position of the principal maxima are as follows—

$$d \sin\theta = m\lambda;\ m = 0, 1, 2, \ldots \tag{3.62}$$

This relation, which is also called the grating equation, can be used to study the dependence of the angle of diffraction θ on the wavelength λ. We will get the zero order principal maximum when $\theta = 0$ of the wavelength. Thus, if, we are using a polychromatic source (e.g. white light) then the central maximum will be of the same colour as the source itself. However, for $m \neq 0$, the angles of diffraction are different for different wavelengths and, therefore, various spectral components appear at different positions. Thus by measuring the angles of diffraction for various colours one can knowing the value of m) determine the values of the wavelengths. For the zeroeth order spectrum where no dispersion occurs the intensity is maximum and it falls according to the value of m increases.

If we differentiate Equation (3.62), we would obtain

$$\frac{\Delta\theta}{\Delta\lambda} = \frac{m}{d\cos\theta} \tag{3.63}$$

From this result we can deduce the following conclusions:

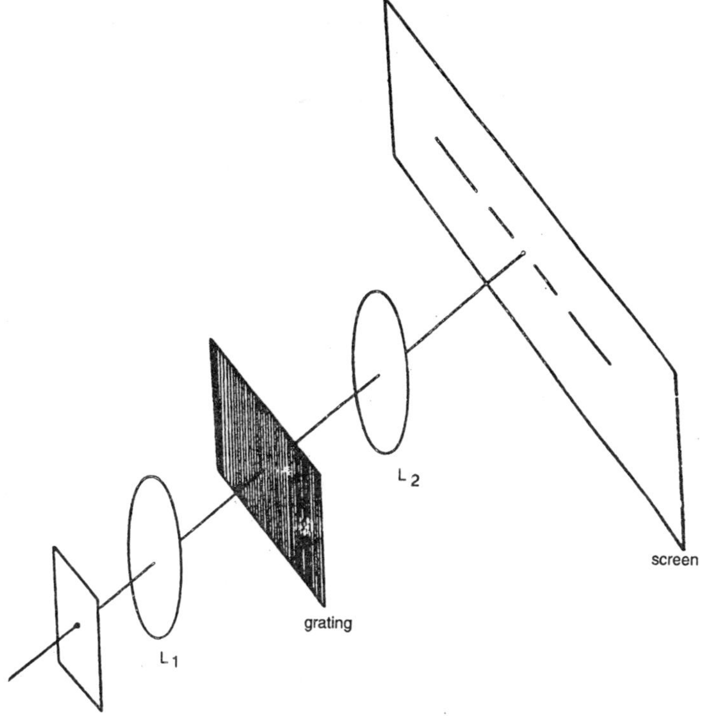

Figure 3.24 : *Fraunhofer diffraction of a plane wave incident normally on a grating.*

(i) Assuming θ to be very small (i.e. cos $\simeq$ 1) we can see that the angle $\Delta\theta$ is directly proportional to the order of spectrum (m) for a given $\Delta\lambda$, so that for a give m, $\Delta\theta/\Delta\lambda$ is a constant. Such a spectrum is known as a normal spectrum and in this the difference in angle for spectral lines is directly proportional to the difference in wavelengths. If θ is large than it is easy to show the dispersion, which is greater at the red end of the spectrum.

(ii) From eq. 3.63 we conclude that $\Delta\theta$ is inversely proportional to d, ánd therefore smaller the grating element, the larger will be the angular dispersion.

We can arrange grating spectrum of a polychromalic source too study which we can see in figure 3.24 and 3.25. In Figure 3.25 we have shown a small hole placed at the focal plane of the lens L_1. A parallel beam of white light emerging from L_1 falls on the grating and the diffraction pattern is observed on the focal plane of the lens L_2. If instead of a hole we have a slit at the focal plane of L_1 (see Figure 3.25)—as it is indeed the case in a typical labo-ratory set up—we would have parallel beams propagating in different directions. As we can see in figure 3.25 the focal plane of the lens L_2 will have a band spectrum.

In a telescope lens L_2 is the objective and we can view the diffraction pattern through an eyepiece. The angles of diffraction for various orders of the grating spectrum can be measured and knowing the value of d, one can calculate the wavelength of different spectral lines.

Resolving Power of a Grating

The resolving power of a grating is the power of distinguish between two nearby spectral lines and is defined by the following equation:

$$R = \frac{\lambda}{\Delta\lambda} \qquad (3.64)$$

where $\Delta\lambda$ is the separation of two wavelengths which the grating can just resolve; the smaller the value of $\Delta\lambda$, the larger the resolving power.

As we discussed, earlier to define the limit of resolution Rayleigh criterian can be used again according to this criterion, if the principal maximum corresponding to the wavelength $\lambda + \Delta\lambda$ falls on the first minimum (on the either side of the principal maximum) of the wavelength λ, then the two wavelengths λ and $\lambda + \Delta\lambda$ are said to be just resolved (see Figure 3.25). If s this common diffraction angle is represented by θ and if we are looking at the *m*th order spectrum, then the two wavelengths λ and $\lambda + \Delta\lambda$ will be just resolved if the following two equations are simultaneously satisfied:

$$d \sin\theta = m(\lambda + \Delta\lambda) \tag{3.65}$$

and

$$d \sin\theta = m\lambda + \frac{\lambda}{N} \tag{3.66}$$

Thus

$$R = \frac{\lambda}{\Delta\lambda} = mN \tag{3.67}$$

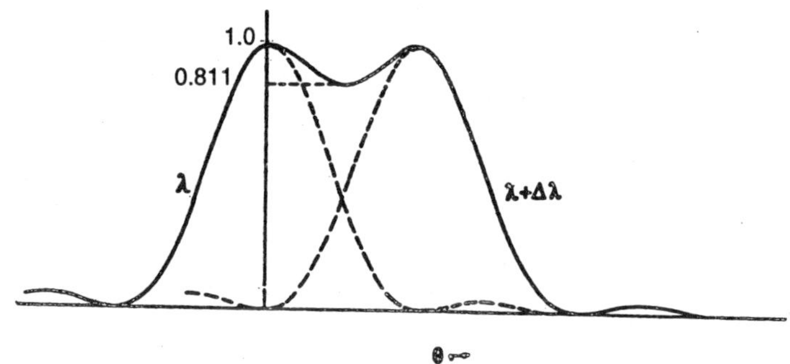

Figure 3.25 : ***The Rayleigh criterion for the resolution of two spectral lines.***

Which implies that the resolving power depends on the total number of lines in the grating—obviously on only those lines which are exposed to the incident beam. In this way we get that in the first order to resolve the D_1 and D_2 lines of Sodium ($\Delta\lambda$ = 6 Å), N must be at least $(5.89 \times 10^{-5})/(6 \times 10^{-8}) \times 1000$.

From eq. (3.67) we can conclude that if N increases than the grating also indefinitely increased; however, for a given width of the grating $D(= Nd)$, as N is increased, d decreases and therefore the maximum value of m also decreases. Thus if d becomes 2.5k, only first and second order spectra will be seen and if it is further reduced to about 1.5* then only the first order spectrum will be seen.

Resolving Power of a Prism

In this section we would come to the result by calculating the resolving power of a prism. Figure 3.26 gives a schematic description of the experimental arrangement for observing the prism spectrum which is determined through the following formula:

$$n(\lambda) = \frac{\sin\frac{A+\delta(\lambda)}{2}}{\sin\frac{A}{2}} \qquad (3.68)$$

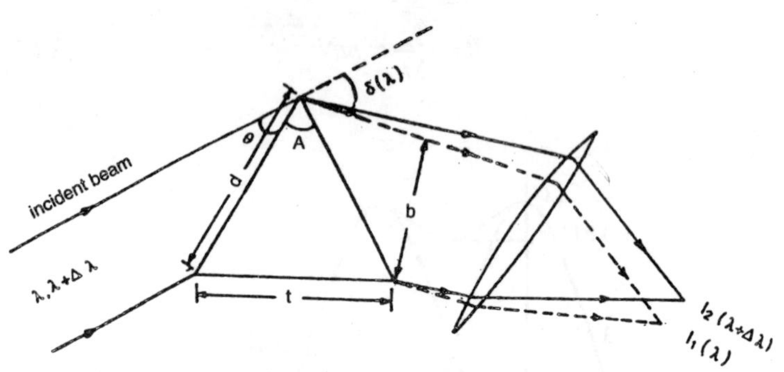

Figure 3.26 : ***The schematic of the experimental arrangement to observe the prism spectrum.*** P_1 ***and*** P_2 ***represent the images corresponding to*** λ ***and*** $\lambda + \Delta\lambda$ ***respectively.***

Here the angle of a prism and angle of minimum deviation is represented by A and δ respectively. We assume that the refractive index decreases with λ (which is usually the case) so that δ also decreases with λ. In Figure 3.26 the points P_1 and P_2 represent the images corresponding to λ and $\lambda + \Delta\lambda$ respectively. We are assuming that $\Delta\lambda$ is small so that the same position of the prism corresponds to the minimum deviation position for both wavelengths. In an actual experiment one usually has a slit source (perpendicular to the plane of the paper) forming line images at

P_1 and P_2. Since the faces of the prism are rectangular, the intensity distribution will be similar to that produced by a slit of width b. If $m = 1$, means the lines resolved the first diffraction minimum of λ should fall at the central maximum of $\lambda + \Delta\lambda$. Now we get—

$$\Delta\lambda \approx \frac{\lambda}{b} \tag{3.69}$$

In order to express $\Delta\delta$ in terms of $\Delta\lambda$, we differentiate Eq. 3.68

$$\frac{dn}{d\lambda} = \frac{1}{\sin\frac{A}{2}} \cos\left[\frac{A+\delta(\lambda)}{2}\right] \frac{1}{2} \frac{d\delta}{d\lambda}$$

Thus

$$\Delta\delta = \frac{2\sin\frac{A}{2}}{\cos\frac{A+\delta(\lambda)}{2}} \frac{dn}{d\lambda} \Delta\lambda$$

From equation (3.26) we get—

$$\theta = \frac{1}{2}\left[\pi - (A+\delta)\right]$$

or

$$\sin\theta = \frac{b}{a} = \cos\frac{A+\delta}{2}$$

where the length a is shown in the figure. Further

$$\sin\frac{A}{2} = \frac{t/2}{a}$$

where t is the length of the base of the prism. Thus

$$\Delta\delta \approx \frac{t}{b} \frac{dn}{d\lambda} \Delta\lambda \tag{3.70}$$

From equation (3.70) , on substitution we would get for resolving power

$$R = \frac{\lambda}{\Delta\lambda} = t\frac{dn}{d\lambda} \tag{3.71}$$

To calculate the dependence of refractive index on wavelength can be properly described by the cauchy formula—

$$n = A + \frac{B}{\lambda^2} + \frac{C}{\lambda^4} + \ldots \tag{3.72}$$

Thus

$$\frac{dn}{d\lambda} = -\left[\frac{2B}{\lambda^3} + \frac{4C}{\lambda^5} + \ldots\right] \tag{3.73}$$

the negative sign implying that the refractive index decreases with increase in wavelength. Suppose for telescope crown glass—

$A = 1.51375$, $B = 4.608 \times 10^{-11}$ cm^2, $C = 6.88 \times 10^{-22}$ cm^4

For $\lambda = 6 \times 10^{-5}$ cm we have

$$\frac{dn}{d\lambda} \simeq \left[4.27 \times 10^2 + 3.54\right]$$

$$\simeq -4.30 \times 10^2 \text{cm}^{-1}$$

Thus, for $t \simeq 2.5$ cm we have

$$R = \frac{\lambda}{\Delta\lambda} \simeq 1000$$

which is an order of magnitude less than for typical diffraction gratings with 15,000 lines.

Oblique Incidence

Now we know that normally waves insident on the grating. For experimental setting it is quite difficult to achieve the condition of normal incidence to a great precision and it is easily seen that slight deviations from normal incidence will introduce considerable errors. It is, therefore, more practical to consider the more general oblique incidence case we can measure wavelenght for the method of minimum deviation similar to prisms.

If the angle of incidence is i, then the path difference of the diffiacted rays from two corresponding points in adjacent slits will be $d \sin \theta + d \sin i$. Thus, principal maxima will occur when

$$d(\sin\theta + \sin i) = m\lambda \tag{3.74}$$

or

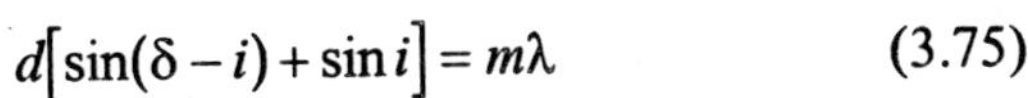

$$d[\sin(\delta - i) + \sin i] = m\lambda \tag{3.75}$$

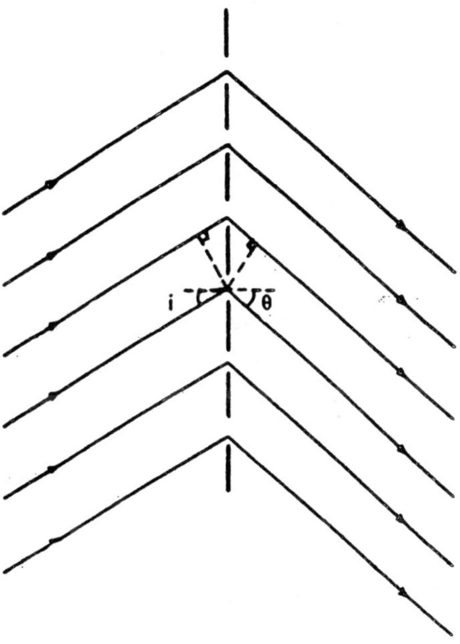

Figure 3.27 : ***Diffraction of a plane wave incident obliquely on a grating.***

If δ is the angle of deviation (i. e. $i + \theta$) than δ minimum would be

$$\frac{d}{di}[\sin(\delta - i) + \sin i] = 0 \tag{3.76}$$

i. e.,

$$-\cos(\delta - i) + \cos i = 0 \tag{3.77}$$

or

$$i = \delta - i = \theta \tag{3.78}$$

Hence, at the position of minimum deviation, the grating condition becomes

$$2d \sin\frac{\delta}{2} = m\lambda \tag{3.79}$$

To get the minimum deviation position in a manner similar to that used in the case of a prism and since the adjustments are relatively simpler, Thus we get a more suitable method for the determination of λ.

X-Ray Diffraction

The light which is visible is an electromagnetic wave with wave length approximately between 40000 Å–7000 Å. X-rays are also electromagnetic waves whose wavelengths are ~1 Å. Obviously, it is extremely difficult to make slits which are narrow enough for the study of X-ray diffraction patterns. Since the interatomic spacings in a crystal are usually of the order of Angstroms, one can use it as a three-dimensional diffraction, grating for studying the diffraction of X-rays. In fact, according to verma and srivastava (1980) X-rays used to study crystal structures. In an ideal crystal, the atoms or molecules arrange themselves in a regular three-dimensional pattern which can be obtained by a three dimensional repetition of a certain unit pattern. This simplest volume which has all the characteristics of the whole crystal and which completely fills space is called the unit cell. Suppose we take many planes in the normal three dimensional periodic arrangement Millerindices are universally used as a system of notation for planes within a crystal. They specify the orientation of planes relative to the crystal axis without giving the position of the plane in space with respect to the origin. These indices are based on the intercepts of a plane with the three crystalxes, each intercept with an axis being measured in terms of unit cell dimensions (*a*, *b* or *c*) along that axis. To get the miller indices of a plane, we have to follow the following procedure–

1. Find the intercepts (of the plane nearest to the origin) on the three axes and express them as multiple or fractions of the unit cell dimension.
2. Take the reciprocals of these numbers and multiply by the LCM of the denominators.
3. Enclose in parentheses.

For example, a (111) plane intercepts all three axes atone unit distance (see Figure 3.28); a (211) plane intercepts the three axes at ½, 1 and 1 unit distances (see Figure 3.28). Similarly, a (110) plane intercepts the z-axis at ∞. Miller indices can also be negative, the minus sign is shown above the digit like (111). Figure 3.29 shows the planes characterized by the Miller indices (111) in a simple cubic lattice.

Suppose *a* monochromatic beam of X-ray falls on crystal. In Figure 3.30 the horizontal dotted lines represent a set of parallel crystal planes with Miller indices (*hkl*). W_1 W_2 and W_3 W_4 represent the incident and reflected wavefronts respectively. Obviously, the secondary wavelets emanating from the points *A*, *B* and *C* are in phase on W_3 W_4 and the waves emanating from the points A_1, B_1 and C_1 will also be in phase on W_3 W_4 if

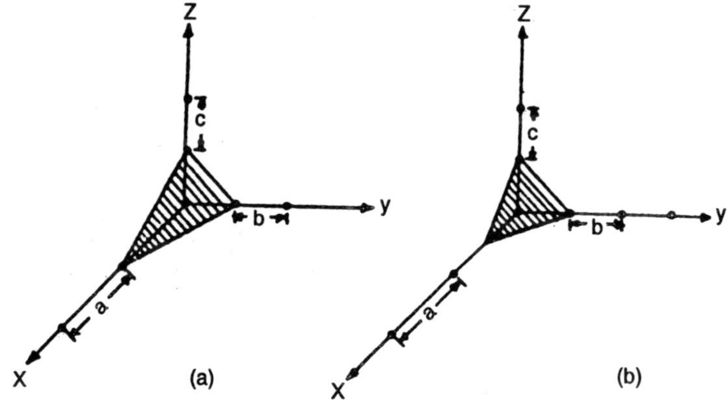

Figure 3.28 : ***(a) The (111) plane intercepts all three axes at one unit distance of each axial dimension. (b) The (211) plane intercepts the three axes at 1/2, 1 and 1 unit distances.***

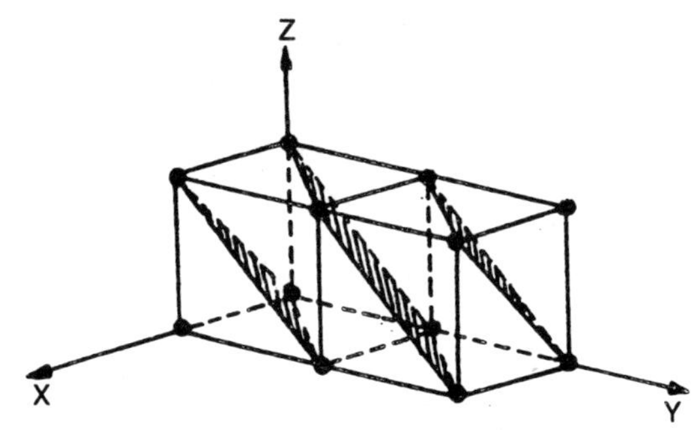

Figure 3.29: ***Planes characterized by the Miller indices (111) in a simple cubic lattice.***

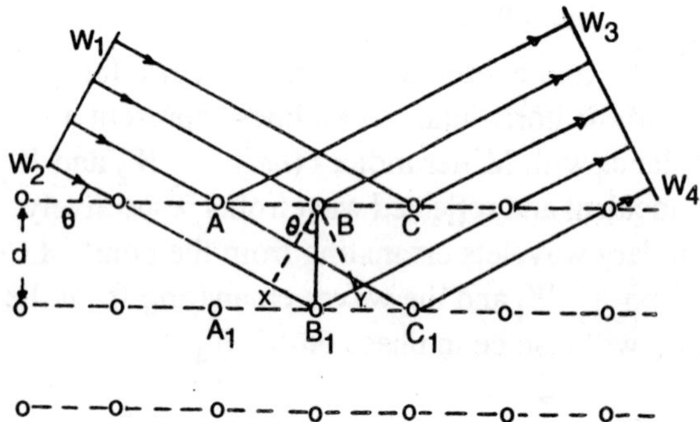

Figure 3.30 : ***Reflection of a plane wave by a set of parallel crystal planes characterized by the Miller indices (hkl). When the Bragg condition 2d sin θ = mλ is satisfied, the waves scattered fro-, different rows will be in phase.***

$$XB_1 + B_1Y = m\lambda, m = 1, 2, 3, \ldots \tag{3.80}$$

or when

$$2d_{hkl} \sin\theta = m\lambda \tag{3.81}$$

In above equation *dhkl* represent the interplanar spacing between crystal planes of indices, m = 1. 2, 3,.... it is known as the order of diffraction. And θ represents the glaing angle. This equation is known as Bragg's law and gives the angular positions of the reinforced diffracted beams in terms of the wavelength λ of the incoming X-rays and of the interplanar spacings d_{hkl} of the crystal planes. When the condition expressed by Equation (3.81) is not satisfied, destructive interference occurs and no reinforced beam will be produced. Constructive interference occurs when the condition given by Equation (3.81) is satisfied leading to peaks in the intensity distribution. In case of solids, which is *a* crystallized cubic structure the interplanar spacing d_{hkl} between two closest parallel planes with Miller indices would be as follows.

$$d_{hkl} = \frac{a}{\sqrt{h^2 + k^2 + l^2}} \tag{3.82}$$

Where a represents the lattice constant. Thus knowing the Miller indices, we can find d_{hkl} and from Bragg's law, we can

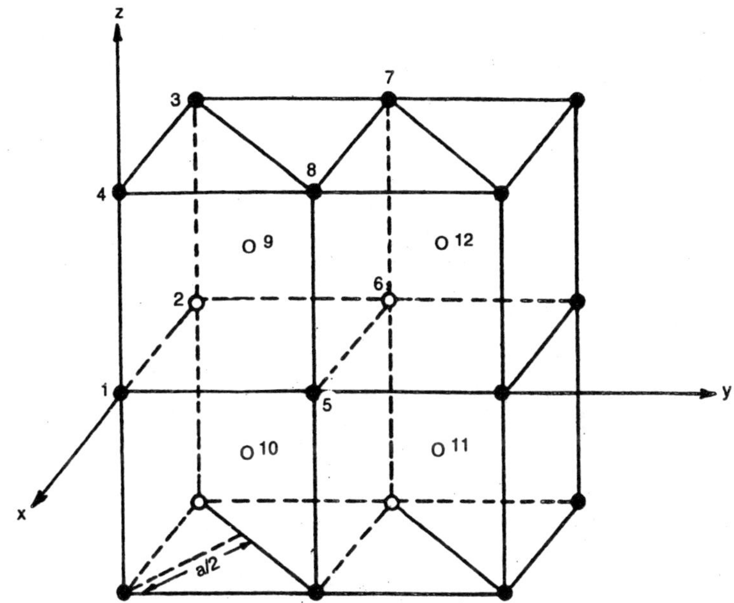

Figure 3.31. A body centred cubic (bcc) lattice. The (110) planes are separated by $a\sqrt{2}$.

determine the value of θ at which Bragg's equation can be satisfied.

If we consider cubic structure we get three types simple cubic, body centred cubic and face centred cubic. Figure 3.39 how a simple cubic structure (abbreviated as SC) in which the atoms are at the corners of a cube which forms what is known as a unit cell. The crystal is built up by the repetition of this unit cell in three dimensions. In addition, if there is an atom at the centre of each cube (shown as 9, 10, 11 and 12 in Figure 3.31 arrangement is known as a body-centred cubic (BCC) structure. The shaded planes shown in Figure3.31 characterized by the Miller indices (110) and using Eq. (3.82) we find that the distance between two adjacent planes is $a / \sqrt{2}$ which can be verified by simple geometry. On the other hand, if instead of having an atom at the centre of the cube there is an atom at the centre of each of the six faces of the cube (see Figure 3.32 we will have the face centred cubic (FCC) structure. Copper, silver and gold crystallize in the FCC form with the lattice parameter a = 3.61 Å, 4.09 Å and 4.08 Å respectively. Some metals are crystallize in the *BBC* form, e. g.- sodium with α = 4.29 Å and barium with α = 5.06 Å and tungsten with α = 3.16 Å.

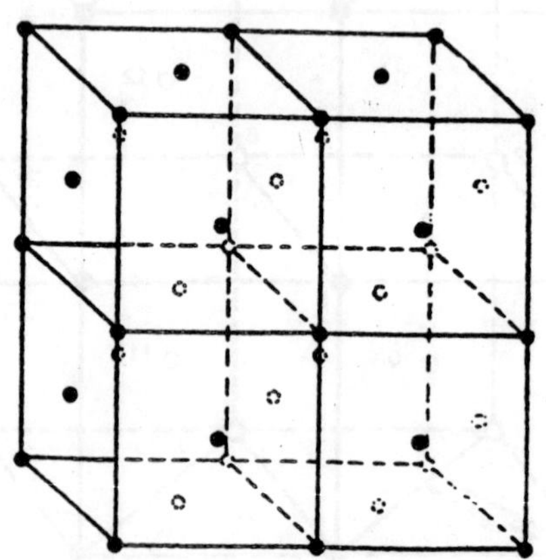

Figure 3.32 : ***A face centred cubic (fcc) lattice.***

Just as there are optical missing orders of a diffraction grating, there are structural extinction of X-ray reflection from a crystal. In case of normal cubic structures all planes are ready for reflection However for the BCC structure, diffraction occurs only on planes whose Miller indices when added together total to an even number. Thus for the BCC structure, the principal diffracting planes for a first order diffraction are (110), (200), (211) (and other similar planes), etc. where $h + k + 1$ is an even number. For *FCC* crystal the all even or all odd Miller indices is the principal diffracting planes.

Experimental methods of X-ray Diffraction

From Bragg's law $2d_{hkl} \sin \theta = m\lambda$ it is clear that there are essentially three methods which can be used so that Bragg's formula can be satisfied:

	λ	θ
Racating crystal method	Fixed	Variable (intentional)

Powder method	Fixed	Variable (inherent)
Lave method	Variable	Fixed

In case of monochomatic X-rays bragg's formula not applicable for an arbitrary value of θ. Hence one rotates the single crystal so that reflection can occur for a discrete set of θ values. This method can only be employed if single crystals of reasonable size are available. If this is not the case, one can still use monochromatic X-rays provided the sample is in the powder form

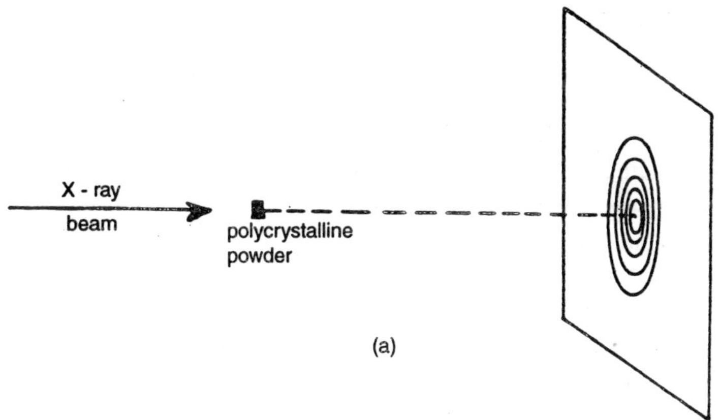

Figure 3.33 : ***When a monochromatic X-ray beam falls on a polycrystalline sample one obtains the Debye-Scherrer rings.***

so that there are always enough crystallites of the right orientation available to satisfy the Bragg relation. A powder will consist of a large number of randomly oriented micro-crystals; each micro-crystal is essentially a single crystal. As the X-ray beam passes through such a polycrystalline material, the orientation of any given set of planes, with reference to the X-ray beam, changes from one micro-crystal to the other. Thus, corresponding to any given set of planes there will be a large number of crystals for which Bragg's condition will be satisfied, and on the photographic plate one will obtain concentric rings [see Figure3.33]; each ring will correspond to a particular value of d_{hkl} and a particular value of m. The appearance of the circular rings can be understood as follows. Consider a set of planes parallel to AB [see Figure 3.34].

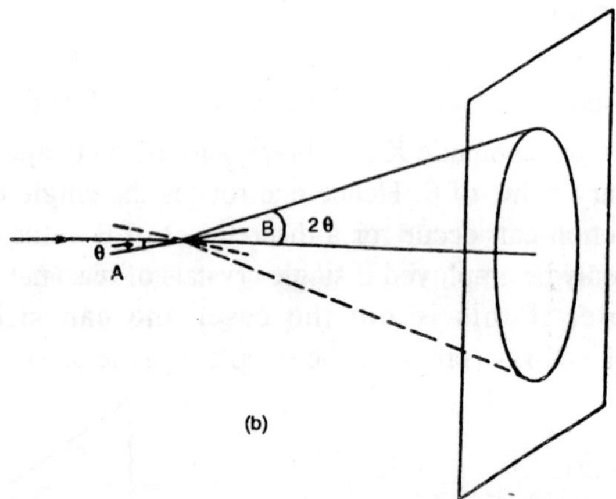

Figure 3.34 : *Diffraction from a polycrystalline sample.*

The glancing angle a is assumed to satisfy the Bragg condition. If the microcrystal is rotated about the direction of the incident X-ray beam, then for all positions of the microcrystal, the glacing angle will be the same for these sets of planes. Thus the direction of the diffracted beam would be different for each position of the microcrystal. But it will always lie on the surface of the cone

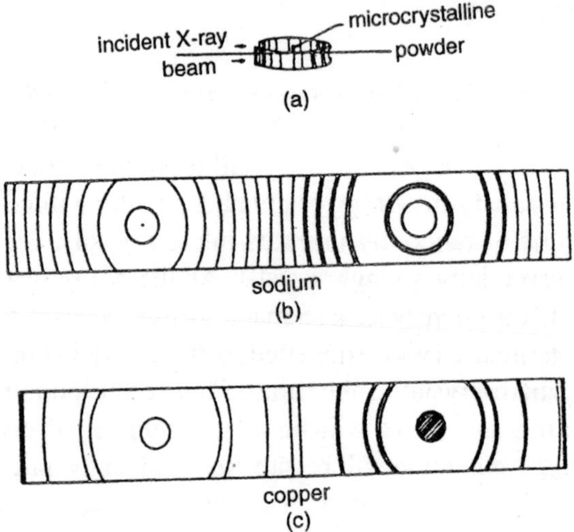

Figure 3.35 : *(a) While using the powder method the photographic film is kept in a cylindrical form as shown in the figure. (b) and (c) represent schematic diffraction patterns for sodium and cooper respectively.*

whose semi-vertical angle will be 2θ. Consequently, one will obtain concentric circular rings on the -photographic plate; these rings are known as Debye-Scherrer rings.

In fig 3.35 power method is shown in which photographic film is in a cylindrical form surrounded with polycrystalline sample. Each Debye-Scherrer ring will produce an arc on the film, and when the film is unrolled, one obtains a pattern as shown in Figure 3.35 (b) and (c). From the position of these arcs one can calculate θ and thus determine the interplanar spacing. From a study of the interplanar spacings one can determine the crystal structure [for more details, one may look up Verma and Srivastava (1982)]. Although a powder camera with an enclosed film strip has been extensively used in the past, modern X-ray crystal analysis uses an X-ray diffractometer which has a radiation counter to detect the angle and intensity of the diffracted beam.

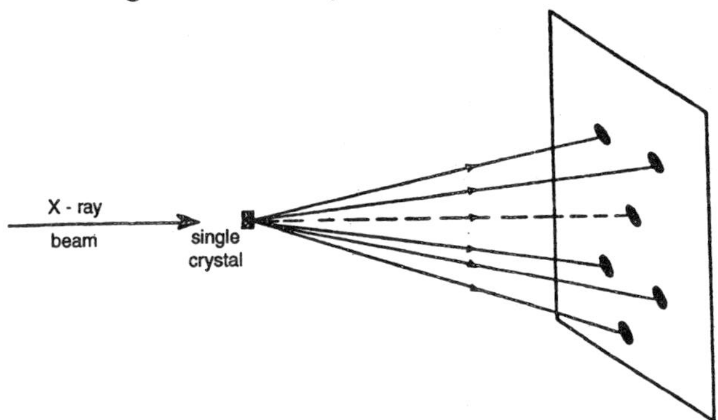

Figure 3.36 : ***When a polychromatic X-ray beam falls on a single crystal, one obtains Laue spots. Each set of planes chooses its own wavelength.***

In fig 3.36 laue method is shown in which we use a single crystal held stationary in a beam of while X-rays. Each set of planes then chooses its own wavelength to satisfy the Bragg relation (see Figure 3.36).

In order to calculate the angles of diffraction we substitute Equation (3.82) in the Bragg's law [Eq. (3.81)] to obtain

$$\frac{2a}{\sqrt{h^2 + k^2 + l^2}} \sin\theta = m\lambda \qquad (3.83)$$

We restrict ourselves only to first order reflections (m = 1); higher order reflections are usually rather weakfinally the equation comes as—

$$\sin\theta = \frac{\lambda}{2a}\sqrt{N} \tag{3.84}$$

where $N = h^2 + k^2 + l^2$

In case of *a* cubie lattice all values of (*hkl*) would be possible as follows—

$$N = 1, 2, 3, 4, 5, 6, 7, \ldots \quad \text{(SC)} \tag{3.85}$$

Similarly, for a BCC lattice $h + k + 1$ must be even implying

$$N = h^2 + k^2 + 1^2 = 2, 4, 6, 8, 10, 12, 14, 16, 18, 20, 22, \ldots \quad \text{(BCC)} \tag{3.86}$$

Finally, for an FCC lattice, Miller indices are either all even or all odd implying

$$N = h^2 + k^2 + l^2 = 3, 4, 8, 11, 12, 16, 19, 20, 24, 27, \ldots \quad \text{(FCC)} \tag{3.87}$$

To calculate the different values of θ we should know the value of λ. For example, if we consider λ = 1.540 Å and 1.544 Å (corresponding to the $CuK_{\alpha 1}$ and $CuK_{\alpha 2}$ lines) then for sodium (which is a BCC structure with a = 4.2906 *), the various values of θ are

(14.70°, 14.74°), (21.03°, 21.09°), (26.08° 26.15°), (30.50°, 30.59°), (34.58°, 34.68°), (38.44°, 38.56°), (42.18°, 42.32°), (45.88°, 46.03°), (49.59°, 49.76°), (53.38°, 53.58°), (57.33°, 57.56°), (61.54°, 61.82°), (66.22°, 66.56°), (79.41°, 80.23°).

The two values inside the parentheses correspond to the two wavelengths 1.540 Å and 1.544 Å respectively. We get double lines for every plane because of the two wavelengths. If is resolvable only at higher scattering angles. Similarly one can consider reflections from other structures. Each value of θ will give rise to a Debye-Scherrer ring shown in Figures 3.34 (a), 3.35 (b) and 3.36 (c).

Finally, we shout mention that the intensity of the diffracted wave depends on the number of atoms per unit area in the plane under consideration. For example, corresponding to the (1 1 0) [see Figure 3.32] and the (2 2 2) planes passing through a BCC lattice, there will be one atom and two atoms, respectively, in an area a^2. *Finally* we get that the intensity of the diffracted wave is higher in first case than second.

The Self-Focusing Phenomenon

We can investigate many non-linear optical phenomena with the help of intense laser beams. One such nonlinear phenomenon is the effect on the propagation of a light beam due to the dependence of the refractive index on the intensity of the beam. Finally it get self focused of the beam.

4

Dispersion

Dispersion is concerns with the speed of light in different material substances and variation of wavelength. Since the speed is *c*/*n*, any change in refractive index *n* entails a corresponding change of speed. As we have studied that the dispersion of colour which occurs upon refraction at a boundary between two different substances is direct evidence of the dependence of the *n*'s on wavelength. The refractive index determination of a prism which measures the deviations of several spectral lines and hence the speed is the function of wave length.

Dispersion of a Prism

As we can see in figure 4.1 of a spectrometer measures the angles of emergence (θ) of the various wavelengths when a ray traverses a prism. The rate of change, $d\theta/d\lambda$, is called the *angular dispersion* of the prism. It can be conveniently represented as the product of two factors, by writing

$$\frac{d\theta}{d\lambda} = \frac{d\theta}{dn}\frac{dn}{d\lambda} \tag{4.1}$$

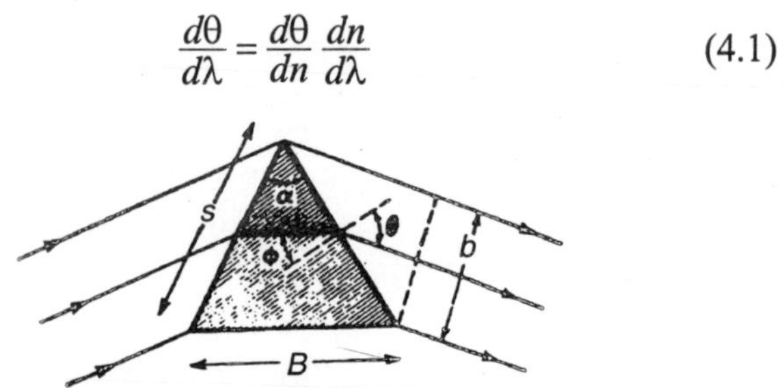

Figure. 4.1 : *Refraction by a prism at minimum deviation.*

The first property of the prism can be detected by geometrical consideration and second property is a characteristic feature of the prism as its dispersion. Before considering the latter quantity, let us evaluate the geometrical factor $d\theta/dn$ for a prism, in the special case of minimum deviation.

For a given angle of incidence on the second face of the prism, we differentiate Snell's law of refraction $n = \sin\theta / \sin\phi$, regarding $\sin\phi$ as a constant, and obtain

$$\frac{d\theta}{dn} = \frac{\sin\phi}{\cos\theta}$$

This is not, however, the value which requires the rate of change of θ for a fixed direction of the rays incident on the first face. However symmetry in the case of minimum deviation will have the two faces, so the total rate of change will be just twice the above value. We then have

$$\frac{d\theta}{dn} = \frac{2\sin\phi}{\cos\theta} = \frac{2\sin(\alpha/2)}{\cos\theta}$$

where α is the refracting angle of the prism. The result becomes still simpler when expressed in terms of lengths rather than angles. As shown in figure 4.1, S B and b represent the lengths. So—

$$\frac{d\theta}{dn} = \frac{2s\sin(\alpha/2)}{s\cos\theta} = \frac{B}{b} \tag{4.2}$$

Hence the required geometrical factor is just the ratio of the base of the prism to the linear aperture of the emergent beam, a quantity not far different from unity. Thus the angular dispersion would be—

$$\frac{d\theta}{dn} = \frac{B}{b}\frac{dn}{d\lambda} \tag{4.3}$$

In connection with this equation, it is to be noted that the equation for the chromatic resolving power follows very simply from it upon the substitution of λ / b for $d\theta$.

Normal Dispersion

Suppose the second factor in equation (4.1) some facts about the variation of n with λ are known facts. Measurements for some typical kinds of glass give the results shown in Tables 4.1 and 4.2. If any set of values of *n* is plotted against wavelength, a curve like one of those in Figure 4.2 is obtained. The curves found for prisms of different optical materials will differ in detail but will all have the same general shape. Normal dispersion is represented by these curves and show following facts—

1. The index of refraction increases as the wavelength decreases.
2. The rate of increase becomes greater at shorter wavelengths.
3. For different substances the curve at a given wavelength is usually steeper the larger the index of refraction.
4. The curve for one substance cannot in general be obtained from that for another substance by a mere change in the scale of the ordinates.

Table 4.1 : Refractive Index for Several Transparent Solids

	Colour wavelength λ, Å					
Substance	**Violet 4100**	**Blue 4700**	**Green 5500**	**Yellow 5800**	**Orange 6100**	**Red 6600**
Crown glass	1.5380	1.5310	1.5260	1.5225	1.5216	1.5200
Light flint	1.6040	1.5960	1.5910	1.5875	1.5867	1.5850
Dense flint	1.6980	1.6836	1.6738	1.6670	1.6650	1.6620
Quartz	1.5570	1.5510	1.5468	1.5438	1.5432	1.5420
Diamond	2.4580	2.4439	2.4260	2.4172	2.4150	2.4100
Ice	1.3170	1.3136	1.3110	1.3087	1.3080	1.3060
Strontium titanate ($SrTi0_3$)	2.6310	2.5106	2.4360	2.4170	2.3977	2.3740
Rutile ($Ti0_2$), E ray	3.3408	3.1031	2.9529	2.9180	2.8894	2.8535

Table 4.2 : Refractive Indices and Dispersions for Several Common Types of Optical Glass

Unit of dispersion 1/Å × 10^{-5}

Wavelength	Telescope crown		Borosilicate crown		Barium flint		Vitreous quartz	
λ, Å	n	$-\frac{dn}{d\lambda}$	n	$-\frac{dn}{d\lambda}$	n	$-\frac{dn}{d\lambda}$	n	$-\frac{dn}{d\lambda}$
C 6563	1.52441	0.35	1.50883	0.31	1.58848	0.38	1.45640	0.27
6439	1.52490	0.36	1.50917	0.32	1.58896	0.39	1.45674	0.28
D 5890	1.52704	0.43	1.51124	0.41	1.59144	0.50	1.45845	0.35
5338	1.52989	0.58	1.51386	0.55	1.59463	0.68	1.46017	0.45
5086	1.53146	0.66	1.51534	0.63	1.59644	0.78	1.46191	0.52
F 4861	1.53303	0.78	1.51690	0.72	1.59825	0.89	1.46318	0.60
G 4340	1.53790	1.12	1.52136	1.00	1.60367	1.23	1.46690	0.84
H 3988	1.54245	1.39	1.52546	1.26	1.60870	1.72	1.47030	1.12

According to these facts it concluded that violet is more deviated than the red in refraction by a transparent substance. The second fact can also be expressed by saying that the dispersion increases with decreasing wavelength. This follows because the dispersion $dn/d\lambda$ is the slope of the curve (its negative sign is usually disregarded), which increases regularly toward smaller λ. An important consequence of this behaviour of the dispersion is that in the spectrum formed by a prism the violet end of the spectrum is spread out on a much larger scale than the red end. The spectrum is therefore far from being a normal spectrum. This will be clear from Figure 4.3 in which the spectrum of helium is shown diagrammatically as given by flint-and crow-glass prism and by a grating used under the proper conditions to give a normal spectrum. If we campare wavelength scale, with the uniform scale of the normal spectrum than we would found wavelength scale is compressed toward the red end.

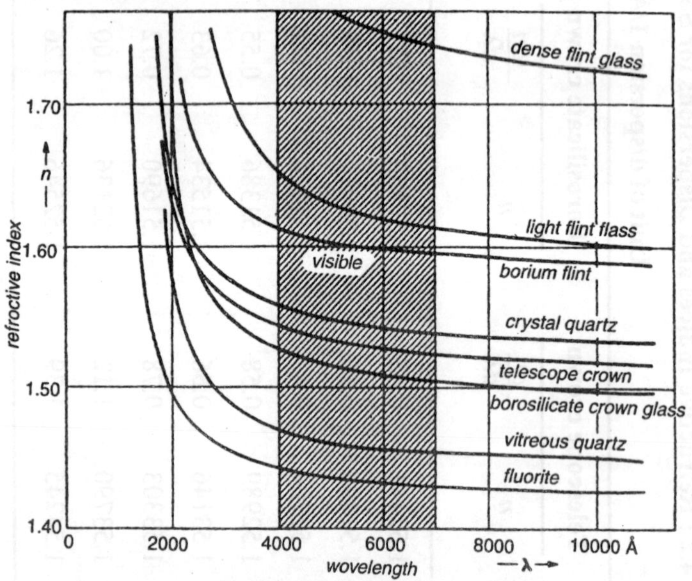

Figure 4.2 : *Dispersion curves for several different materials commonly used for lenses and prisms.*

According to third fact, substance of higher index of refraction require greater dispersion ($dn/d\lambda$). Thus, comparing (a) and (*b*) in Figure 4.3, the flint glass has the higher index of refraction and gives a longer spectrum because of its greater dispersion. To compare the *relative* spacing of the lines in (*b*) with those in (a), the spectrum from crown glass has been enlarged, in (*c*), to have the same overall length between the two lines λ3888 and λ6678. When this is done, it is seen that there is not complete agreement with the lines of (*a*). In fact, the spectra from prisms of different substances will never agree exactly in the relative spacing of their spectrum lines. This is a consequence of the fourth of the above

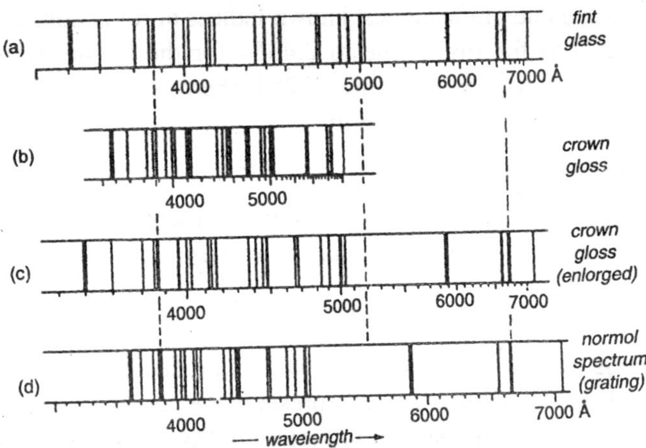

Figure. 4.3 : *Comparison of the helium spectrum produced by flint-glass and crown-glass prism spectrographs with a normal spectrum.*

facts, according to which the shape of the dispersion curve is different for every substance. The curve for flint glass in Figure 4.2 has a greater slope at the violet end, relative to that in the red, than does the curve for crown glass. As a result the dispersion of different substances is irrational, because the relation between the curves is not easy.

Every colourless substance show normal dispersion in the visible region. The magnitude of the index of refraction may be quite different in various substances, but its change with wavelength always shows the characteristics described above. In general, the greater the density of the substance the higher its index of refraction and its dispersion. As the density of flint glass is 2.8 which is higher than 2.4 than ordinary crown glass. Water has a smaller n and $dn/d\lambda$, while in a very light substance like air n is practically unity and $dnld\lambda$ very nearly zero. For air n = 1.000276 for red light (Fraunhofer's C line), rising to only 1.000279 for blue light (F line). This rule relating density to index of refraction is only a qualitative one, and many exceptions are known. For instance, ether has a higher index than water (1.36 as compared with 1.33), yet it is less dense, as is shown by the fact that ether floats on the surface of water. As we discussed above the relation between high dispersion and high index is only rough. Diamond has a density of 3.52 and one of the highest known indices of refraction, varying from 2.4100 for the C line to 2.4354 for the F line. Here the difference in dispersion is only 0.0254 where as a dense flint glass gives 0.05 for the same quantity.

Cauchy's Equation

In 1836 cauchy made a equation to represent the curve of normal dispersion. His equation may be written

$$n = A + \frac{B}{\lambda^2} + \frac{C}{\lambda^4}$$

where A, B, and C are constants which are characteristic of any one substance. This equation represents the curves in the visible region, such as those shown in Figure (4.2) with considerable accuracy. To find the values of the three constants, it is necessary to know values of n for three different λ's. Then three equations may be set up which, when solved as simultaneous equations, give A, B, and C. For some purposes it is sufficiently accurate to include only the first two terms and the two constants can be found from values of n at only two λ's. Thus we have two equation given by cauchy—

$$n = A + \frac{B}{\lambda^2} \tag{4.4}$$

from which the dispersion becomes, by differentiation

$$\frac{dn}{d\lambda} = -\frac{2B}{\lambda^3} \tag{4.5}$$

from above we can say that the variation in dispersion is approximate to the inverse cube of the wave length. At 4000 Å it will be about 8 times as large as at 8000 Å. The decreasing slop of the dispersion curve is like minus sign.

The theoretical reasoning on which Cauchy based his equation was later shown to be false, so that it is to be considered essentially as an empirical equation. Nevertheless it holds very satisfactorily for case of normal dispersion and is a useful equation from a practical standpoint. Later we will study about the special case of a more complete equation with sound theoretical foundation.

Anomalous Dispersion

Cauchy's equation is not applicable on infrared region of the spectrum, in measuring index of refraction as a transparent substance like quartz show deviation in disperse curve from the cauchy equation. The deviation is always of the type illustrated in Figure 4.4 where, starting at the point *R*, the index of refraction is seen to fall off more rapidly than required by a Cauchy equation that represents the values of n for visible light (between *P* and *Q*) quite accurately. This equation predicts a very gradual decrease of *n* for large values of λ (broken curve) the index approaching the limiting value *A* as λ approaches infinity [Eq. 4.4] (In contrast to this, the measured value of *n* first decreases more and more rapidly as it approaches a region in the infrared where light ceases to be transmitted at all. This is an absorption band a region of selective absorption, the position of which is characteristic of the material. Within the absorption band, *n* cannot usually

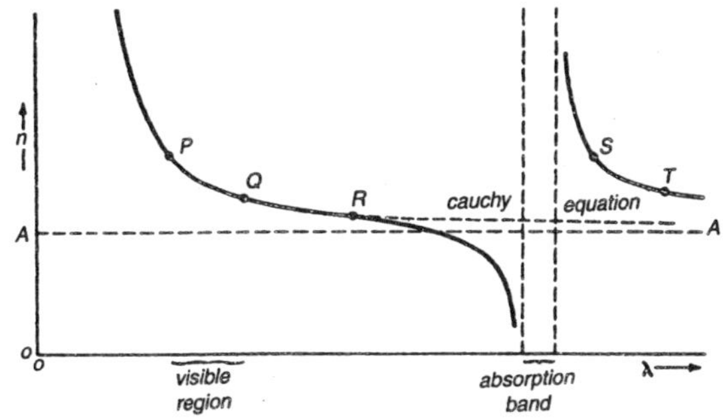

Figure. 4.4 : ***Anomalous dispersion of a transparent substance like quartz in the infrared.***

be measured because the substance will not transmit radiation of this wavelength. On the long-wavelength side of the absorption band the index is found to be very high, decreasing at first rapidly and then more slowly as we go farther beyond the absorption band. Over the range from *S* to *T*, the Cauchy equation will again represent the data, but with different constants. As we can see in fig the constant *A* is larger.

If dispersion curve has deviations and it crosses absorption band, than it is called anomalous dispersion. The dispersion is anomalous because in this neighbourhood the longer wavelengths have a higher value of *n* and are more refracted than the shorter ones. The dye fuchsin and iodine vapor having absorption bands in visible region. A prism formed of such a substance will deviate the red rays more than the violet, giving a spectrum which is very different from that formed by a substance having normal dispersion. When it was later discovered that transparent substances like glass and quartz possess regions of selective absorption in the infrared and ultraviolet, and therefore show anomalous dispersion in these regions, the term "anomalous" was seen to be inappropriate. Selective absorption at same wavelength is necessary to exist for any substance and hence the phenomenon, far from being anomalous, is perfectly general. The so-called normal dispersion is found only when we observe those wavelengths which lie between two absorption bands, and fairly far removed from them. Nevertheless, the term "anomalous dispersion" has been retained, although it has little more than historical significance.

In 1904 R. W. Wood did a experiment to show the anomalous dispersion of sodium vapor in the neighbourhood of the yellow *D* lines. White light when passed through sodium vapor undergoes strong selective absorption at these lines, which form a close doublet of wavelengths 5890 and 5896 Å. At wavelengths far removed from these values, the index of refraction is only very slightly greater than unity, as we expect for a gas. With sodium vapor of appreciable density, the index of refraction in the neighbourhood of the *D* lines passes through a region of anomalous dispersion (strictly speaking, two regions very close together) of the type show in Figure 4.4. As the *D* lines are approached from the side of shorter wavelengths, n begins to decrease rapidly, becoming much less than unity as we get very close to them. In other case it starts from high and finally comes to λ and than increases further.

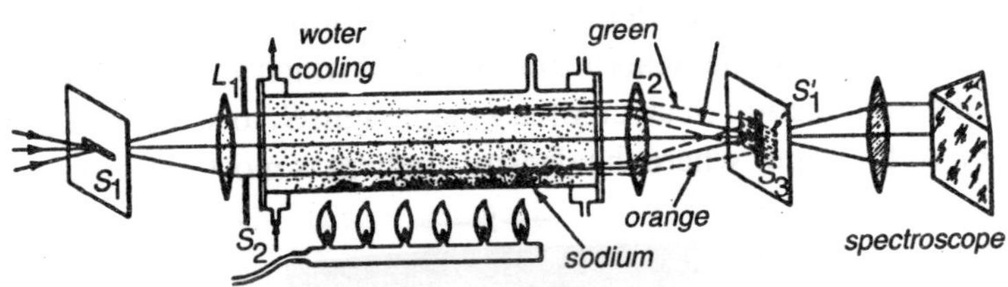

Figure. 4.5 : *Experimental arrangement for observing the anomalous dispersion of sodium vapor.*

We can produce sodium vapor similar to the prism, by vaporizing the metal is a evaluated tube, is heated from the bottom. The arrangement is shown in Figure 4.5. A number of lumps of metallic sodium are placed along the bottom of a steel tube provided with water-cooled glass windows at the ends and an outlet for pumping. White light from a narrow horizontal slit S_1 is rendered parallel by the lens L_1 and after passing through the tube, forms a horizontal image S_1' *on* the vertical slit S_3 of an ordinary prism spectroscope. When the sodium tube is cold, S_1' will be a sharp, white image, illuminating one point of the spectroscope slit, and this will be spread out into a narrow horizontal continuous spectrum in the focal plane of the spectroscope camera. If the tube is evacuated to about 2 cm pressure and the sodium is heated by the row of gas burners, it will vaporize slowly, the vapor diffusing upward through the residual gas in the tube. A density gradient is set up, the vapor being densest at the bottom and rarest at the top of the tube. This is equivalent to a prism of vapor, the refracting edge of the prism being perpendicular to the plane of the figure and its thickness increasing downward. This prism will form an anomalous spectrum on S_3, in which the wavelengths shorter than the yellow, i.e., on the green side, are deviated upward since heir n is less than 1, and the longer ones (on the orange side) will be deviated downward. Thus, we observe by spectroscope that the deviation of spectrum is on the green side of the *D* line in case of direction and red side in case of downward direction. (The directions are actually reversed because the spectroscope inverts the image of the slit.) Three actual photographs of the resulting spectra with different densities of the vapor are shown in Figure. 4.6. As a consequence of the inversion mentioned above, the photographs form qualitatively a plot of *n* against λ as in Figure In the practical performance of this experiment, several refinements are desirable, of which an important one is the introduction of an auxiliary diaphragm S_2 to select that portion of the vapor where the density gradient is most uniform.

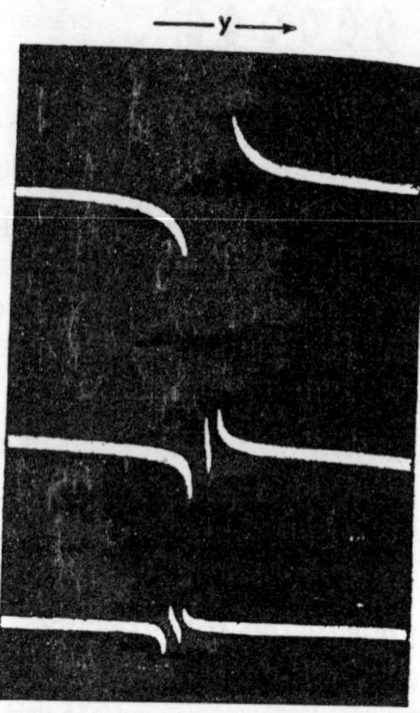

Figure. 4.6 : ***Anomalous dispersion of sodium vapor at three different gas densities.***

Sellmeier's Equation

Now we know that the equation of cauchy is not applicable on dispersion curve in a region of anomalous dispersion. The first success in deriving a formula of more general applicability was obtained by postulating a mechanism by which the medium could affect the velocity of the light wave. It was assumed that the medium contains particles bound by elastic forces, so that they are capable of vibrating with a certain definite frequency ν_0. This is the so-called *natural frequency, i.e.,* one with which the particles will vibrate in the absence of any periodic force, and is identical with the natural frequency in connection with absorption and selective reflection. Passage of the light waves through the medium is then assumed to exert a periodic force on the particles, which causes them to vibrate. If the frequency ν of the light wave does not agree with ν_0, the vibrations will be forced vibrations of relatively small amplitude and of frequency ν. As the frequency of the light approaches ν_0, the response of the particles will be greater and a very large amplitude will be built up by resonance when $\nu = \nu_0$ exactly. These vibrations will in turn react upon the light wave and alter its velocity. In 1871 sellmeier investigated mathematically a equation and i.e—

$$n^2 = 1 + \frac{A\lambda^2}{\lambda^2 - \lambda_0^2} \tag{4.6}$$

This equation contains two constants, A and λ_0, the latter being related to the natural frequency of the particles by the equation $\nu_0\lambda_0 = c$. Hence λ_0 is the wavelength in vacuum corresponding to ν_0. Because the several natural frequencies have different existence than the equation can be written as follows—

$$n^2 = 1 + \frac{A_0\lambda^2}{\lambda^2 - \lambda_0^2} + \frac{A_1\lambda^2}{\lambda^2 - \lambda_1^2} + \ldots = 1 + \sum_t \frac{A_i\lambda^2}{\lambda^2 - \lambda_i^2} \tag{4.7}$$

in which $\lambda_0, \lambda_1, \ldots$ correspond to the possible natural frequencies. The constants A_i are proportional to the number of oscillators capable of vibrating with these frequencies.

Figure 4.7 shows a plot of n against λ according to Equation (4.7), showing two frequencies. As λ approaches λ_0 or λ_1, n goes to $-\infty$ or $+\infty$ on the short-wavelength or long-wavelength side, since the denominator of one of the terms in Equation (4.7) goes to zero. Other important characteristics of the curve to be noted are that n approaches unity as λ approaches zero, and that at $\lambda = \infty$, n^2 takes the value $+\sum_i A_i$.

Equation 4.9 is sellmeier's equation is modified cauchy's equation and it is derived from the electromagnetic theory. Not only does it take account of anomalous dispersion, but it also gives a more accurate representation of *n* in regions far from absorption bands than a Cauchy equation with the same number of constants. That Cauchy's equation is an approximation to Sellmeier's can be seen by writing Equation (4.6) in the form

$$n^2 = 1 + \frac{A}{1-\left(\lambda_0^2/\lambda^2\right)}$$

On expanding by the binomial theorem, we find

$$n^2 = 1 + A\left(1 + \frac{\lambda_0^2}{\lambda^2} + \frac{\lambda_0^4}{\lambda^4} + \ldots\right)$$

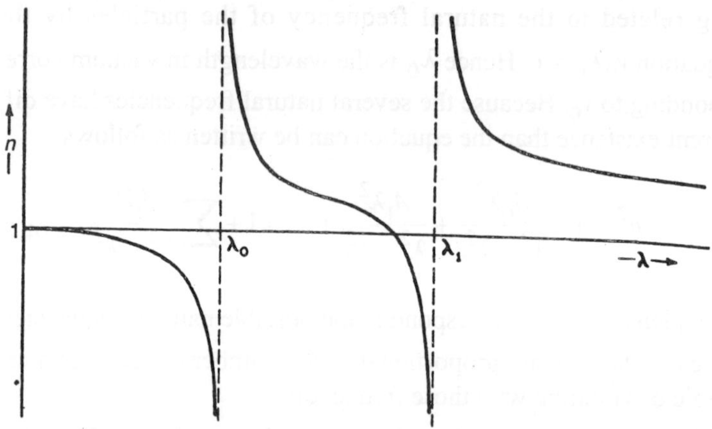

Figure. 4.7 : *Theoretical dispersion curves given by Sellmeier's equation for a medium having two natural frequencies.*

For that part of the dispersion curve where λ is considerably greater than λ_0, the higher powers of λ_0/λ will be small and may be neglected. This gives

$$n^2 = 1 + A + A\frac{\lambda_0^2}{\lambda^2}$$

Writing M for $1 + A$ and N for $A\lambda_0^2$, we have

$$n = \left(M + N\lambda^{-2}\right)^{1/2}$$

On further explanation we get—

$$n = M^{1/2} + \frac{N}{2M^{1/2}\lambda_2} + \frac{N^2}{8M^{3/2}\lambda^4} + \ldots$$

and neglecting higher powers of $1/\lambda$ leads to

$$n = P + \frac{Q}{\lambda^2} + \frac{R}{\lambda^4}$$

This is Cauchy's equation.

We can show the origin of dispersion with a simple pendulum, to the both of which is attached a sight rubber band. If the end of the rubber band is held in the hand and moved to and fro, a periodic force is exerted on the pendulum similar to the action of the light wave on one of the oscillators in the medium. If the frequency motion of the hand is very high compared to the natural frequency of the pendulum, the bob will remain practically motionless. This corresponds to a wave of high frequency and short wavelength,, the velocity which is practically uninfluenced by the presence of the oscillators. Fig 4.7 shows the approaches of n and λ to unity and zero respectively, but the velocity is like free space.

If we will move our hands faster than pendulum, we get that the pendulum swings 180° out of phase with the motion of the hand. Under these conditions, the rubber band is considerably stretched when the displacements of the hand and bob are in opposite directions and so exerts its maximum force on the hand, tending to pull it back to the central position. This corresponds to an increased restoring force on the "ether" which propagates the wave, and hence to an *increase* in the velocity of the wave. Thus in Figure 4.7 n becomes considerably less than 1 at a wavelength slightly less than λ_0. Finally when the frequency of motion of the hand is made less than the natural frequency, the pendulum will follow the hand, practically in phase with it. In this case the rubber band exerts only small forces on the hand, since the displacements of the pendulum are in the same direction. The forces are less than if the pendulum were at rest, so this corresponds to a decreased restoring force on the ether. Thus velocity of the wave decreases and on the long wavelenght side of λ_0 the value of *n* is greater than 1.

The large discontinuity in the dispersion curve at λ_0 is thus seen to be a consequence of the abrupt change of phase by 180° of the oscillator relative to the impressed vibration as the latter passes through resonance frequency. This effect can be demonstrated directly by hanging three pendulums side by side from a horizontal rod clamped atone end. The center pendulum is a heavy one and corresponds to the, ether wave while the other two are very light, one being slightly longer and the other slightly shorter than the heavy pendulum. If center pendulum is swinging, two light swing in opposite phases, the shorter one nearly agreeing in phase with the impressed vibration.

Effect of Absorption on dispersion

As we know that sellmeier's equation is better than cauchy's equation for close absorption bands bed it not applicable for wavelengths with medium of appreciable absorption. This can be seen directly from the fact that the curve of Figure 4.7 goes to infinity on either side of each λ_i. Not only is this physically impossible, but the form of the curve near λ_i does not agree with experiment. It has been possible to measure the dispersion curve right through an absorption band, although this is a difficult matter because practically all the light is absorbed. By using prisms of very small refracting angle, or thin films of the material with a Michelson interferometer, the indices of refraction of a few dyes, such as cyanine, which have an absorption band in the visible, have been carefully m assured. The resulting curve resembles one of those shown by a heavy solid line in Figure 4.8. In comparison with λ the curve is different from, that required by sellmeier's equation.

Sellmeier's equation take no account of the absorption of energy of the wave first shown by Helmholtz. In the above discussion, and in the suggested mechanical analogy, it was assumed that the oscillator does not experience any frictional resistance to its vibration. Such a resistance is necessary if energy is to be taken continuously from the wave by the oscillator. Helmholtz assumed a frictional force directly proportional to the velocity of the oscillator, and he therefore derived an equation for the index of refraction which takes account of absorption. As a measure of the strength of the absorption, we could use the absorption coefficient α, To express in terms of *a* constant k_0 the equation are simpler, concerned with α which is as follows—

$$\kappa_0 = \frac{\alpha\lambda}{4\pi} \qquad (4.8)$$

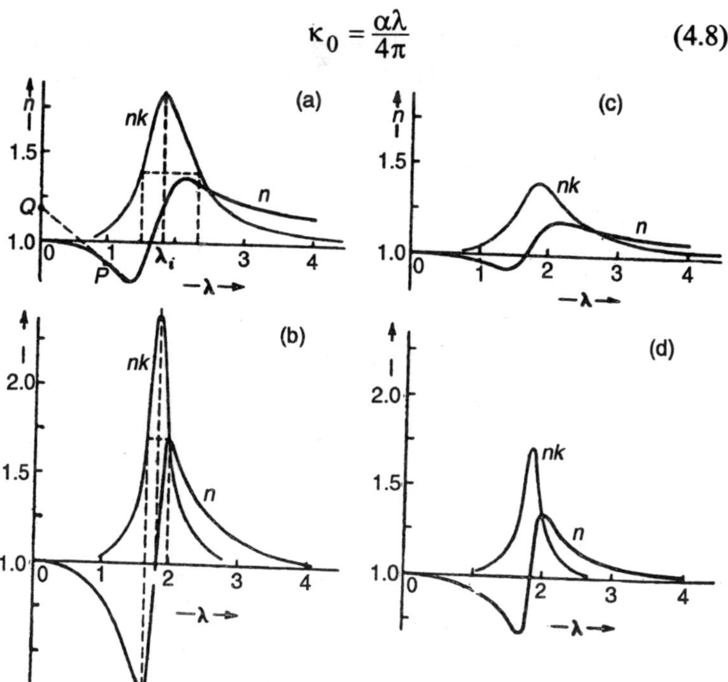

Figure. 4.8 : *Ideal dispersion curves for an oscillator with different amounts of friction and absorption: (a) strong absorption, strong friction; (b) strong absorption, weak friction; (c) weak absorption, strong friction; (d) weak absorption, weak friction.*

Here λ is the wavelength measured in vacuum. The physical significance of κ_0 is best expressed by the fact that the intensity falls to $1/e^{4\pi\kappa_0}$ of its initial value in going the distance λ through the medium. The dispersion equations resulting from this purely mechanical theory of Helmholtz may be written

$$n^2 - \kappa_0^2 = 1 + \sum_i \frac{A_i\lambda^2}{\left(\lambda^2 - \lambda_i^2\right) + g_i\lambda^2 / \left(\lambda^2 - \lambda_i^2\right)}$$

$$2n\kappa_0 = \sum \frac{A_i\sqrt{g_i}\lambda^3}{\left(\lambda^2 - \lambda_i^2\right)^2 + g_i\lambda^2} \tag{4.9}$$

The strength of the frictional force is constant and denoted by $g_{i.}$ These equations should now hold for all wavelengths, including those within an absorption band. In regions far from absorption bands, x_0 and g_i are both essentially zero and the first of the equations reduces to Sellmeier's equation (4.7). Figure 4.8 (a) is a plot of n and of $n\kappa_0$ the latter of which by Equation (4.8) is a measure of the absorption coefficient a, for a case of large friction ($g = 1.96 \times 10^{-3}$). At $\lambda_1 = 0.1732$ μm we can see the quantitative course of dispersion and absorption curves at maximum. It will be seen that n no longer goes to infinity, as in Figure 4.7 but remains finite at $\lambda = \lambda_i$. The other curves of Figure 4.8 are drawn to show the effects of changing both the strength of the absorption and the frictional damping. The former is determined by the total number of oscillators causing the absorption, while the latter depends on the magnitude of the various effects responsible for the breadth of spectral lines. Figure (b) and (d) shows refractive index curve which are come at the point where the absorption is half of its maximum value.

The above experiment of pendulum can include the effect of friction after modification and to give some insight into the physical reason for the resulting change in the form of the dispersion curve. Thus if the smaller pendulum, which represents the oscillator, has a wire attached to it which dips in water or oil, we have the desired condition. Pendulam shows two charges in response of impressed vibration. In the first place, the amplitude will not become nearly as large when the impressed frequency is exactly equal to the natural frequency of the pendulum. With no friction, the amplitude produced by resonance is theoretically infinite (in the final equilibrium state), and the corresponding value of n goes to infinity also. The effect of friction, however, limits this maximum amplitude, and this accounts for the fact that only moderate variations of n are actually observed. In the second place, the change of relative phase between the pendulum and the impressed vibrations when the latter pass through the natural frequency is no longer abrupt but takes place more or less gradually. This accounts for the fact that there is no longer a sharp continuity in the dispersion curve. Which is rounded off into a continuous curve. If a wire dipped into the water (or any other more viscous liquid), than its phase change becomes more and more gradual with increase in friction.

Wave and Group Velocity in the Medium

In the curves of Figures 4.7 and 4.8, the abscissas are wavelengths in vacuum $v = c/v$ and the ordinates are the ordinary indices of refraction $n = c/v$, where v is the wave velocity in the medium. Here wave velocity are greater than that of light in vacuum at part of the curve where n is smaller than one. This is at first sight a contradiction to one of the fundamental results of the theory of relativity, according to which c is the highest attainable velocity. There is actually no contradiction here, however. Since relativity applies to the velocity with which energy (a light signal) is transmitted, and this is always less than c. Remembering that the energy travels with the group velocity u, we require that it is c/u that shall be greater than unity, rather than c/v. For transform we can relate u and v together as—

$$\frac{c}{u} = n - \lambda \frac{dn}{d\lambda} \tag{4.10}$$

where λ is the wavelength in vacuum. Thus the geometrical construction of * 12.8 may also be applied to refractive indices. If in Figure 4.8 (a) we draw a tangent to the dispersion curve, it will intersect the axis of n at a point Q whose ordinate is c/u. That is, while the ordinate of P is n, or c/v, for that wavelength, the ordinate of Q is the corresponding value of c/u for the same wavelength.

This geometrical construction shows, then, that for any point on the curve where it is descending toward the right, the corresponding *c/u* is greater than unity even though *n* itself may be less than unity. Thus velocity less than *c* is not violation of the principle of relativity. An exception to this statement appears to occur in the region within the absorption band, where the curve slopes up steeply to the right. However, in this region we have strong absorption, so that the amplitude of the wave drops practically to zero in a fraction of a wavelength. Here the wave velocity and group velocity are almost useless but the requirement of relativity is fulfilled by other considerations.

The Complete Dispersion Curve of a Substance

Although the curve of the refractive index against wavelength is different for every different substance, the curves for all optical media, i.e., substances more or less transparent in the visible region, possess certain general features in common. To illustrate these, let us consider the schematic curve of Figure 4.9, which represents the variation of *n* from $\lambda = 0$ to several kilometers for an ideal substance. If $\lambda = 0$ than the refraction of index would be unity. For the very short waves (γ rays and hard X rays), the index is slightly less than 1. Siegbahn proved this fact experimentally by refracting X rays through a prism. It was found that the beam was deflected very slightly *away from* the base of the prism, as would be the case if the waves travel faster in the prism than in air. It has also been demonstrated that X rays can be totally reflected from a solid substance by using grazing incidence so that the X rays strike the surface at an angle greater than the critical angle. A. H compton and some other, used this property to measure the wavelength of X-rays by diffracting them from an ordinary ruled grating (used at grazing incidence).

The first absorption is encountered in the X-ray region at a wavelength depending upon the atomic weight of the heaviest element in the material. The maximum values for silicon and uranium is 6.731 Å and 0.1075 Å respectively This absorption rises rapidly to a maximum, and then falls off sharply at the *K-absorption limit* of the element. It gives rise to a relatively narrow region of strong anomalous dispersion, marked *K* in Figure 4.9. Beyond this will lie other absorption discontinuities of this element, called *L, M,...* limits, as well as the *K, L, M....* limits of other elements present There fore any actual optical medium there will be many of these sharp discontinuities. In figure three are simplified.

From the X-ray region the curve descends more rapidly toward longer wave eventually reaching the broad region λ_1 of strong absorption and anomalous dispersion in the ultraviolet. The region between soft X-rays and near ultraviolet is completely covered The descending course of—

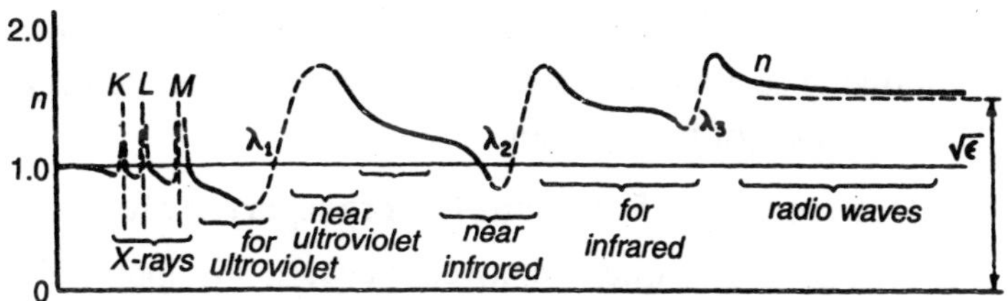

Figure. 4.9 : ***Schematic diagram of a complete dispersion curve for a substance transparent to the visible spectrum.***

the curve in the visible region, characteristic of normal dispersion, is seen to be connected with the existence of this ultraviolet absorption. In general the curve will have a steeper slope in the visible region, so that the dispersion $dn/d\lambda$. is greater, the nearer this absorption band lies to the visible. As we can see in the figure 4.2 and table 4.1 the ability dispersion of light is in the sequence from smaller to greater is as—fluorite < quartz < glass. Dense flint glass, which gives the highest dispersion, frequently as a yellowish colour, owing to the fact that the absorption band encroaches slightly on the violet end of the visible spectrum.

Curve descend steeply near infrared and runs in to another absorption band at λ_2. The center of this band is at 8.5 μm for quartz, but the absorption begins to become strong at 4 or 5 μm. Beyond this first absorption band there usually exist one or more others. In passing each of these bands, the index of refraction increases. Thus the index will be higher for certain infrared wavelengths than for any part of the visible. For example, Rubens measured values of *n* for quartz varying from 2.40 to 2.14 in the region λ = 51 to

63 μm. An. interesting method of isolating radiation of very long wavelengths, called the method of *focal isolation,* is based on this fact. Owing to the high value of *n*, a convex lens will have a much smaller focal length for these long waves than for the shorter waves and the latter can be screened off with suitable diaphragm. Nichols an tear measured the longlet infrared rays.

At wavelengths beyond all the infrared bands, the index decreases slowly and more or less uniformly through th ion of radio waves, approaching a certain limiting value for infinitely long wave In radio frequencies some narrow regions of absorption are there but its always weak. The limiting value will be shown in the following section to be the square root of ε, the ordinary dielectric constant of the medium.

The Electromagnetic Equations for Transparent Media

As we discussed earlier. Moxwell's equation is applicable on empty space, and how predicted electromagnetic wave of velocity *C.* It is now of interest to investigate the characteristics and velocity of such waves in a material substance. For the present we shall consider only nonconducting media, and the more difficult case of conductors will be taken up later. When a steady electric field acts upon a nonconducting dielectric, there is a small displacement of the bound charges in the atoms, and we say they become *polarized.* The charges do not move continuously along, as in a conductor, but are merely displaced through minute distances and come to rest again in a fashion analogous to the stretching of a spring electric displacement is denoted by vector quantity *D* and it is directly proportional to the impressed electric field *E* in an isotropic medium and we get—

$$\mathbf{D} = \varepsilon \mathbf{E} \tag{4.11}$$

Here ε is the dielectric constant. To apply Maxwell's equations to such a medium it now becomes necessary to replace **E** by **D** wherever it occurs in the equations for empty space. Hence Maxwell's equations for a nonconducting isotropic medium are written:

$$\frac{\varepsilon}{c}\frac{\partial E_x}{\partial t}=\frac{\partial H_z}{\partial y}-\frac{\partial H_y}{\partial z} \qquad -\frac{1}{c}\frac{\partial H_x}{\partial t}=\frac{\partial E_z}{\partial y}-\frac{\partial E_y}{\partial z}$$

$$\frac{\varepsilon}{c}\frac{\partial E_y}{\partial t}=\frac{\partial H_x}{\partial z}-\frac{\partial H_z}{\partial z} \qquad -\frac{1}{c}\frac{\partial H_y}{\partial t}=\frac{\partial E_x}{\partial z}-\frac{\partial E_z}{\partial x}$$

$$\frac{\varepsilon}{c}\frac{\partial E_z}{\partial t}=\frac{\partial H_y}{\partial x}-\frac{\partial H_x}{\partial y} \quad (4.12) \qquad -\frac{1}{c}\frac{\partial H_z}{\partial t}=\frac{\partial E_y}{\partial x}-\frac{\partial E_x}{\partial y} \quad (4.13)$$

$$\varepsilon\left(\frac{\partial E_x}{\partial x}+\frac{\partial E_y}{\partial y}+\frac{\partial E_z}{\partial z}\right)=0 \tag{4.14}$$

$$\frac{\partial H_x}{\partial_x}+\frac{\partial H_y}{\partial y}+\frac{\partial H_z}{\partial z}=0 \tag{4.15}$$

In case of plane waves as we did earlier, from equations (4.12) and (4.13) it would be as—

$$\frac{\partial^2 H_z}{\partial_x}=\frac{c^2}{\varepsilon}\frac{\partial^2 H_z}{\partial x^2} \text{ and } \frac{\partial^2 E_y}{\partial t^2}=\frac{c^2}{\varepsilon}\frac{\partial^2 Hy}{\partial x^2}$$

Comparison with the general wave equation shows the new velocity to be $c/\sqrt{\varepsilon}$. The index of refraction becomes

$$n=\frac{c}{v}=\sqrt{\varepsilon} \tag{4.16}$$

The solution of Equations (4.12) to (4.15) for monochromatic plane waves, analogous to Equation (20n), is now to be written

$$E_y = A\sin(\omega t-kx) \qquad H_z=\sqrt{\varepsilon}\,A\sin(\omega t-kx)$$

and the magnitude of the electric and magnetic vectors at any instant is such that

$$H_z=\sqrt{\varepsilon}\,E_y$$

Therefore in the usual case $\varepsilon > 1$ the amplitude of the magnetic wave is greater than that of the electric wave in a ratio equal to the index of refraction [Eq. (4.10)].

In case of dielectric field the electromagnetic energy can be determined by principles used earlier, only replacing the *E* by *D*.

The instantaneous energy densities of the above electric and magnetic waves becomes $\varepsilon E_y^2/8\pi$ and $H_z^2/8\pi$, and are therefore again equal. Their sum may be written as $\sqrt{\varepsilon}E_yH_z/4\pi$, and when this is multiplied by v from Eq. (4.16) to obtain the intensity, one finds

$$I = \frac{c}{\varepsilon}\frac{\varepsilon E_y^2}{4\pi} = \frac{cn}{4\pi}E_y^2 = \frac{cn}{8\pi}A^2 \tag{4.17}$$

If the flow of energy is averaged over a time long compared with the period than the E_4 represents the root mean square (*rms*) in above equation. The result may also be written $cE_yH_z / 4\pi$. In this form it represents the expression of a general law of electromagnetism known as *Poynting's theorem,* according to which the direction and magnitude of the energy flow are given by the *Poynting vector* $(c/4\pi)[\mathbf{E}\times\mathbf{H}]$, the quantity in brackets being the vector product.

The values of *n* in equation (4.16) are more applicable on gases in comparison to the denser medium. Thus the dielectric constant for water, measured by placing it between the plates of a condenser charged to a steady potential, is 81, indicating a value of 9 for the index of refraction. For sodium light, the measured index of water is 1.33. For various kinds of glass, ε varies from 4 to 9, which would require *n* to vary from 2 to 3. This again is higher than the observed values for visible light.

We can find cause of this discrepancy easily. It lies in the fact that the electric field of a light wave is not a steady field but a rapidly alternating one. For yellow light the frequency is $5 \times 10^{14}\text{s}^{-1}$. If the dielectric constant of a substance is measured using an alternating potential on the plates in place of a steady one, the result is found to vary with the frequency. From this we see that the index of refraction must also vary with frequency, or wavelength. As the wavelength becomes very large and approaches infinity, the frequency approaches zero. The limiting case of a steady field thus corresponds to zero frequency, and we are led to expect the index of refraction to approach the square root of the dielectric constant for steady fields. That this is in fact the case is shown by the measurements of the index of refraction of water for electromagnetic waves quoted in Table 4.3. The value of $\sqrt{\varepsilon}$ measured for a steady field is shown for comparison. The value of n finded exactly the same as predicted for infinitely long waves.

Theory of dispersion

From electromagnetic theory the variations of n can be explained. One must take account of the molecular structure of matter. When an electromagnetic wave is incident on an atom or molecule, the periodic electric force of the wave sets the bound charges into a vibratory motion having the frequency of the wave.

In this phase in which motion is related to the impressed electric force, depends upon the impressed frequency and will vary with the difference between the impressed frequency and the natural frequency of the bound charges in the way discussed earlier. As the wave traverses the empty space between molecules, it will, of course, have the velocity c, and we must now inquire how it is possible that the presence of the oscillating charges in the molecules produces an effective alteration in the rate at which the wave progresses through the medium.

The dispersion can be explained by secondary wave due to the induced oscillations of the bound charges. These secondary waves are identical with those which give rise to molecular scattering as in the explanation of the blue colour of the sky. When a light beam traverses a transparent liquid or solid, the amount of light scattered laterally is extremely small, even though the concentration of scattering centers is much greater than that in the air which gives the sky light. This is due to the fact that the scattered wavelets travelling out laterally from the beam have their phases so arranged that there is practically complete destructive interference. But the secondary waves travelling *in the same direction* as the original beam do not cancel out but combine to form sets of waves moving parallel to the original waves. Now the secondary waves must be added to the primary ones according to the principle of super position, and the results will depend on the phase difference between the two sets. The phase of the primary wave can be modified to get equivalent to a change in their velocity of wave. That is, since the wave velocity is merely the rate at which a condition of equal phase is propagated, an alteration of the phase by interference changes the velocity. We have seen that the phase of the oscillators, and hence of the secondary waves, depends on the impressed frequency, so it becomes clear that the velocity in the medium varies with the frequency of light. Here the physical interpretations of dispersion is explained.

The foundations for the mathematical treatment of the above mechanism were laid by Rayleigh, who considered the case of mechanical waves, and the theory was later extended to cover the case of electromagnetic waves by Planck, Schuster, and others.

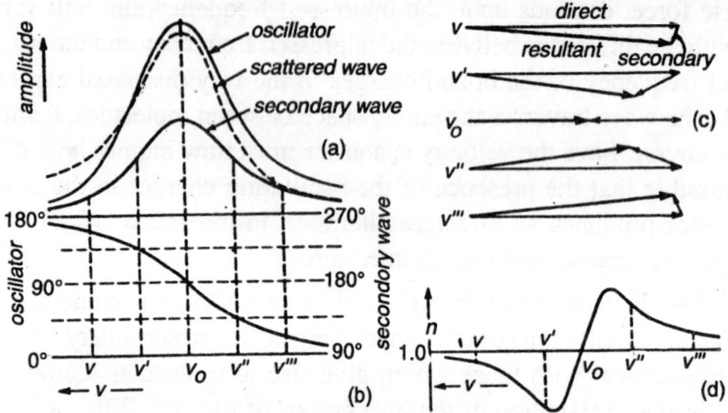

Figure. 4.10 : ***The interpretation of dispersion as the result of interference of the secondary wave with the direct wave.***

We would not discuss here about this development. It leads to dispersion formulas similar to that of Helmhotz [Eq. 4.9]. In fact, there is a close analogy throughout between the electromagnetic and mechanical pictures of the phenomenon. The oscillations of the bound charges must be regarded as damped by a frictional force, just as the particles were in Helmholtz's theory. In electromagnetic theory the nature of the damping forces postulates will be discussed later.

Table 4.3 : Variation of n with λ for water

Wavelength,	cm	Frequency, Hz	n
5.89×10^{-5}	5.1	$\times 10^{14}$	1.333
12.56×10^{-5}	2.9	$\times 10^{14}$	1.3210
258	$\times 10^{-5}$	0.116×10^{14}	1.41
800	$\times 10^{-5}$	0.0375×10^{14}	1.41
0.40	750	$\times 10^{8}$	5.3
1.75	171	$\times 10^{8}$	7.82.
8.1	37	$\times 10^{8}$	8.10
65	4.6	$\times 10^{8}$	8.88
∞	0	$\times 10^{8}$	$(9.03 = \sqrt{\varepsilon})$

In fig 4.10 we can see the relative amplitudes and phases of the incident wave, oscillator and secondary wave. The first curve in (a) shows the response of a damped *oscillator* of natural frequency v_0 to an impressed vibration of frequency v, the amplitude becoming a maximum when $v = \gamma_0$. The broken curve shows the amplitude radiated by the oscillator, i.e., of the *scaltered wave.* As a consequence of Rayleigh's law that the shorter waves are scattered more effectively, his curve is higher on the sides of higher v but drops to zero at low frequencies. The scattered wavelets builds the third curve which gives the amplitude of the secondary waves. Curve (b), in conjunction with the left-hand scale of ordinate gives the phase difference between the oscillator and the impressed wave this changes from 0 to 180° in passing through the natural frequency but not abruptly because of the damping. At v_0 it is 90° behind that of the impressed wave. Theory shows, furthermore, that the phase of the scattered waves, and therefore of the secondary waves as well, lags 90° behind that of the oscillators. This is because electromagnetic radiation is proportional to the rate of change of current, or to the acceleration of a charge. The oscillation of charges is same as the phase of current and velocity. Therefore, since in a simple harmonic motion the acceleration is one-quarter period behind the velocity, the phase of the radiated waves is retarded this much behind that of the oscillating source taking account of this additional 'retardation, it will be seen that the-right-hand scale of ordinates in Figure 4.10 (*b*) applies to the phase lag of the secondary waves behind the impressed waves.

In figure (4.10) we can see the further development in to compound vectorial the amplitude of the direct and secondary waves. For the frequency ν, the amplitude of the secondary waves is small [curve (*a*)] and lags in phase behind the direct waves by nearly 270° [curve (*b*)]. The vector diagram at the top in (*c*) shows that the resultant amplitude is nearly the same but that the phase is slightly *advanced,* corresponding to a rotation of the vector in a clockwise sense. An advance of phase means an increase in velocity, since it will be remembered that the phase increases as we move backward along a wave. Thus in the dispersion curve (*d*), the index of refraction at ν is slightly less than 1. The second vector diagram, for ν', gives a greater advance of phase, and a considerably smaller resultant amplitude. At $\nu = \nu_0$ there is no change of phase or velocity produced, but merely a reduction of intensity. The energy removed from the resultant forward wave appears in other directions as resonance radiation. Beyond ν_0 there is a retardation instead of an advance of phase, and the velocity of the wave is decreased finally in curve (*d*) we get the form for anomalous dispersion in qualitative way.

Nature of the vibrating Particles and Frictional forces

finally in the various discontinuties of the typical dispersion curve is due to the charged particles and damping forces (see fig 4.9). The X-ray absorptions are attributed to the innermost electrons in the atoms, which are assigned to the "shells" *K*, *L*, *M*, etc., of decreasing energy and increasing distance from the nucleus. Because they are deep in the atom, these electrons are shielded from the effects of collisions and electric fields due to neighbouring atoms. These two causes of line breadth in spectrum lines are not important for X-rays, and the absorption discontinuities are sharp, even in solids. It is only in this region th. radiation damping makes any appreciable contribution to the line widths.

The outer electrons in the atoms and molecules of the material more broad absorption in the far ultraviolets. These are not shielded, and consequently in solids and liquids an extensive region of continuous absorption is produced. For molecular gases the bands may consist of individual rotational lines that are quite sharp, but these are so numerous that they are usually unresolved. In this region the damping due to collisions begins to become more important than that due to radiation, and at still longer wavelengths it usually predominates. The near-infrared absorption bands represent the various natural frequencies of the atoms as a whole, or even of molecules. Since these vibrators are much heavier than electrons, it is clear why they possess lower vibration frequencies. In the far infrared, other molecular vibrations of lower frequency may be involved. In that case frequencies of rotation of molecules plays an important part e.g., in gases.

5

Absorption and Scattering

If a beam of light propagates from solid, liquid or gaseous state than its propagation is affected by two ways which are as—(1) the intensity will always decrease to a greater or lesser extent as the light penetrates farther into the medium, and (2) the velocity will be less in the medium than in free space. The loss of intensity is chiefly due to absorption, although under some circumstances scattering may play an important part. In this chapter we shall discuss the consequences of absorption and scattering, while the effect of the medium on the velocity, which comes under dispersion, we shall consider in the following chapter. The term *absorption* as used in this chapter refers to the decrease of intensity of light as it passes through a substance. This definition is different from that of absorptance which has been discussed in last chapter. The two terms refer to different physical quantities, but there are certain relations between them, as we shall now see.

General and Selective Absorption

If a substance reduces the intensity of all wavelengths of light in same amount than it is called a general absorption. For visible light this means that the transmitted light, as seen by the eye, shows no marked colour. There is merely a reduction of the total intensity of the white light, and such substances therefore appear to be gray. There is no perfect substance which can absorb all wavelengths equally, but some substances as—suspensions of lamp black or their semitransparent films of platinum almost do it. Practically all coloured substances owe their colour to the existence of selective absorption in some part or parts of the visible spectrum. Thus a piece of green glass absorbs completely the red and blue ends of the spectrum, the remaining portion in the transmitted light giving a resultant sensation of green to the eye. The colours of most natural objects such as paints, flowers, etc., are due to selective absorption. These objects are said to show pigment or *body colour,* as distinguished from *surface colour,* since their colour is produced by light which penetrates a certain distance into the substance. Then, by scattering or reflection, it is deviated and escapes from the surface, but only after it has traversed a certain thickness of the medium and has been robbed of the colours which are selectively absorbed. In all such cases the absorptance of the body will be proportional to its true absorption and will depend in the same way upon wavelength. Surface colour, on the other hand, has its origin in the process of reflection at the surface itself. Some substances, particularly metals like gold or copper, have a higher reflecting power for some colours than for others and therefore show colour by reflected light. The transmitted light here has the complementary colour, whereas in body colour the colour is the same for the transmitted and reflected light. A thin gold foil, for example, looks yellow by reflection and blue-green by transmission. As we know the body absorption of these materials is very high. Thus the reflectance is higher and absorptance is low.

Distinction Between Absorption and Scattering

Figure 5.1 shows the enterance of a light of intensity I_0 in a long glass cylinder filled with smoke. The intensity I of the beam emerging from the other end will be less than I_0. For a given density of smoke, experiment shows that I depends on the length d of the column according to the exponential law:

$$I = I_0 e^{-\alpha d} \tag{5.1}$$

In above equation α is the coefficient of absorption, which is the rate of loss of light from the direct beam. However, most of the decrease of intensity of I is in this case not due to a real disappearance of the light but results from the fact that some light is *scattered* to one side by the smoke particles and thus removed from the direct beam. A considerable intensity I_s of scattered light may easily be detected by observing the tube from the side in a darkened room even with a very dilute smoke. We can see rays of sunlight passing through a window with the help of fine dust particles suspended in the air.

True *absorption* represents the actual disappearance of the light, the energy of which is converted into *heat motion of* the molecules

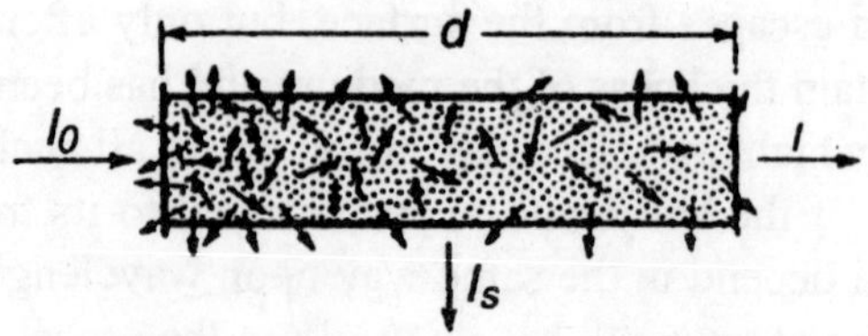

Figure 5.1 : ***Scattering of light by finely divided particles as in smoke.***

of the absorbing material. In the above experiment this will occur to only a small extent, so that the name "absorption coefficient" for α is not appropriate in this case. In general, we can regard α as being made up *of* two parts, α_a due to true absorption and α_s due to scattering. Thus we get the following equation from equation (5.1)—

$$I = I_0 e^{-(\alpha_a + \alpha_s)d} \tag{5.2}$$

α_a or α_s can be negligible for many cases but existence of these two different processes is important.

Absorption by Solids and Liquids

A thick solid or a liquid covered by a transparent cell on passing monochromatic light give small incident light than others, owing to absorption. The amount of absorption will also change to a greater or less extent if the wavelength of the incident light is changed. A simple way of investigating the amount of absorption for a wide range of wavelengths simultaneously is illustrated in Figure 5.2 S_1 is a source which emits a continuous range of wavelengths, such as an ordinary tungsten-filament lamp. From this source the light is rendered parallel by the lens L_1 and traverses a certain thickness of the absorbing medium *M*. It is then focused by L_2 on the slit S_2 of a prism spectrograph, and the spectrum photographed on the plate *P*. If *M* is a "transparent" substance like glass or water, the part of the spectrum on *P* representing visible wavelengths will be perfectly continuous, as if *M* were not present. If *M* is coloured, part of the spectrum will be blotted out corresponding to the wavelengths removed by *M,* and we call this an *absorption band*. In case of solids and liquids these bands continue in character to fading off slowly and constantly at the end.

A transparent visible region do absorption if its observations are extended far enough into the infrared or the ultraviolet region. Such an extension includes considerable experimental difficulty when a prism spectrograph is used, because the material of the prism and lenses (usually glass) may itself have strong selective absorption in these regions. Thus flint glass cannot be used much beyond 25,000 Å (or 2.5 μm) in the infrared or beyond about 3800 Å in the ultraviolet. Quartz will transmit somewhat farther in the infrared and much farther in the ultraviolet. Prisms can use various transparent substances to transmit an appreciable amount of light, regions of which are shown in Table 5.1.

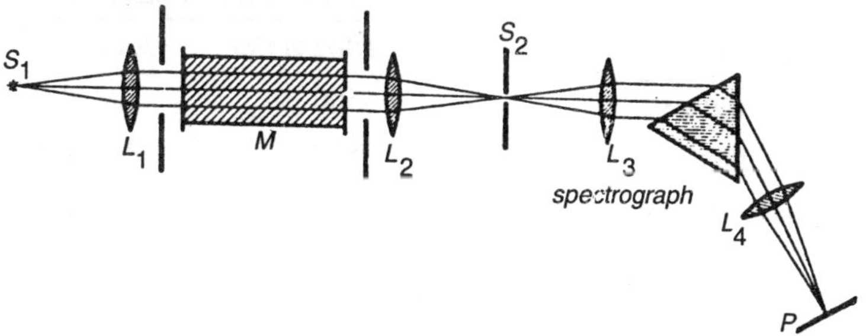

Figure 5.2 : *Experimental arrangement for observing the absorption of light by solids, liquids, or gases.*

For infrared and ultraviolet investigations prisms are of rock salt and quartz respectively. There is no advantage in using fluorite unless air is completely removed from the light path in an ultraviolet spectrograph. This is because, below 1850 Å air begins to absorb strongly. Also, specially prepared photographic plates must be used below this wavelength, since the gelatin of the emulsion by its absorption renders ordinary plates insensitive below about 2300 Å. The innovative methods of sensitzing plates is used in the infrared photography (as far as 13000 Å). Beyond this, an instrument based upon measurement of the heat produced, such as a thermopile, is usually used, although as far as 6 μm the *photoconductive cell,* utilizing the change of electrical resistance upon illumination, gives greater sensitivity.

No substances exists when absorption measurements are extended over the whole electromagnetic spectrum. This does not show strong absorption for some wavelengths. The metals exhibit general absorption, with a very minor dependence on wavelength in most cases. The pronounced 'transmission band' (near 3160 Å) of silver is exception case. A film of silver which is opaque to visible light may be almost entirely transparent to ultraviolet light of this wavelength. Dielectric materials, which are poor conductors of electricity, exhibit pronounced selective absorption which is most

Table 5.1

	Limit of transmission, Å	
Substance	***Ultraviolet***	***Infrared***
Crown glass	3500	20,000
Flint glass	3800	25,000
Quartz (SiO_2)	1800	40,000
Fluorite (CaF_2)	1250	95,000
Rock sali (NaC1)	1750	145,000
Sylvin (KCI)	1800	230,000
Lithium fluoride	1100	70,000

easily studied when scattering is avoided by having them in a homogeneous condition such as that of a single crystal, a liquid, or an amorphous solid. Similarly some substances are more or less transparent to X rays and Y rays as light waves of wavelength below about 10 Å. Proceeding toward longer wavelengths, we encounter a region of very strong absorption in the extreme ultraviolet, which in some cases may extend to the visible region, or beyond, and in others may stop somewhere in the near ultraviolet (see Table 5.1). In the infrared, Further absorption bands are encountered in the infrared but these eventually give way to almost complete transparency in the region of radio waves. Thus for dielectrics we may usually expect three large regions of transparency in the region of radio waves. Thus for dielectrics we may usually expect three large regions of transparency one at the shortest wavelengths, one at intermediate wavelengths (perhaps including the visible), and one at very long wavelengths. In different substances the limits of these regions vary a great deal and one substance, such as water, may be transparent to the visible but opaque to the near infrared, besides some substances like rubber, etc. are opaque to the visible but transparent to the infrared.

Absorption by Gases

In case of gases the absorption spectra show narrow, dark lines. In certain cases it is also possible to find regions of continuous absorption but the outstanding characteristic of gaseous spectra is the presence of these sharp lines. The spectrum will be a true line spectrum if the gas is momatomic like helium or mercury vapor frequently showing clearly defined series. In comparison to emission spectrum the number of lines in the absorption spectrum is invariably less. For instance, in the case of the vapors of the alkali metals, only the lines of the principal series are observed under ordinary circumstances. The absorption spectrum is therefore simpler than the emission spectrum. The sharp lines form the rotational structure of the absorption bands characteristic of molecules if the gas consists of diatomic or polyatomic molecules. In same gas the less bands are observed in absorption than in emission, because the absorption spectrum is simple.

Resonance and Fluorescence of Gases

Now we discuss about the energy of incident light, removed by the gas. If true absorption exists, according to the definition. This energy will all be changed into beat and the gas will be somewhat warmed. Unless the pressure is very low, this is generally the case. After an atom or molecule has taken up energy from the light beam, it may collide with another particle, and an increase in the average velocity of the particles is brought about in such collisions. The length of time that an energized atom can exist as such before a

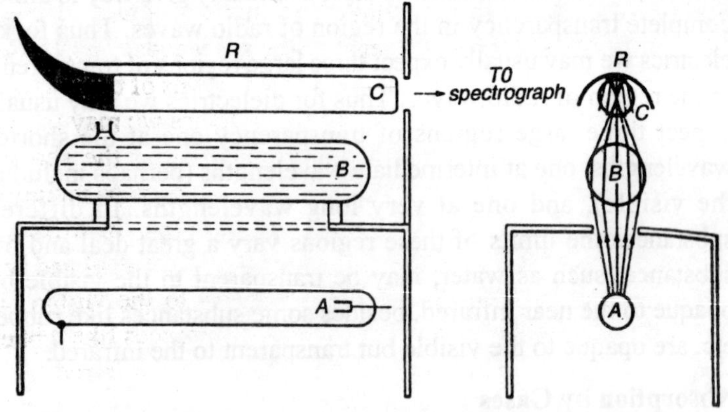

Figure 5.3 : ***Experimental arrangement for observing the fluorescence of iodine vapor with excitation by monochromatic light.***

collision is only about 10^{-7} or 10^{-8} s, and unless a collision occurs before this time, the atom will get rid of its energy as radiation. At low pressures, where the time between collisions is relatively long. the gas will become a secondary source of radiation, and we do not have true absorption. In such cases the wavelength of reemitted light is equal to that of incident light and called resonance radiation. This radiation was discovered and extensively investigated by R. W. Wood. The origin of its name is clear, since as has been mentioned the phenomenon is analogous to the resonance of a tuning fork. The reemitted light may have a longer wavelength than the incident light under some circumstances. This effect is called *fluorescence*. In either resonance or fluorescence, Some of the light is removed from the direct beam in either resonance or fluorescence. Dark lines will be produced in the spectrum of the transmitted light. Resonance and fluorescence are not to be classed as scattering. This will be discussed later.

By using a sodium-arc lamp, rasonance radiation from a gas can readily be demonstrated. A small lump of metallic sodium is placed in a glass bulb connected to a vacuum pump. By heating with a bunsen burner the sodium is distilled from one part of the bulb to another. This liberates the large quantities of hydrogen always contained in this metal. After a high vacuum is attained, the bulb is sealed off and the light of the arc is focused by a lens on the bulb. The bulb must of course be observed from the side in a dark room. On gently warming the sodium with the flame, a cone of yellow light defining the path of the incident light will be seek. Cones glow shorter at higher temperature, so on the inner surface of the glass it look like their bright skin.

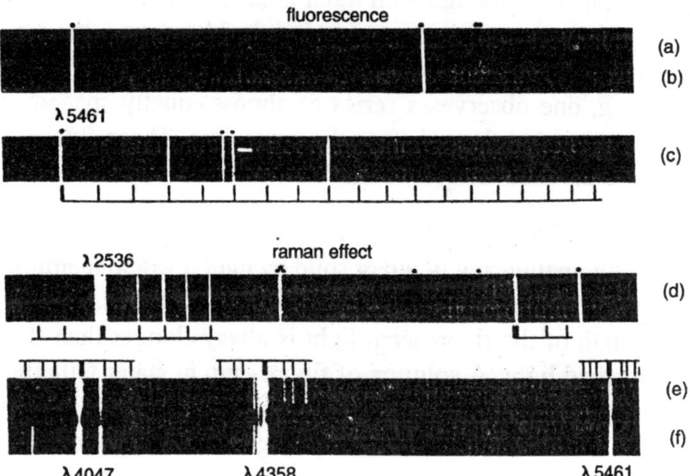

Figure 5.4 : ***Photographs of (a) mercury-arc spectrum; (b) fluorescence spectrum of iodine; (c) enlarged section of (b); (d) Raman spectrum of hydrogen; (e) Raman spectrum of liquid carbon tetrachloride; (f) mercury arc.***

Fluorescence of a gas can be seen through a iodine vapor because of diatomic molecules (I_2). When a white light from a carbon arc is focused in a bulb containing iodine vapor in vacuum at room temperature it produces a greenish cone of light. A still more interesting experiment can be performed by using monochromatic light from a mercury arc, as shown in Figure 5.3. The source of light is a long horizontal arc *A*, which is enclosed in a box with a long slot cut in the top parallel to the arc. Immediately above this is a glass tube *B* filled with water. This acts as a cylindrical lens to concentrate the light along the axis of tube *C*, containing the iodine vapor in vacuum. At the plane window on the end of *C* a spectroscope is pointed which observe the fluorescent light from the vapors. The other end is tapered and painted black to prevent reflected light from entering the spectroscope; and a screen with a circular hole placed close to the window helps in this respect. A polished reflector *R* laid over C increases the intensity of illumination. If *B* contains a solution of potassium dichromate and neodymium sulfate, only the green line of mercury, λ5461, is transmitted. Figure 5.4 (b) and (c) were reproduced from a spectrogram taken in this way, though with water in the tube *B*. Beside the lines of the ordinary mercury spectrum (marked by dots in the figure) which are present as a result of ordinary reflection or Rayleigh scattering, one observes a series of almost equally spaced lines extending toward the red from the green line. These fluorescent lights are the modified wave lengths.

Fluorescence of Solids and Liquids

On illuminating any liquid or solid by light it would reemit fluorescent light inspite of absorbtion. According to *Stokes' law,* the wavelength of the fluorescent light is always longer than that of the absorbed light. A solution of fluorescein in water will absorb the blue portion of white light and will fluoresce with light of a greenish hue. Thus a beam of white light traversing the solution becomes visible through emission of green light when observed from the side but is reddish when looked at from the end. Certain solids show a persistence of the reemitted light, so that it lasts several seconds or even minutes after the incident light is turned off. Thus, it is known as phosphorescence.

A mercury arc with ultraviolet light can illuminate various objects which produce striking fluorescent effect. A special nickel oxide glass can be obtained which is almost entirely opaque to visible light but transmits freely the strong group of mercury lines near 13650. If only this light from the arc comes through the glass, many organic as well as inorganic substances are rendered visible almost exclusively by their fluorescent light. Teeth illuminated by ultraviolet light appear unnaturally bright, but artificial teeth look perfectly black. The ruby gem stone also show good demonstration of fluorescence.

Selective Reflection. Residual Rays

If certain wavelengths are reflected more strongly than others than they show selective reflection. This usually occurs at those wavelengths for which the medium possesses very strong absorption. We are speaking now of dielectric substances, i.e., those which are nonconductors of electricity. We will discuss this case later on. That there is an intimate connection between selective reflection, absorption, and resonance radiation can be seen from an interesting observation made by R. W. Wood with mercury vapor. At a pressure of a small fraction of a millimeter, mercury vapor shows the phenomenon of resonance radiation when illuminated by λ2536 from a mercury arc. As the pressure of the vapor is increased, the resonance radiation becomes more and more concentrated toward the surface of the vapor where the incident radiation enters, i.e., on the inner wall of the enclosing vessel. Finally, at a sufficiently high pressure, the secondary radiation ceases to be visible except when viewed at an angle corresponding to the law of reflection. At this angle fully 25 percent of the incident light is reflected in the 'ordinary way, the remainder having been absorbed and transformed into heat by atomic collisions. However, this high reflection, which is comparable to that of metals in this region, exists only for the particular wavelength λ2536. Other wavelengths are freely transmitted. This experiment show us continuous transition from resonance radiation to selective reflection.

Some solids also snow selective reflection because of the strong absorption bands in their visible region. The dye fuchsine is an example. Such substances have a peculiar metallic sheen by reflected light and are strongly coloured. Their colour is due to the very high reflection of a certain band of wavelengths-so high that it is frequently termed "metallic" reflection. It is this type of reflection that is responsible for surface colour.

The characteristic feature of selective reflection is to locate the absorption bands which present far in the infrared. For example, quartz is found to reflect 80 to 90 percent of radiation having a wavelength of about 8.5 μm, or 85,000 Å. The method of *residuals* for isolating a narrow band of wavelengths is based upon is fact. In figure 5.5 *S* is a thermal source of radiation, giving a continuous spectrum. After reflection from the four quartz plates Q_1 to Q_4, the radiation is analyzed by means of a wire grating *G* and thermopile *T*. It is found to consist almost entirely of the wavelength 8.5 μm. Supposing this wavelength to be 90 percent reflected, each quartz surface and other wavelengths 4 percent reflected, we have, after four reflections, $(0.9)^4 = 0.66$ of the former remaining, but only $(0.04)^4 = 0.0000026$ of the latter. The wavelengths of the residual rays of many substances have been measured in this way. By the measurement of wavelengths we get substance which have longest are sodium chloride, potassium chloride, and rubidium chloride with 52, 63 and 74 μm respectively.

Theory of the Connection between Absorption and Reflection

In the electromagnetic theory for the production of resonance radiation it is assumed that light waves are incident upon matter which contains *bound charges* capable of vibrating with a natural frequency equal to that of the impressed wave. Thus a charge e is acted upon by the electric field E with a force *eE,* and if E varies with a frequency exactly matching that with which the charged particle would normal vibrate, a large amplitude may be produced. Finally charged particle reradiates an electromagnetic wave of the same frequency. In a gas at low pressure. Where the atoms are relatively far apart, the frequency which can be absorbed will be sharply defined and there will be no systematic relation between the phases of the light reemitted from different particles. The observed intensity from *N* particles will then be just *N* times that due to one particle. These are concerned with resonance radiation.

If, on the other hand, the particles are close together and interacting strongly with each other, as in a liquid or solid, the absorption will not be limited to a spar, defined frequency but will spread over a considerable range. The conclusion comes that the phases of the reemitted light from adjacent particles are agree.

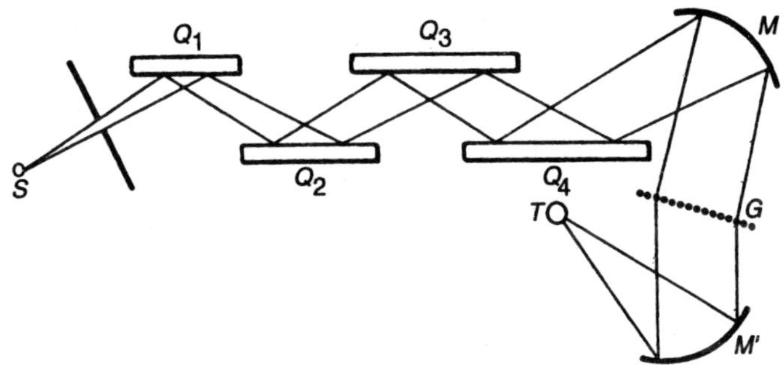

Figure 5.5 : *Experimental arrangement for observing residual rays by selective reflection.*

This will give rise to regular reflection, since the various secondary waves from the atoms in the surface will cooperate to produce a reflected wave front traveling off at an angle equal to the angle of incidence. In fact, this is just the conception used in applying Huygens' principle to prove the law of reflection. Hence selective reflection is also a phenomenon of resonance, and occurs strongly near those wavelengths corresponding to natural frequencies of the bound charges in the substance. The substance will not transmit light of these wavelengths; instead it reflects strongly. True absorption, or the conversion of the light energy into heat, may also occur to a greater or less extent because of the large amplitudes of the vibrating charges which are here involved. We would get 100 percent reflecting power if absorption is entirely absent.

Scattering by Small Particles

As we studied equation the beam of light after scattering traverse a cloud of fine suspended matter. That this phenomenon is closely connected both with reflection and with diffraction can be seen by consideration of Figure 5.6. In (a) is shown a parallel beam consisting of plane waves advancing toward the right and striking a small plane reflecting surface. The successive wave fronts drawn are one wavelength apart, so that here the size of the reflector is somewhat greater than a wavelength. The light coming off from the surface of the reflector is produced by the vibration of the electric charges in the surface with a definite phase relation, and the spherical wavelets produced by these vibrations cooperate to produce short segments of plane wave fronts. These are not sharply bounded at their edges by the reflected rays from the edges of the mirror (dotted lines) but spread out somewhat, owing to diffraction. In fact, the distribution of the intensity of the reflected light with angle is for the light- transmitted by a single slit. In this case the width of the reflector is in place of the slit width for generator spreading and the reflector relative to the wavelength would be smaller.

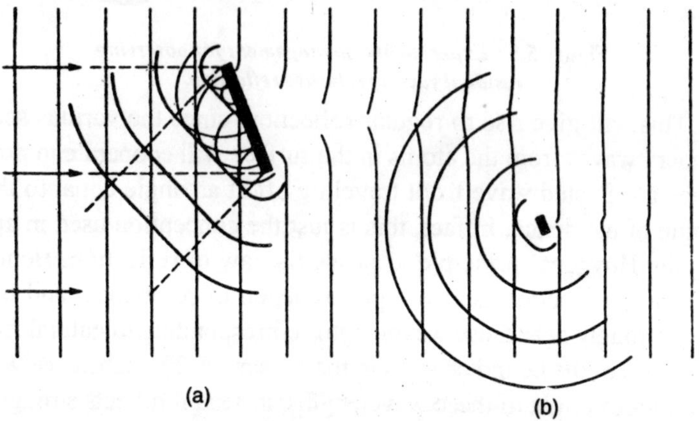

Figure 5.6 : ***The reflection and diffraction of light by small objects comparable in size with the wavelength of light.***

As we can see in figure 5.6 (b) the larger wavelength than reflector, the reflected waves different from uniform spherical waves due to large spreading. In this case the light taken from the primary beam is said to be scattered, rather than reflected, since the law of reflection has ceased to be applicable. Scattering is therefore a special case of diffraction. The wave scattered from an object much smaller than a wavelength of light will be spherical regardless of whether or not the object has the plane form assumed in Figure 5.6 (b). This follows from the fact that there can be no interference between the wavelets emitted by the several points on the surface of the scattering particle, inasmuch as the extreme points are separated by a distance much less than the wavelength.

In 1871 Rayleigh first time study quantitative laws of scattering by small particles and such scattering is frequently called *Rayleigh scattering.* The mathematical investigation of the problem gave a general law for the intensity of the scattered light, applicable to any particles of index of refraction different from that of the surrounding medium. The only restriction is that the linear

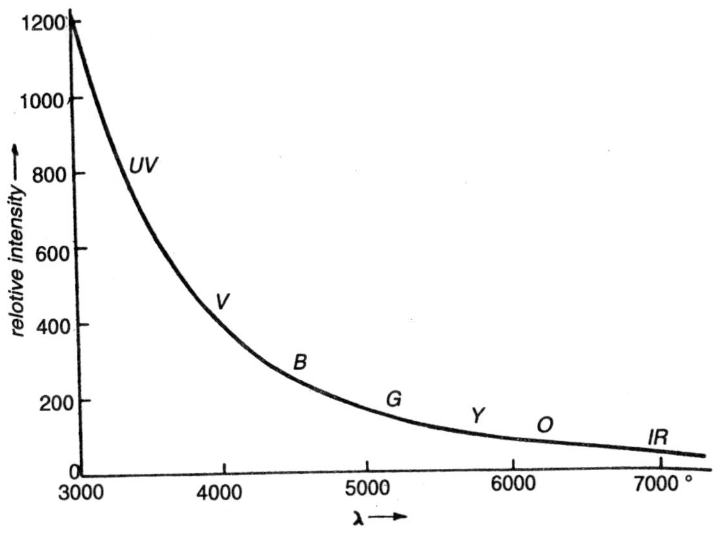

Figure 5.7 : *Intensity of scattering versus wavelength according to Rayleigh's law.*

dimensions of the particles be considerably smaller than the wavelength. *As* we might expect, the scattered intensity is found to be proportional to the incident intensity and to the square of the volume of the scattering particle. The most interesting result, however, is the dependence of scattering on wavelength. With a given size of the particles long waves would be expected to be less effectively scattered than short ones, because the particles present obstructions to the waves which are smaller *compared with the wavelength* for long waves than for short ones. Indeed, the intensity is directly proportional to $1/\lambda^4$ will be proved later.

$$I_s = k\frac{1}{\lambda^4}$$

Since red light, λ7200, has a wavelength 1.8 times as great as violet light, λ4000, the law predicts $(1.8)^4$ or 10 times greater scattering the violet light from particles much smaller than the wavelength of either colour. In figure 5.7 we can see about quantitative study of this relation.

The scattering of white light can be best seen in tobacco smoke, which always has a bluish colour. If the size of the particles is increased until they are no longer small compared with the wavelength, the light becomes white, as a result of ordinary diffuse reflection from the surface of the particles. The blue colour seen with very small particles and its dependence on the size of the particles were first studied experimentally by Tyndall,* and his name is often associated with the phenomenon. The white light scattering can be seen from large particles of chalk dust from an eraser, falling across a beam of light from a carbon arc.

Molecular Scattering

If in a pure liquid, all suspended particles of dust are free than a strong beam of sunlight traverse in it. Observation in a dark room will show that there is a small amount of bluish light scattered laterally from the beam. Although some of this light is still due to microscopic particles in suspension, which seem to be almost impossible to eliminate entirely, a certain amount appears to be attributable to the scattering by individual molecules of the liquid. At first sight it is surprising to find that the scattering from liquids is so feeble, in view of the large concentration of molecules present. It is, in fact, much weaker than the scattering from the same number of molecules of a gas. In the latter, the molecules are randomly distributed in space, and in any direction except the forward one the waves scattered by different molecules have perfectly random phases. Thus for N molecules the resultant intensity is just N times that scattered from any individual one. In a liquid, and even more so in a solid, the spatial distribution has a certain degree of regularity. Furthermore, the forces between molecules act to destroy the independence of phases. The result is that the scattering from liquids and solids in directions other than forward is very weak indeed. To determine the velocity of light in the medium forward scattered waves are used.

Because of the smaller number of scattering centers, the lateral scattering from gases is weak. When a great thickness of gas is available, however, as in our atmosphere, the scattered light is easily observed. It has been shown by Rayleigh that practically all the light that we see in a clear sky is due to scattering by the molecules of air. If it were not for our atmosphere, the sky would look perfectly black. Actually, molecular scattering causes a considerable amount of light to reach the observer in directions making an angle with that of the direct sunlight, and thus the sky appears bright. Its blue colour is the result of the greater scattering of short waves. Rayleigh measured the relative amount of light of different wavelengths in sky light and found rather close agreement with the $1/\lambda^4$ law. The same phenomenon is responsible for the red colour of the sun and surrounding sky at sunset. In this case, the scattering removes the blue rays from the direct beam more effectively than the red, and the very great thickness of the atmosphere traversed gives the transmitted light its intense red hue. We can prove these blue of the sky and the red of the sun set by examples.

Raman Effect

Raman effect is all about the scattering with change of wavelength like fluorescence but, it differs from fluorescence, however, in two important respects. In the first place, the light which is incident on the scattering material must have a wavelength that does not correspond to one of the absorption lines or bands of the material. Otherwise we obtain fluorescence, the green line of mercury is absorbed by the iodine vapor. In the second place, the intensity of the light scattered in the Raman effect is much less intense than most fluorescent light. Thus Raman effect is quite difficult to detect, and it is trapped by photography.

The Raman effect can be observed by the apparatus shown in figure 5.3. For this purpose, a liquid or gas which is transparent to the incident light must be used in the tube *C*. It is convenient to fill tube *B* with a saturated solution of sodium nitrite, since this absorbs the ultraviolet lines of the mercury arc but transmits the blue-violet line λ4358 with great intensity. Figure 5.4 shows the *S* Raman spectrum of CCl_4. It will be seen that the same pattern of Raman lines is *E* excited by each of the strong mercury lines. Figure 5.4 (*d*) illustrates the Raman spectrum of gaseous hydrogen, showing two sets of lines on the side toward the red DS' of the exciting line, which in this case was λ2536. Occasionally still fainter lines are seen on the violet side, two of which are visible in (*d*) and three in (*e*). This is also sometimes observed in the case of fluorescence. Because in these lines the wavelength of modified light is shorter than the that of incident light, and they show violation of strokes law which is known as anti-stokes lines.

Theory of Scattering

A small elastically bound charged particle start motion, if an electromagnetic wave passes over it, because of electrified. We considered the case where the frequency of the wave was equal to the natural frequency of free vibration of the particle. We then obtain resonance and fluorescence under certain conditions, and selective reflection under others. In both cases there may exist a considerable amount of absorption. Scattering, on the other hand, takes place for frequencies not corresponding to the natural frequencies of the particles. The resulting motion of the particles is then one of *forced vibration.* If the particle is bound by a force obeying Hooke's law, this vibration will have the same frequency and direction as that of the electric force in the wave. Its amplitude, however, will be very much smaller than that which would be produced by resonance. Hence the amplitude of the scattered wave will be much less, and this accounts for the relative faintness of molecular scattering. The phase of the forced vibration will differ from that of the incident wave, and this fact is responsible for the difference of the velocity of light in the medium from that in free space. Further in this chapter we discuss about the scattering forms of dispersion.

The electromagnetic theory show the changes of wavelength in Raman effect and in fluorescence. If the charged oscillator is bound by a force which does not obey Hooke's law, but some more complicated law, it will be capable of reradiating not only the impressed frequency but also various combinations of this frequency with the fundamental and overtone frequencies of the oscillator. For the complete explanation of these phenomena, however, the electromagnetic theory alone is not adequate. It cannot explain the actual magnitudes of the changes in frequency nor the fact that these are predominantly toward lower frequencies. To analyse this quantum theory is necessary. Rayleigh scattering yields a characteristic distribution of intensity in different directions

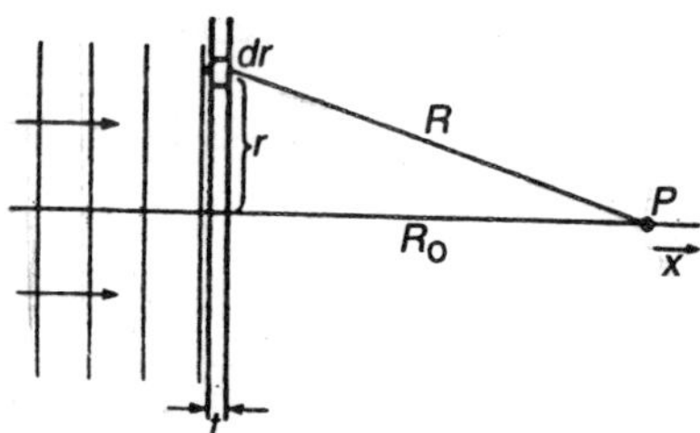

Figure 5.8 : ***Geometry of scattering by a thin lamina.***

with respect to that of the primary beam. According to electromagnetic theory scattered light is strongly polarized. These features are in general agreement with the predictions of the electromagnetic theory. We shall not discuss them, however, until we have taken up the subject of polarization

Scattering and Refractive Index

Due to the effect of scattering the velocity of light is different in matter and vacuum. The individual molecules scatter a certain part of the light falling on them, and the resulting scattered waves *interfere* with the primary wave, bringing about a change of phase which is equivalent to an alteration of the wave velocity. In this chapter we will discuss about the connection between scattering and refractive index in detail.

Figure 5.8 shows a plane wave striking an infinitely wide sheet of a transparent material, which has small thickness compared to the wavelength. Let the electric vector in this incident wave have unit amplitude, so that in the exponential notation. It may be represented at a particular time by $E = e^{ikx.}$ If the fraction of the wave that is scattered is small, the disturbance reaching some point P will be essentially the original wave, plus a small contribution due to the light scattered by all the atoms in the thin lamina. To evaluate the latter, we note that its intensity is proportional to the coefficient α_s of Eq. (5.2). This measures the fractional decrease of intensity by scattering in traversing the small thickness t, to which the scattered intensity must be proportional. We therefore have

$$-\frac{dI}{I} = \alpha_s t \approx I_s \tag{5.3}$$

The intensity scattered by a single atom, since there are Nt atoms per unit area of the lamina, becomes

$$I_1 \approx \frac{\alpha_s t}{Nt} = \frac{\alpha_s}{N}$$

and the amplitude

$$E_1 \approx \sqrt{\frac{\alpha_s}{N}}$$

If waves scattering non coherently from the different centers than this relation works as in smoke particles. The present case of Rayleigh scattering in the forward direction must be taken as *coherent,* however, so that all waves leave the scatterer in phase with each other we should consider amplitude in place of intensities, than the total scattered amplitude would be as—

$$E_s \approx Nt\sqrt{\frac{\alpha_s}{N}} = t\sqrt{\alpha_s N}$$

The complex amplitude at P is obtained by integrating this quantity over the surface of the lamina, and adding it to the amplitude of the primary wave. Thus the resultant would be as—

$$E + E_s = e^{ikR_0} + t\sqrt{\alpha_s N}\int_0^\infty \frac{2\pi r\, dr}{R} e^{ikR}$$

where the factor $1/R$ enters because of the inverse-square law.

Now since $R^2 = R_0{}^2 + r^2$, we have $r\,dr = R\,dR$, and the integral may be written

$$\int_0^{\infty} \frac{2\pi}{r} e^{ikR} r \; dr = 2\pi \int_{R_0}^{\infty} e^{ikR} \; dR = \frac{2\pi}{ik}\left[e^{ikR}\right]_{R_0}^{\infty}$$

Since the wave trains always have a finite length, the scattering as $R \to \infty$ can contribute nothing to the coherent wave. On integration the lower limit the substitution would be—

$$\begin{aligned} E + E_s &= e^{ikR_0} - t\sqrt{\alpha_s N}\frac{\lambda}{i} e^{ikR_0} \\ &= e^{ikR_0} + t\sqrt{\alpha_s N}\, i\,\lambda e^{ikR_0} \\ &= e^{ikR_0}\left(1 + i\lambda t\sqrt{\alpha_s N}\right) \end{aligned}$$

By our original assumption, the second term in parentheses is small compared with the first. These will be recognized as the first two terms in the expansion of $e^{i\lambda t\sqrt{\alpha_s N}}$, and may here be equated to it, giving

$$E + E_s = \exp ik\, R_0 \exp\left(i\lambda t\sqrt{\alpha_s N}\right) = \exp\left[i\left(kR_0 + \lambda t\sqrt{\alpha_s NM}\right)\right]$$

At P the phase of the wave can be altered by the amount $\lambda t\sqrt{\alpha_s N}$. But we know that the presence of a lamina of thickness t and refractive index n gives a phase retardation of $(2\pi/\lambda)$ $(n - 1)\, t$. Hence

$$\lambda t\sqrt{\alpha_s N} = \frac{2\pi}{\lambda}(n-1)t$$

and finally
$$n - 1 = \frac{\lambda^2}{2\pi}\sqrt{\alpha_s N} \qquad (5.4)$$

Thus we get Rayleigh's law of scattering. Since, by Equation (5.3), I_s is proportional to a_s, this scattered intensity varies as $1/\lambda^4$, assuming n to be independent of wavelength. In our derivation no absorption has been considered, so that the equation is valid only for wavelengths well away from any absorption bands. Further we discuss about the behaviour of refractive index for wavelength of an absorption band.

6

Polarization

Polarization

The shape of VHF (very high frequency) television antennas is vertical in England and horizontal in North America. The difference is due to the direction of oscillation of the electromagnetic waves carrying the TV signal. In England, the transmitting equipment is designed to produce waves that are **polarized** vertically; that is, their electric field oscillates vertically. So, for the electric field of the incident television waves to drive a current along an antenna and thus provide a signal to a television set the antenna must be vertical. The waves are polarized horizontally in North America.

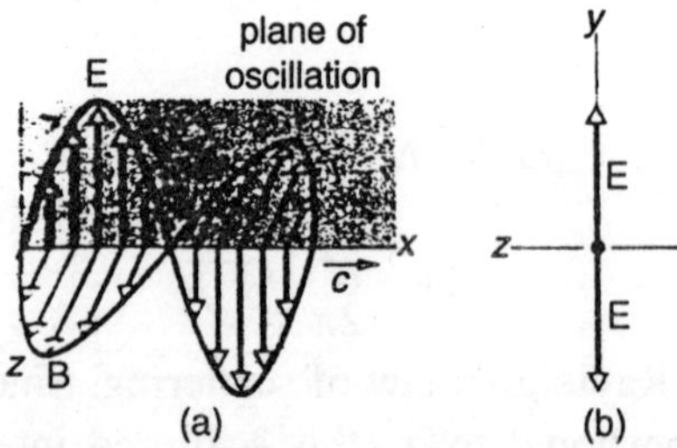

Figure 6.1 : *(a) The plane of oscillation of a polarized electromagnetic wave. (b) To represent the polarization, we view the phase of oscillation "head-on" and indicate the amplitude of the oscillating electric field.*

An electromagnetic wave with its electric field oscillating parallel to the vertical y axis. The plane of oscillation of the wave is that plane which having E vector and in this case waves are plane polarized in the *y* direction. We can represent the wave's **polarization** (state of being polarized) by showing the extent of the electric held oscillations in a "head-on" new of the plane of oscillation, as in Figure (6.1 b).

Polarized Light

If electromagnetic waves are emitted from both, a television station and common source (as bulb) of light than in case of television station it get same polarization and in case of common source it get polarized randomly. That is, the electric field at any given point is always perpendicular to the direction of travel of the waves but changes directions randomly. So if we try to represent a head-on view of the oscillations over some time period, we do not have a simple drawing like that of Figure 6.1 (b) instead we have a mess like that in Figure 6.2 (a).

From figure 6.2 (a) we can simplify the mass by resolving each electric field according to the principle into *y* and *z* components and then finding the net fields along the *y* axis and the *z* axis separately, as shown in Figure 6.2. In doing so, we mathematically change unpolarized light into the superposition of two polarized waves whose planes of oscillation are perpendicular to each other. The result is the double

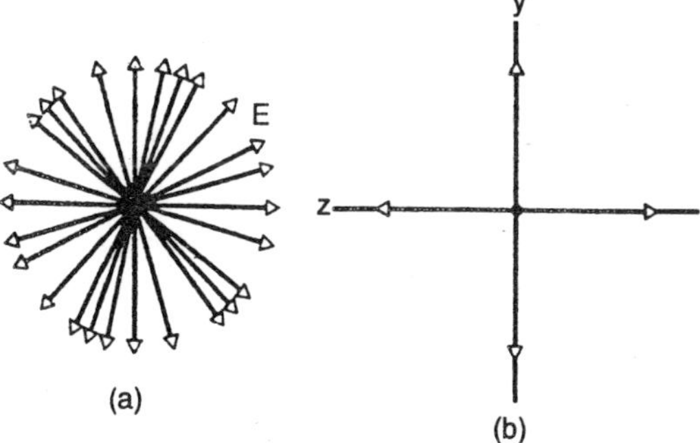

Figure 6.2 : ***(a) Unpolarized light consists of waves with randomly directed electric fields. Here the waves are all traveling along the same axis, directly out of the page, and all have the same amplitude*** E. ***(b) A second way of representing unpolarized light: the light is the superposition of two polarized waves whose planes of oscillation are perpendicular to each other.***

Arrow representation of Figure 6.2 (b) which simplifies drawings of unpolarized light. Similarly we can represent light that is partially polarized (its field oscillations are not completely random as in Figure 6.2 (a) nor are they parallel to a single axis as in Figure 6.1 (b). In that case we would use the longer arrow than other in double arrows.

In figure 6.3 we can see the transformation of unpalorized visible light in to polarized light by passing it through a polarizing sheet. Such sheets, commercially known as Polaroids or Polaroid filters, were invented in 1932 by Edwin Land while he was an undergraduate student. A sheet consists of certain long molecules embedded in plastic. When the sheet is manufactured, it is stretched to align molecules in parallel rows, like rows in a plowed field. If we will pass the light through the sheet the components of electric field of one direction pass through the sheet and components perpendicular to that direction are absorbed by the molecules and disappear.

We shall not dwell on the molecules but, instead, shall assign to the sheet a *polarizing direction,* along which electric field components are passed:

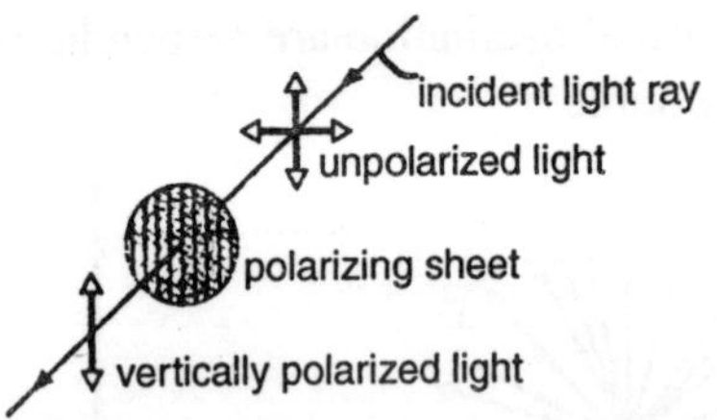

Figure 6.3 : ***Unpolarized light becomes polarized when it is sent through a polarizing sheet. Its direction of polarization is then parallel to the polarizing direction of the sheet, which is represented here by the vertical lines drawn in the sheet.***

An electric field component parallel to the polarizing direction is passed *(transmitted)* by a polarizing sheet; a component perpendicular to it is absorbed.

Thus the light emerging from the sheet consists of only the components that are parallel to the polarizing direction of the sheet; hence the light must be polarized in that direction. In Figure 6.3 the vertical components are transmitted by the $_t$ sheet; the horizontal components are absorbed. The waves which are transmitted now polarized vertically.

We next consider the intensity of the transmitted light. We start with unpolarized light, whose electric field oscillations we can resolve into y and z components as in Figure 6.1 (b). Further, we can arrange for the y axis to be parallel to the polarizing direction of the sheet. Then only the y components of the light are passed by the sheet; the z components are absorbed. As suggested by Figure 6.1 (b), if the original waves are randomly oriented, the sum of the y components and the sum of the z components are equal. When the z components are absorbed, half the intensity I_0 of the original light is lost. The emerging polarized light which represent intensity by I would be—

$$I = \frac{1}{2} I_0 \tag{6.1}$$

Let us call this the *one-half rule;* we can use it *only* when the light reaching a polarizing sheet is unpolarized.

Assuming that, a polarizing sheet getting light which is already polarized. Figure 6.4 shows a polarizing sheet in the plane of the page and the electric field **E** of such a polarized light wave traveling toward the sheet (and thus prior to any absorption). We can resolve **E** into two components relative to the polarizing direction of the sheet: parallel component E_y is transmitted by the sheet, and perpendicular component E_Z is absorbed.

Because in between **E** and the polariging direction of the sheet θ is the angle than the parallel component of transmission would be as

$$E_y = E \cos\theta \tag{6.2}$$

From the equation which show that the intensity of an electromagnetics wave is proportional to the square of the electric filed's magnitude. In our present case then, the intensity I of the emerging wave is proportional to E_y^2 and the intensity I_0 of the original wave is proportional to E^2. Hence, from Equation 6.2 we can write $I/I_0 = \cos^2\theta$, or

$$I = I_0 \cos^2 \theta \qquad (6.3)$$

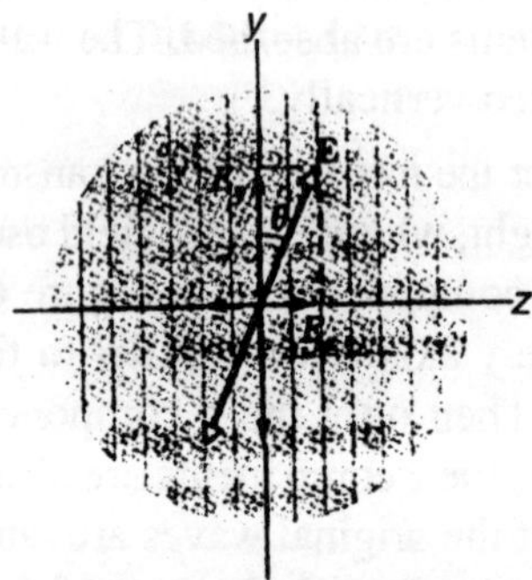

Figure 6.4 : *Polarized light approaching a polarizing sheet. The electric field E of the light can be resolved into components* Ey (parallel *to the polarizing direction of the sheet) and* E_z *(perpendicular to that direction). Component* E_y *will be transmitted by the sheet: component* E_z will *be absorbed.*

Let us call this the *cosine-squared rule;* we can use it *only* when the light reaching a polarizing sheet is already polarized. The transmitted intensity I_o is a maximum and is equal to the original intensity I_0 when the original wave is polarized parallel to the polarizing direction of the sheet (when θ in Eq. 6.3 is 0° or 180°). If original wave is polarizing with 90° angle means perpendicular to the sheet, than I would be zero.

Figure 6.5 shows an arrangement in which initially unpolarized light is sent through two polarizing sheets P_1 and P_2. (Often, the first sheet is called the *polarizer,* and the second the *analyzer.)* Because the polarizing direction of P_1 is vertical, the light transmitted by P_1 to P_1 is polarized vertically. If the polarizing direction of P_2 is also vertical, then all the light transmitted by *P1 is* transmitted by P_2. If the polarizing direction of P_2 is horizontal, none of the light transmitted by P_1 is transmitted by *P2. We* reach the same conclusions by considering only the *relative ori*entations of the two sheets: if their polarizing directions are parallel, all the light passed by the first sheet is passed by the second sheet. If wave is polarizing in parallel direction than no light is passed by the second sheet.

Finally, if the two polarizing directions of Fig 6.5 make an angle between 0° and 90°, some of the light transmitted by *P1 will* be transmitted by P_2. From eq. 6.3 we can find the intensity of that light.

The polarization can also be done by other than polarizing sheets such as by reflection and by scattering from atoms or molecules. In *scattering,* light that is intercepted by an object, such as a molecule is sent off in many, perhaps random, directions. An example is the scattering of sunlight by molecules to the atmosphere, which gives the sky its general glow.

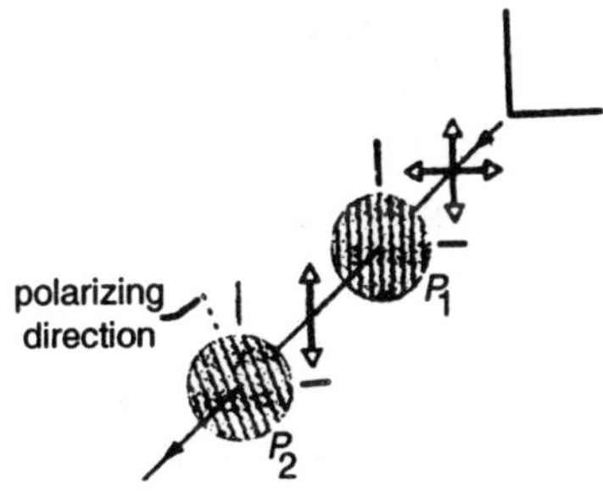

Figure. 6.5 : ***The light transmitted by polarizing sheet vertically polarized, as represented by the vertical arrow. The amount of that light that is then transmitted by polarizing depends on the angle between the polarization direction of that light and the polarizing direction of P_2 (indicated by the drawn in the sheet and by the dashed line).***

Indeed light directly from seen is unpolarized but light from sky is partially polarized due to scattering. Bees use the polarization of sky light in navigating to and from their hives. Similarly, the Vikings used it to navigate across the North Sea when the daytime Sun was below the horizon (because of the high latitude of the North Sea). These early seafarers had discovered certain crystals (now called cordierite) that changed colour when rotated in polarized light. If we look through a crystal rotating about their line of sight than it can shown as the hidden sun, thus determine the south.

Amplitude of oscillation is variable as in case of one direction it has greater amplitudes than other directions. However, when At light is incident at a particular incident angle, called the *Brewer angle* θ_B, the reflected light has only perpendicular components.

The reflected light is then fully polarized perpendicular to the plane of incidence The parallel components of the incident light do act disappear; they and perpendicular components form die light that is refracted through the glass surface.

As we know dielectrie materials as glass, and water etc. can polarize light partically or fully by reflaction. When You intercept sunlight reflected from such a surface, you' see a bright spot (the glare) on the surface where the reflection takes place If the surface is horizontal the reflected light is partially or fully polarized horizontally. In polarizing sunglasses if these lenses are keep in vertical polarizing direction, than such glare can be eliminated from the horizontal surface.

Brewster's Law

If the angle of incident of a light is Brewster (θ_B) than the reflected and refracted rays would be perpendicular to each other it, has proved. Because the reflected ray is reflected at the angle θ_B and the refracted ray is at angle θ_r, we have

$$\theta_B + \theta_r = 90°$$

Arbitrarily assigning subscript 1 ipso the material through which the incident and reflected rays travel, we have, from that equation,

$$n_1 \sin\theta_B = n_2 \sin\theta_r$$

Combining these equations leads to

$$n_1 \sin\theta_B = n_2 \sin(90° - \theta_B) = n_2 \cos\theta_B$$

which gives us

$$\theta_B = \tan^{-1}\frac{n_2}{n_1} \text{ (Brewster angle)} \qquad (6.4)$$

(Note carefully' that the subscripts in Equation 6.4 are not arbitrary because of our decision as to their meanings.) If the incident and reflected rays travel :in air, wee can approximate n_1 as unity and let n represent n_2 in order. to write Equation 6.4 as

$$\theta_B = \tan^{-1} n \text{ (Brewster's law)} \tag{6.5}$$

the modified form of equation 6.4 is called Brewsler's law. Like θ_B, it is named after Sir David Brewster who found both experimentally in 1812.

The direction of S (and thus of the wave's travel and the energy transport) is perpendicular to the directions of both **E** and **B**. The time-averaged rate per unit area at which energy is transported is $\bar{S}$, which is called the intensity I of the wave:

$$I = \frac{1}{c\mu_0} E_{rms}^2, \tag{6.6}$$

in which $E_{rms} = E_m / \sqrt{2}$. A point source of electromagnetic waves units the waves isotropically, that is with equal intensity in all direction. A point source of power P_s would have intensity of the waves at distance is polarization.

$$I = \frac{P_s}{4\pi r^2} \tag{6.7}$$

Radiation Pressure

When a surface intercepts electromagnetic radiation, a force and a pressure are exerted on the surface. If surface obsorbs all the radiation than the force would be—

$$F = \frac{IA}{c} \text{ (total absorption)} \tag{6.8}$$

in which l is the intensity of the radiation and A is the area of the surface perpendicular to the path of the radiation. If the radiation is totally reflected back along its original path, the force is

$$F = \frac{2IA}{c} \text{ (total reflection back along path).} \tag{6.9}$$

The radiation pressure p_r is the force per unit area:

$$p_r = \frac{I}{c} \text{ (total absorption)} \tag{6.10}$$

and

$$p_r = \frac{2I}{c} \text{ (total reflection back along path).} \tag{6.11}$$

Polarization

If the electric field vectors of electro magnetic waves are in a single plane than this palne is called oscillation. Light waves from common sources are not polarized, that is, they are **unpolarized** or **randomly polarized.**

Polarizing Sheets

When a polarizing sheet is placed in the path of light, only electric field components of the light parallel to the sheet's **polarizing direction** are transmitted by the sheet ; components perpendicular to the polarizing direction are absorbed. If light emerges from a polarizing sheet than the direction of the polarization would be parallel to the sheet.

If the original light is initially unpolarized, the transmitted intensity I is half the original intensity I_0:

$$I = \frac{1}{2} I_0$$

If the original light is initially polarized, the transmitted intensity depends on the angle θ between the polarization direction of the original light and the polarizing direction of the sheet:

$$I = I_0 \cos^2 \theta \qquad (6.12)$$

Geometrical Optics

Geometrical optics is a medium to represent the light waves as straight line rays.

Reflection and Refraction

When a light ray encounters a boundary between two transparent media, a reflected ray and a refracted ray generally appear. Both rays remain in the plane of incidence. The angle of reflection is equal to the angle of incidence, and the angle of refraction is related to the angle of incidence by

$$n_1 \sin\theta_1 = n_2 \sin\theta_2 \qquad (6.13)$$

where n_1 and n_2 are the indices of refraction of the media in which the incident and refracted rays travel.

Total Internal Reflection

A wave encountering a boundary across which the index of refraction decreases will experience total internal reflection if the angle of incidence exceeds a critical angle θ_c, where

$$\theta_c = \sin^{-1} \frac{n_2}{n_1} \tag{6.14}$$

Polarization by Reflection

If the direction os polarization is perpendicular to the plane of incidence than the wave will be fully polarized if it strikes a boundary as the Brewster angle θ_B, where

$$\theta_B = \tan^{-1} \frac{n_2}{n_1} \text{ (Brewster angle)} \tag{6.15}$$

Polarization by Reflection

Suppose a wave incidence on a diclectrie plane. We assume that the electric vector associated with the incident wave lies in the plane of incidence as shown in Figure 6.6 the angle of incidence (θ) is such that

$$\theta = \theta_p = \tan^{-1}\left(\frac{n_2}{n_1}\right) \tag{6.16}$$

then the reflection coefficient is zero. Thus, if an unpolarized beam is incident at this angle, then the reflected beam will be linearly polarized with its electric vector perpendicular to the plane of incidence [see Figure 6.6 (b)]. Equation 6.16 is referred to as the Brewster's law and at this angle of incidence, the reflected and the transmitted rays are at right angles to each other, θ_p represents the angle of polarization or brewster angle.

For the air-glass interface, $n_1 = 1$ and $n_2 \simeq 1.5$ giving $\theta_p \simeq 57°$. The transmitted beam is partially polarized and if one uses a large number of reflecting surfaces, one would obtain a nearly plane polarized beam (see Figure 6.6)].

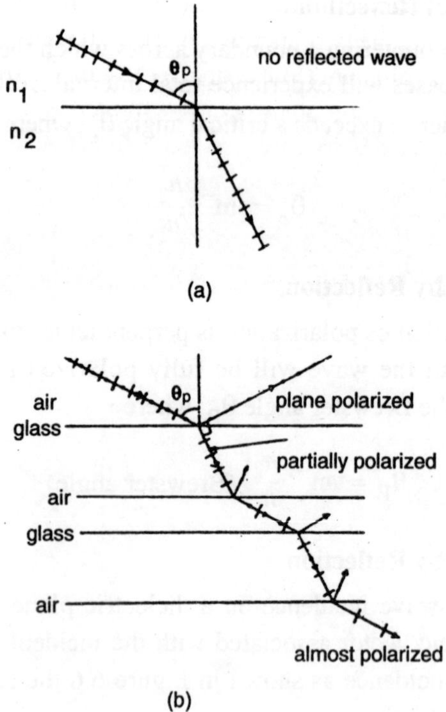

Figure 6.6 : ***(a) If a linearly polarized wave (with its E in the plane of incidence) is incident on the interface of two dielectrics with the angle of incidence equal to θ_p (= $\tan^{-1} n_2/n_1$) then the reflection coefficient is zero. (b) Thus; if an unpolarized beam is incident at this angle, the reflected beam is plane polarized whose electric vector is perpendicular to the plane of incidence. The transmitted beam is partially polarized and if this beam is made to undergo several reflections, then the emergent beam is almost plane polarized with its electric vector in the plane of incidence.***

Polarization by Double Refraction

We will discuss the phenomenon of double refraction and will show that when an unpolarized beam enters an anisotropic crystal, it splits up into two beam each of them being characterized by a certain state of polarization. To get a linear polarized beam we would have to eliminate the one beam.

A simple method for eliminating one of the beams is through selective absorption; this property of selective absorption is known as dichroism. A crystal like tourmaline has different coefficients of absorption for the two linearly polarized beams into which the incident beam splits up. Consequently, one of the beams gets absorbed quickly and the other component passes through without much attenuation. Accordingly to get a linear polarized be from a unpolarized beam we have to pass it through a tourmaline crystal.

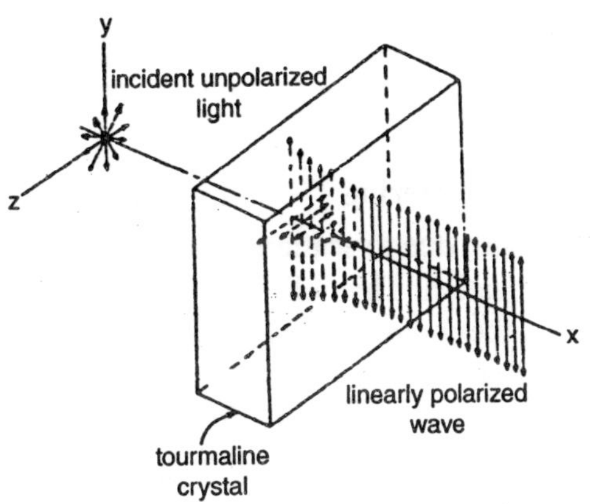

Figure 6.7 : ***When an unpolarized beam enters a dichroic crystal like tourmaline, it splits up into two linearly polarized components. One of the components gets absorbed quickly and the other component passes through without much attenuation.***

We can also eliminate a polarized beam by total internal reflection. The two beams have different ray velocities and as such the corresponding refractive indices will be different. If one can sandwich a layer of material whose refractive index lies between the two, then for one of the beams, the incidence will be at a rarer

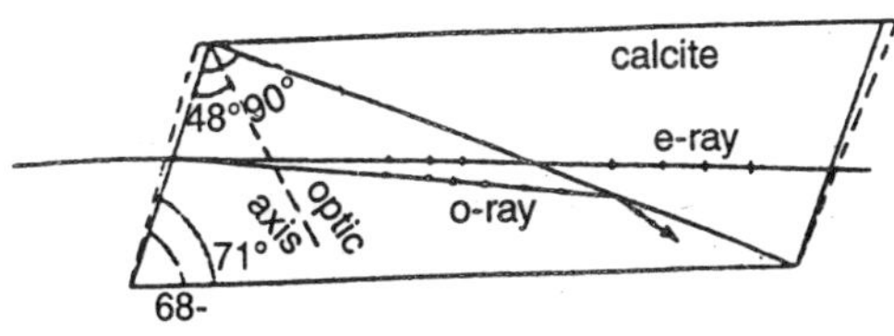

Figure 6.8 : ***The Nicol prism. The dashed outline corresponds to the natural crystal which is cut in such a way that the ordinary ray undergoes total internal reflection at the Canada balsam layer.***

medium and for the other it will be at a denser medium. This principle is used in a Nicol prism which consists of a calcite crystal cut in such a way that for the beam, for which the sandwiched material is a rarer medium, the angle of incidence is greater than the critical angle. Thus this particular beam will be eliminated by total internal reflection. In properly cut calcite crystal, due to a layer of Canada Balsam in between a ray undergoes total internal reflection (see figure 6.8). The extraordinary component passes through and the beam emerging from the crystal is linearly polarized.

Polarization by Scattering

In case of a gas, due to scattering at 90° an unpolarized beam get linear polarization. This follows from the fact that the waves propagating in the *y*-direction produced by the *x*-component of the dipole oscillations (see Figure 6.9). The *x*-component of the dipole oscillations will produce no field in the *y*-direction. It may be of interest to mention that it was through scattering experiments that Barkla could establish the transverse character of X-rays. Clearly, if the incident beam is linearly polarized with its electric vector along the *x*-direction, then there will be no scattered light along the *x*-axis. Now allow to undergo a scattered wave to undergo a further scattering to analyse this. (see fig 6.9)

Malus's Law

Let us consider a polarizer which has a pass-axis parallel to the *y*-axis i.e. if a beam propagating in the z-direction is incident on the polarizer, then the electric vector associated with the emergent wave will oscillate along the *y*-axis. Consider the incidence of a linearly polarized beam with its electric vector oscillating along a direction which makes an angle θ with *y*-axis (see Figure 6.10). In a electric field having amplitude E_0, the amplitude of the wave emerging from the polaroid wile be $E_0 \cos \theta$ and thus the intensity of the emerging beam would be as—

$$I = I_0 \cos^2 \theta \tag{6.17}$$

In this equation I_0 is the intensity of the incident beam. Equation (10) presents the Malus' Law. Thus, if a linearly polarized beam is incident Polaroid and if the Polaroid is rotated about the z-axis, then the intensity of the emergent wave will vary according to the above law. For example, if the polaroid shown in Figure 6.10 is rotated in the clockwise direction, then the intensity will increase till the pass-axis is parallel to OE; a further rotation will result in a decrease in intensity till the pass axis is perpendicular to OE, where the intensity will be almost zero. On further rotating it , to reach its original position it passes through a maximum and a minimum position.

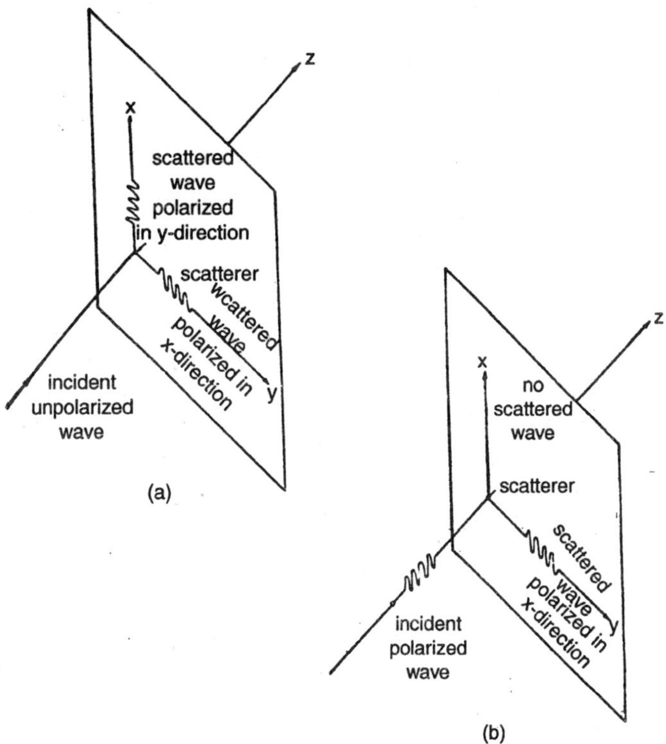

Figure 6.9 : ***(a) If the electromagnetic wave is propagating along the z-direction, then the scattered wave along any direction perpendicular to the z-axis will be linearly polarized. (b) If a linearly polarized wave (with its E oscillating along the x-direction) is incident on a dipole, then there will be no scattered wave in the x-direction.***

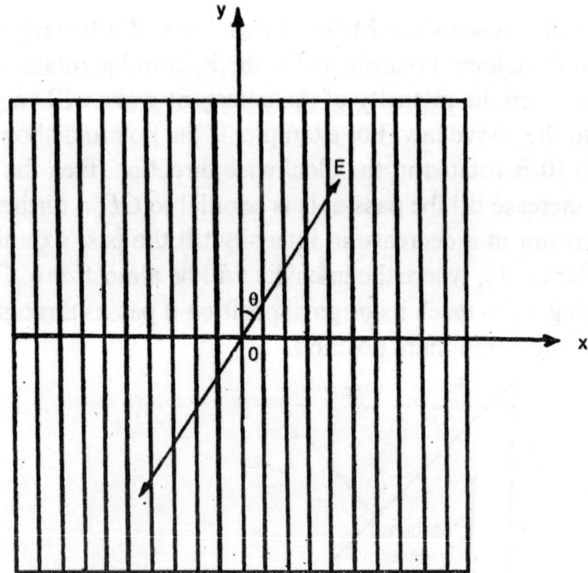

Figure 6·10 : *Incidence of a linearly polarized plane wave on a polaroid whose pass-axis is along the y-direction.*

Superposition of Tow disturbances

Let us consider the propagation of two linearly polarized electromagnetic waves (both propagating along the z-axis) with their electric vectors oscillating along the x-axis. The electric field of waves are as follows—

$$\mathbf{E}_1 = \hat{\mathbf{x}} a_1 \cos(kz - \omega t + \theta_1) \quad (6.18)$$

$$\mathbf{E}_2 = \hat{\mathbf{x}} a_2 \cos(kz - \omega t + \theta_2) \quad (6.19)$$

where a_1 and a_2 represent the amplitudes of the waves; $\hat{\mathbf{x}}$ represents the unit vector along the x-axis and θ_1 and θ_2 are phase constants resultant of these two waves would be given by

$$\mathbf{E} = \mathbf{E}_1 + \mathbf{E}_2 \quad (6.20)$$

The Mathematical Analysis

From above discussion as will obtain equation for an ellips which is as—

$$E_x = a_1 \cos \omega t \quad (6.21)$$

$$E_y = a_2 \cos(\omega t - \theta) = a_2 \cos\omega t \cos\theta + a_2 \sin\omega t \sin\theta$$

$$= a_2 \cos\omega t \cos\theta \pm a_2 \sqrt{1 - \cos^2 \omega t} \sin\theta$$

$$= \frac{a_2}{a_1} E_x \cos\theta \pm \sqrt{1 - (E_x / a_1)^2}\, a_2 \sin\theta \tag{6.22}$$

or $$\left(E_y - \frac{a_2}{a_1} E_x \cos\theta\right)^2 = a_2^2 \sin^2\theta \left[1 - \left(\frac{E_x}{a_1}\right)^2\right]$$

Rearranging, we get

$$\frac{E_x^2}{a_1^2} + \frac{E_y^2}{a_2^2} - \frac{2E_x E_y}{a_1 a_2} \cos\theta = \sin^2\theta \tag{6.23}$$

i.e.

$$\frac{E_y}{E_x} = (-1)^n \frac{a_2}{a_1} \tag{6.24}$$

which represents the equation of a straight line, implying a linearly polarized wave. For $\theta = \left(n + \frac{1}{2}\right)\pi, (n = 0, 1, 2, \ldots)$, Eq. (6.23) becomes

$$\frac{E_x^2}{a_1^2} + \frac{E_y^2}{a_2^2} = 1 \tag{6.25}$$

which represents an ellipse with its minor and major axes along the x- and y-axes. If we further have $a_1 = a_2$, then Eq. (6.25) will represent a circle.

In general, Eq (6.23) represents an ellipse and if ϕ is the angle between the coordinate axes and the axes of the ellipse, then

$$\tan 2\phi = \frac{2a_1 a_2 \cos\theta}{a_1^2 - a_2^2} \tag{6.26}$$

Geometry of Calcite Crystal and Optical Axis

If we will mark a point in paper and put a calcite cyrstal over it than we get two images of that point. The appearance of the two dots is due to the phenomenon of double refraction. In general, a ray entering such a crystal will split up into two rays and for a crystal like calcite, one of the rays will follow the Snell's law of refraction and the other will not. The former is referred to as the ordinary ray and the latter as the extraordinary ray. The velocity of the ordinary ray is the same in all directions whereas the velocity of the extraordinary ray is different in different directions; a substance which exhibits different properties in different directions is called an anisotropic substance. If the two velocities are equal for one direction than this direction would called the optic axis of the crystal. Indeed, if we imagine a point source imbedded in as crystal like calcite, then we will have two different sets of wavefronts, one corresponding to the ordinary ray which is spherical and the other corresponding to the extraordinary ray which is an ellipsoid of revolution. An ellipsoid of revolution is a surface obtained by rotating an ellipse about either the major or the minor axis; this axis of rotation is along the optic axis (see Figure 6.11). If the ellipsoid of revolution lies completely outside the sphere (i.e. if the velocity of the extraordinary ray is every-

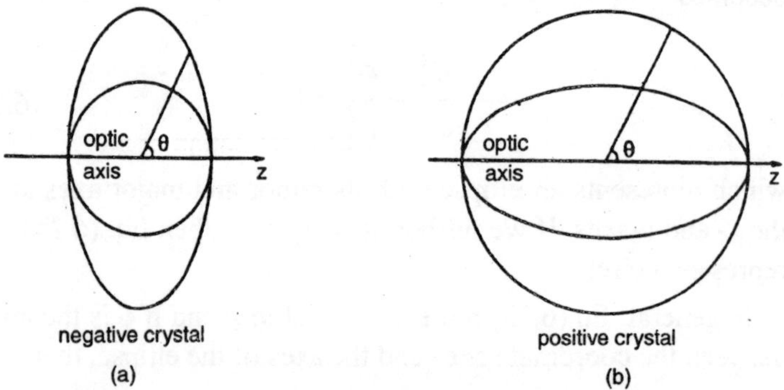

Figure 6.11 : ***The ellipsoid of revolution (which corresponds to the extraordinary ray) lies outside the sphere (which corresponds to the ordinary ray) in a negative crystal as shown in (a); in a positive crystal, the converse is true—see (b).***

where greater than the ordinary ray [see Figure 6.11 (a)] then the crystal is known as a negative crystal. This is indeed the case for calcite. On the other hand, if the ellipsoid of revolution lies completely inside the sphere, i.e., if the velocity of the extraordinary ray is everywhere less than the ordinary ray [see Figure 6.11] then the crystal is known as a positive crystal which is the case for quartz. The velocities of rays either ordinary or extraordinary, can be calculated by the following equation.

$$v_{ro} = v_x = c / n_o \text{ (ordinary ray)} \tag{6.27}$$

$$\frac{1}{v_{re}^2} = \frac{\sin^2\theta}{v_z^2} + \frac{\cos^2\theta}{v_x^2} \text{ (extraordinary ray); } v_z = \frac{c}{n_e} \tag{6.28}$$

Here θ is the angle between ray and optic axis which is parallel to the z-axis, v_x $(= c / n_o)$ and v_z $(= c / n_e)$ are constants of the crystal. If we plot v_{re} as a function of θ we will obtain an ellipsoid of revolution; on the other hand, since v_{ro} is independent of θ, if we plot v_{ro} (as a function of θ), we will obtain a sphere. Along the optic axis, $\theta = 0$ and $v_{ro} = v_{re}$ When $v_z < v_x$, v_r will be greater than v_{ro} (i.e. it will be negative crystal) and conversely.

Both the ordinary and the extraordinary waves are linearly polarized, the vibration corresponding to the ordinary wave is always at right angles to the direction of the ray as well as to the optic axis. On the other hand, for the extraordinary wave, the direction of vibration is at right angles to the propagation vector **k** and lies in the plane containing the optic axis and the ray direction. In this section, we will consider the refraction of a plane electromagnetic wave incident on a negative crystal like calcite. This can also be alone with positive crystals.

Normal incidence

suppose a uniaxial crystal refract a plane electromagnetic wave. The direction of the optic axis is shown as a dashed line 6.12 In order to determine the ordinary ray we draw a sphere of —radius v_x from the point *B*. Similarly, we draw another sphere from the point *D*. The common tangent plane to these spheres is shown as *OO′* which represents the wavefront corresponding to the ordinary refracted ray. In order to determine the extraordinary ray, we draw an ellipse (centred at the point B) with its minor axis along the optic axis. The ellipsoid of revolution is obtained by rotating the ellipse about the optic axis. Similarly, we draw another ellipsoid of revolution from the point *D*. The common tangent plane to these ellipsoids (which will be perpendicular to **k**) is shown as *EE'* in the figure 6.12. If we join the point *B* to the point of contact **0**, then corresponding the incident ray *AB*, the direction of the ordinary ray will be along *BO*. In this way matching point *B* with *E* (the ellipsoid of revolution and the tangent plane) the direction of the extraordinary ray would be along *BE*.

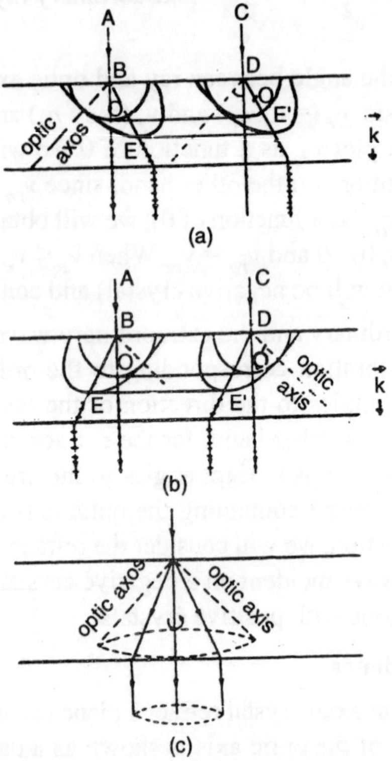

Figure 6.12 : ***(a) and (b) The refraction of a plane wave incident on a negative crystal whose optic axis is along the dashed line. (c) For a ray incident normally, if the crystal is rotated about a line which is normal to the surface, the extraordinary ray will rotate on the periphery of a cone.***

In fact, if the direction of optic axis is different then the direction of ordinary ray will be same but extraordinary would have different direction (see figure 6,12) Thus if a ray is incident normally on a calcite crystal and if the crystal is rotated about the normal, then the extraordinary ray will also rotate (about the normal) on the periphery of a cone; each time the ray will lie in the plane containing the normal and the optic axis [see figure 6.10(c)]

As mentioned earlier, the direction of vibrations for the ordinary ray is normal to the optic axis and the ordinary ray; as such, in this case, they will be normal to the plane of the paper, Therefore, the directions of these vibrations have been shown as dots in 6.12. Similarly, since the direction of vibrations for the extraordinary ray is perpendicular to **k** and lies in the plane containing the extraordinary ray and the optic axis, they are along the small straight lines drawn on the extraordinary ray in Figure 6.12.

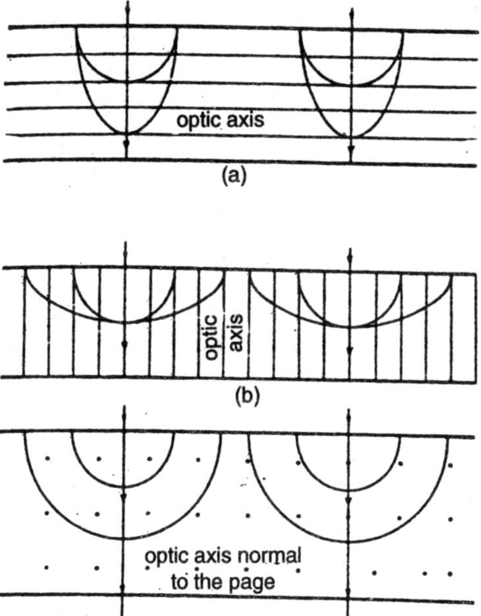

Fig 6.13 : ***Propagation of a plane wave incident normally on a negative uniaxial crystal. In (a) and (b) the optic axis is shown as parallel straight lines and in (c) the optic axis is perpendicular to the plane of the figure and is shown as dots. In each case, the extraordinary and the ordinary rays travel in the same direction.***

In this way incident ray splits in to two rays which travel in different direction and these came out from the crystal we get two linear polarized beams. In the above case, we have assumed the optic axis to make an arbitrary angle α with the normal to the surface.

In the special case of $\alpha = 0$ and $\alpha = \pi / 2$, the ordinary and the extraordinary rays travel along the same direction as shown in Figs 6.13 (a) and 6.13 (b). Figure 6.13 (c) corresponds to the case when the optic axis is normal. to the plane of the paper; and as such, the section of the extraordinary wavefront in the plane of the paper will be a circle. Once again, both the ordinary and the extraordinary rays travel along the same direction. It may be mentioned that Figures 6.13 (a) and 6.13 (c) indeed correspond to the same configuration in fact, they represent two different cross-sections of the same set of spherical and ellipsoidal wavefronts.

In fig 6.13 : show the configuration that both the rays travel with same direction but propagate with different velocities. Fig 6.14 show. This phenomenon is used in the fabrication of quarter and half-wave plates. On the other hand, in the configuration shown in Figure 6.13 (b), both the rays not only travel in the same direction but they also propagate with the same velocity.

Oblique incidence

Now we see in fig (6.14) (a) oblique incident on a negative uniaxial crystal, of a plane wave. Once again we use Huygens' principle to determine the shape of the refracted wavefronts. Let *BD* represent the incident wavefront. If the time taken for the disturbance to reach the point *F* from the point *D* is t, then with *B* as centre we draw a sphere of radius $v_x t$ and an ellipsoid of revolution of semi-minor and semi-major axes $v_x t$ and $v_z t$ respectively; the semi-minor axis is along the optic axis. We draw tangent planes *FO* and *FE* from *a* point *F* to the sphere and the ellipsoid of resolution respectively. These planes would represent the refracted wavefronts corresponding to the ordinary and the extraordinary rays respectively. If the points of contacts are *0* and *E*, then the ordinary and extraordinary refracted rays will propagate along *BO* and *BE* respectively; this can also be shown by using Fermat's

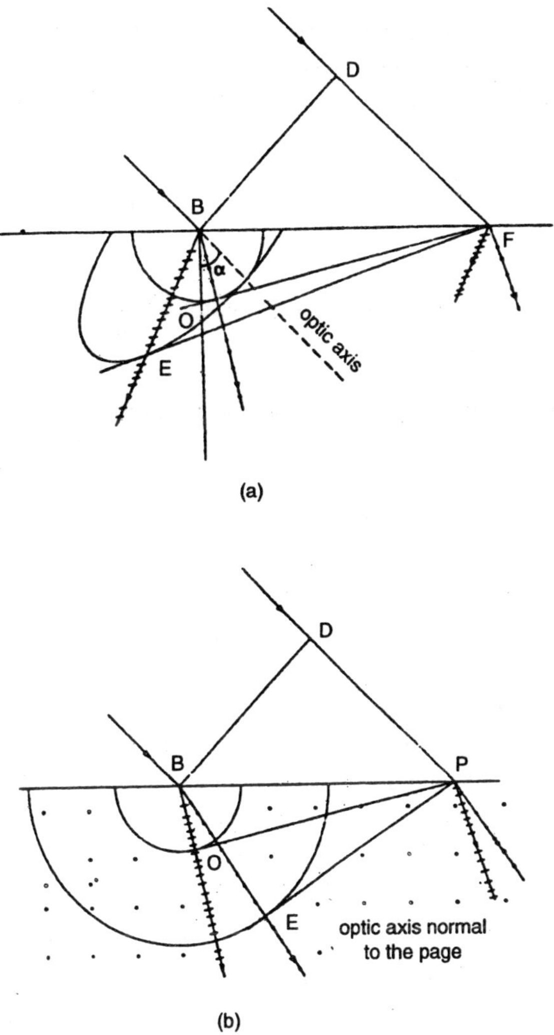

Figure 6.14 : *Refraction of a plane wave incident obliquely on a negative uniaxial crystal. In (a), the direction of the optic axis is along the dashed line. In (b), the optic axis is perpendicular to the plane of the paper.*

principle. The directions of vibration of these rays are shown by dots and small lines respectively and are obtained by using the general rules discussed earlier. In case of $\alpha = 0$ and $\alpha = \frac{\pi}{2}$ the shape of the refracted wavefrants can so easily measure

Figure 6.14 (b) corresponds to the case when the optic axis is normal to plane of incidence. The sections of both the wavefronts will be circles; consequently, the extraordinary ray will also satisfy Snell's law and we will have

$$\frac{\sin i}{\sin r} = \frac{c}{c_z} \tag{6.29}$$

Indeed the refractive index of the extraordinary ray for a negative crystal is defined by

$$n_e = \frac{c}{\text{maximum velocity of the extraordinary ray}}$$

$$= \frac{c}{v_z}$$

Similarly for a positive crystal, the refractive index is given by

$$n_e = \frac{c}{\text{minimum velocity of the extraordinary ray}}$$

$$= \frac{c}{v_z}$$

The refractive indices in above equations are the principal refractive indices. For example, for calcite, which is a negative crystal, the values of n_o and n_e are 1.65836 and 1.48641 respectively; for quartz, which is a positive crystal, n_o and n_e are 1.54425 and 1.55336 respectively; here n_o represents the refractive index for the ordinary ray.

Interference of Polarized Light

As we studied, when *a* ray incident on *a* doubly refracting crystal it splits in to two rays. (each having ceraine state of polarization). The direction of vibration associated with the ordinary ray is at right angles to the direction of the ray as well as to the optic axis; and the direction of vibration for the extraordinary ray lies in the plane containing the optic axis and the ray direction and is perpendicular to **k**. The considerations assumed that the incident beam was unpolarized. Here we shall see the normal incidence of *a* plane polarized beam and sate of polarization of the beam emerging from the crystal.

Suppose, optic axis is parallel to the surface of the crystal in figure 6.16. If the direction of vibration of the incident ray is perpendicular to the plane of the paper [see Figure 6.15 (a)], then the ray will pass through as an ordinary ray and the extraordinary component will be absent; the plane of the paper is assumed to contain the optic axis. In this way incident ray is in the plane of the paper, of the direction of vibration,_then the ray will pass through as an extraordinary ray and the ordinary component will be absent. For any other state of polarization (of the incident beam), both the extraordinary and the ordinary components will be present. Let the electric vector associated with the incident polarized beam make an angle ϕ with the z-axis; the z-axis is chosen to be parallel to the optic axis and the direction of propagation is along the x-axis [see Figure 6.15 (c)] Such a beam can be assumed to be a superposition of two linearly polarized beams (vibrating in phase), polarized along the y- and z-directions with amplitudes $E \sin \phi$ and $E \cos \phi$ respectively. The z-component (whose amplitude is $E \cos \phi$) passes through as an extraordinary beam propagating with speed c/n_e. The y-component (whose amplitude is $E \sin \phi$) passes through as an ordinary beam propagating with speed c/n_o (these are now wave velocities. Since $n_e \neq n_o$, the two beams will propagate with different velocities is as such, when they come out of the crystal, they will not be in phase. There for, the beam, superposition of these two beams polarized elliptically.

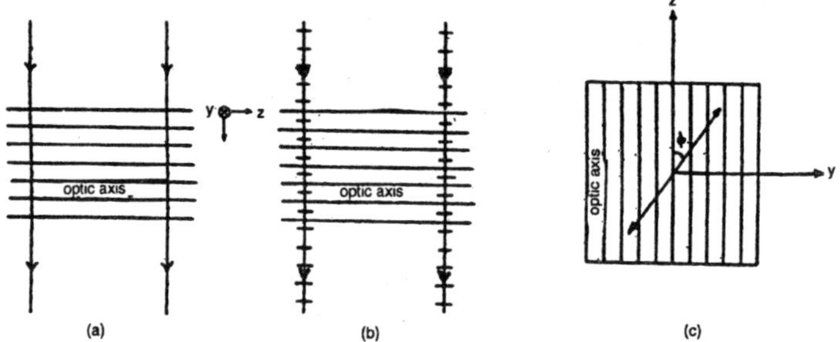

Figure 6.15 : A linearly polarized wave is incident normally on a crystal whose optic axis is parallel to the surface which is shown as parallel straight lines. The dots and dashes, in (a) and (b), denote the direction of polarization and in (c) the incident wave is linearly polarized at an angle ϕ with the z-axis.

Let the plane $x = 0$ represent the surface of the crystal on which the beam is incident. The y- and z-components of the incident beam can be written in the form

$$\left.\begin{aligned} E_y &= E\sin\phi\cos(kx-\omega t) \\ E_z &= E\cos\phi\cos(kx-\omega t) \end{aligned}\right\} \qquad (6.30)$$

where k (= ω/c) represents the free-space wave number. Thus, at $x = 0$, we will have

$$\left.\begin{aligned} E_y(x=0) &= E\sin\phi\cos\omega t \\ E_z(x=0) &= E\cos\phi\cos\omega t \end{aligned}\right\} \qquad (6.31)$$

Inside the crystal, the two components will be given by

$$\left.\begin{aligned} E_y &= E\sin\phi\cos\left(n_o kx-\omega t\right) \\ E_z &= E\cos\phi\cos\left(n_o kx-\omega t\right) \end{aligned}\right\} \qquad (6.32)$$

If the thickness of the crystal is d, then at the emerging surface, we will have

$$\left.\begin{aligned} E_y &= E\sin\phi\cos\left(\omega t-\theta_o\right) \\ E_z &= E\cos\phi\cos\left(\omega t-\theta_o\right) \end{aligned}\right\} \qquad (6.33)$$

where $\theta_o = n_o kd$ and $\theta_o = n_e kd$. By appropriately choosing the instant $t = 0$, the components may be rewritten as

$$\left.\begin{aligned} E_y &= E\sin\phi\cos(\omega t-\theta) \\ E_z &= E\cos\phi\cos\omega t \end{aligned}\right\} \qquad (6.34)$$

where

$$\theta = \theta_o - \theta_e = kd\left(n_o - n_e\right) = \frac{\omega}{c}\left(n_o - n_e\right)d \qquad (6.35)$$

It shows the difference of extraordinary and ordinary beam. Clearly, if the thickness of the crystal is such that $\theta = 2\pi, 4\pi, \ldots$, the emergent beam will have the same state of polarization as the-incident beam; if $\theta = \pi, 3\pi, \ldots$, emergent beam will again be linearly polarized but the electric vector will be along a direction. For any other value of θ the emergent beam will be elliptically polarized; [of course, there the propagation was along the z-axis and here it is along the axis].

Suppose $\phi = \frac{\pi}{4}$ and $\theta = \frac{\pi}{2}$ means y and z-components of the incident wave have equal amplitudes with *a* phase difference of $\frac{\pi}{2}$. Thus, for the emergent beam we have

$$E_y = \frac{E}{\sqrt{2}} \sin \omega t$$

$$E_z = \frac{E}{\sqrt{2}} \cos \omega t \tag{6.36}$$

which represents a circularly polarized wave because

$$E_y^2 + E_z^2 = \frac{E^2}{2} \tag{6.37}$$

In order to determine the direction of rotation of the electric vector, we note that at $t = 0$,

$$E_y = 0, E_z = \frac{E}{\sqrt{2}} \tag{6.38}$$

and at $t = \Delta t$,

$$E_y \simeq \frac{E}{\sqrt{2}} \omega \Delta t, E_z \simeq \frac{E}{\sqrt{2}} \tag{6.39}$$

Above equation conclude that with increase in time, the direction of electric vector rotates in clockwise direction and beam is left circularly polarised. In order-to introduce a phase difference of π/2, the thickness of the crystal should have a value given by the following equation:

$$d = \frac{c}{\omega(n_o - n_e)} \frac{\pi}{2} = \frac{1}{4} \frac{\lambda}{(n_o - n_e)} \tag{6.40}$$

where λ is the free-space wavelength. For calcite,

$$n_o = 1.65836, n_e = 1.48641 \tag{6.41}$$

which correspond to λ = 5893 Å and at 18°C. Substituting these values, we obtain

$$d = \frac{5893 \times 10^{-8}}{4 \times 0.17195} \text{ cm} \simeq 0.000857 \text{ mm} \qquad (6.42)$$

Thus a calcite crystal of thickness 0.000857 mm with its optic axis parallel to the surface will introduce a phase difference of $\pi/2$ (and hence a path difference of $\pi/4$) between the ordinary and the extraordinary components at λ = 5893 Å. Such a plate is known as a quarter wave plate the thickness is an odd multiple of above quantity than it would be as—

$$d = (2m+1)\frac{\lambda}{4(n_o - n_e)}; m = 0, 1, 2, \ldots \qquad (6.43)$$

it will still produce a path difference of $\pi/4$. In the example considered above (i.e. when $\theta = \pi/4$), it can easily be shown that the emergent beam will be left circularly polarized for m = 0, 2, 4, . . . and right circularly polarized form m = 1, 3, 4, . . .

For a positive crystal (like quartz), $n_e > n_o$ and Eq. 6.34 should be written in the form

$$\left.\begin{aligned} E_y &= E \sin\phi \cos(\omega t + \theta') \\ E_z &= E \cos\phi \cos\omega t \end{aligned}\right\} \qquad (6.44)$$

where $\theta' = \frac{\omega}{c} d(n_e - n_o)$. For a quarter wave plate,

$$d = (2m+1)\frac{\lambda}{r(n_e - n_o)}; m = 0, 1, 2, \qquad (6.45)$$

However, in this case, for $\phi = \pi/4$, the emergent beam will be right circularly polarized for m = 0, 2, 4, . . . and left circularly polarized for m = 1, 3, 5, . . .

If thickness is concerned with *a* path difference of $\lambda/2$ then it would be as a half wave plate. Thus, for a half-wave plate, the thickness (for a negative crystal) would be given by

$$d = (2m+1)\frac{\lambda}{2(n_o - n_e)} \qquad (6.46)$$

Analysis of Polarized Light

In the previous sections we have seen that a plane wave can be characterized by different states of polarizations, which may be any one of the following: (a) linearly polarized, (b) circularly polarized, (c) elliptically polarized, (d) unpolarized, mixture of linearly polarized and unpolarized, (f) mixture of circularly polarized and unpolarized, or (g) mixture of elliptically polarized and unpolarized light. All the states of polarizations can be seen by the naked eyes. In this section, we will discuss the procedure for determining the state of polarization of a light beam.

If we introduce a polaroid in the path of the beam and rotate it about the direction of propagation, then either of the following three possibilities can occur:

(i) If there is complete extinction at two positions of the polarizer, then the beam is linearly polarized.

(ii) If there is no variation of intensity,, then the beam is either unpolarized or circularly polarized or a mixture of unpolarized and circularly polarized light. We now put a quarter wave plate on the path of the beam followed by the rotating polaroid. If there is no variation of intensity then the incident beam is unpolarized. If there is complete extinction at two positions, then the beam is circularly polarized (this is due to the fact that a quarter wave plate will transform a circularly polarized light into a linearly polarized light. A beam would be *a* mixture of unxolarized and circularly polarized light, if there is a variation of intensity.

(iii) If there is a variation of intensity (without complete extinction) then the beam is either elliptically polarized or a mixture of linearly polarized and unpolarized or a mixture of elliptically polarized and unpolarized light. We now put a quarter wave plate in front of the polaroid with its optic axis parallel to the pass-axis of the polaroid at the position of maximum intensity. Linear polarized light originate from elliptical polarization. Thus, if one obtains two positions of the polaroid where complete extinction occurs, then the original beam is elliptically polarized. If complete extinction does not occur, and the position of maximum intensity occurs at the same orientation as before the beam is a mixture of unpolarized and linearly polarized light. The beam would be the mixture of elliptically polarized and unpolarized light due to the position of maximum intensity of different orientations.

Optical Activity

When a linearly polarized light beam propagates through an "optically active" medium like sugar solution then—as the beam propagates—Its plane of polarization rotates. This rotation is directly proportional to the distance traversed by the beam and also to the concentration of sugar in the solution. Basically by measuring the angle of rotation (polarization) we can detect right concentration of sugar in the solution.

The rotation of the plane of polarization is due to the fact that the "modes" of the optically active substance are left circularly polarized (LCP) and right circularly polarized (RCP) which propagate with slightly different velocities. By "modes" we imply that if an LCP light beam is incident on the substance then it will propagate as an LCP beam; similarly, an RCP light beam will propagate as an RCP beam but with a slightly different velocity. On the other hand, if a linearly polarized light beam is incident, then we must express the linear polarization as a superposition of an RCP and an LCP beam and then consider the independent propagation of the two beams. This can be explained by following example—

We consider an RCP beam propagating in the +*z* direction

$$\left.\begin{aligned} E_x^r &= E_0 \cos(k_r z - \omega t) \\ E_y^r &= -E_0 \sin(k_r z - \omega t) \end{aligned}\right\} \tag{6.47}$$

where $k_r = \frac{\omega}{c} n_r$ and the superscript (and the subscript) *r* signify that we are considering an RCP beam. In this way to explain *a* beam of same amplitude propagating in the f_2 direction we should use following equation—

and

$$\left.\begin{aligned} E_z^l &= E_0 \cos(k_l z - \omega t) \\ E_y^l &= E_0 \sin(k_l z - \omega t) \end{aligned}\right\} \tag{6.48}$$

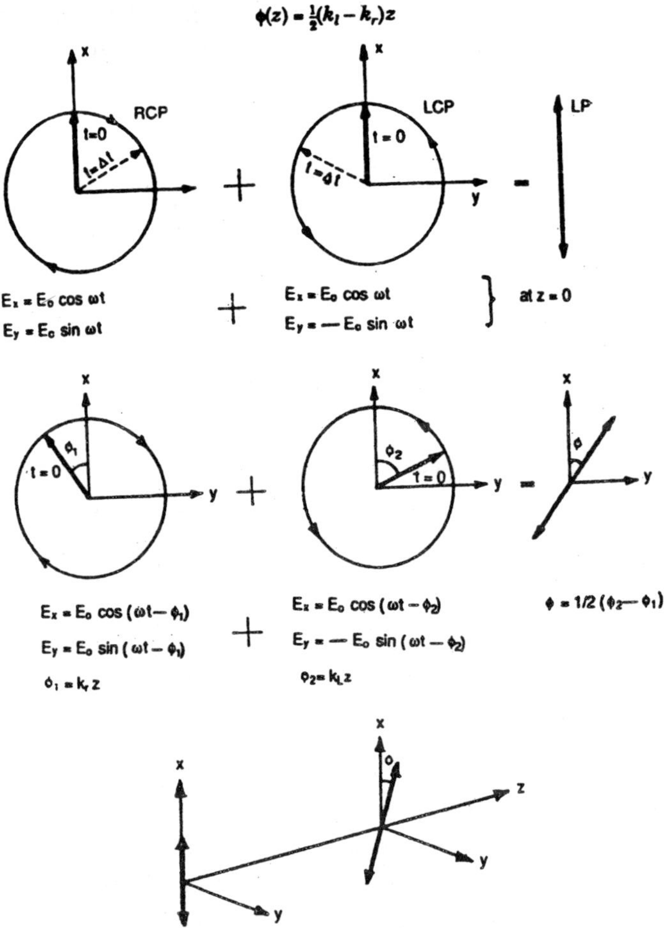

Figure (6.16) : ***The "clockwise" rotation of the plane of polarization for a "right-handed" optically action medium.***

where $k_l = \frac{\omega}{c} n_l$; n_r and n_l are the refractive indices corresponding to the RCP and LCP beams respectively. Suppose two beams propagating simultaneous then the resultant-fields of x and y components would be as follows—

$$E_x = E_0\left[\cos(k_r z - \omega t) + \cos(k_l z - \omega t)\right]$$

or

$$E_x = 2E_0 \cos\left[\frac{1}{2}(k_l - k_r)z\right]\cos[\omega t - \theta(z)]$$

Similarly $$E_y = 2E_0 \sin\left[\frac{1}{2}(k_l - k_r)z\right]\cos[\omega t - \theta(z)]$$

where $$\theta(z) = \frac{1}{2}(k_l + k_r)z$$

Finally because the plane of polarization rotating with z than the resultant wave would be linearly polarized. If the direction of the oscillating electric vector makes an angle ϕ with the x-axis then [see figure 6.16]

$$\phi(z) = \frac{1}{2}(k_l - k_r)z$$

or $$\phi(z) = \frac{\pi}{\lambda_0}(n_l - n_r)z = \frac{\omega}{2c}(n_l - n_r)z \tag{6.49}$$

where λ_0 is the free space wavelength. Now, if

$n_l > n_r \Leftrightarrow$ the "optically active" substance is said to be right-handed or dextro-rotatory

and, if

$n_r > n_l \Leftrightarrow$ the optically active substance is said to be left-handed or laevo-rotatory.

For example, for turpentine,

$$\phi = +37° \text{ for } z = 10 \text{ cm}$$

As we know sugar solution has optical activity due to the helical structure of sugar molecules. The method of determining the concentration of sugar solutions by measuring the rotation of the plane of polarization is a widely used method in industry. It may be noted that if $n_l = n_r$ (as is indeed the case in an isotropic substance) then $\phi(z) = 0$ and a linearly polarized light remains linearly polarized along the same orientation.

Crystals also show optical activity. For example, for a linearly polarized light propagating along the optic axis of a quartz crystal, the plane of polarization gets rotated. Indeed

$$|n_l - n_r| \simeq 7 \times 10^{-5} \Rightarrow \phi \simeq \frac{7}{60}\pi = 21° \text{ for } z = 0.1 \text{ cm}$$

for $\lambda_0 = 6000$ Å.

Change in the Sop (State of Polarization) of a light Beam Propagating through an Elliptic core Single Mode Optical Fibre

Propagation of polarized light through an elliptic core optical fibre is *a* interesting phenomenon. We will have a brief discussion on optical fibres earlier; it will suffice here to say that in an ordinary optical fibre we have a cylindrical core (of circular cross-section) cladded with a medium of slightly lower refractive index. The guidance of the light beam takes place through the phenomenon of total internal reflection (see Figure). Because of the circular symmetry of the problem, the incident beam can have any state of polarization which will be maintained as the beam propagates through the fibre. Now, if we have an elliptic core fibre [see Figure 19.20 (a)] then the "modes" of the fibre are (approximately) x and Polarization of y means *a* polarized beam x on incident will propagate without any change in the state of polarization with phase velocity ω/β_x. Similarly, a y-polarized beam will propagate as a y-polarized beam with velocity ω/β_y. Now, let circularly polarized beam be incident on the input face of the fibre at z = 0. On resolving incident beam in to x and y polarized beam these would go with different velocities as—

$$\mathbf{E}(x, y, z) = \Psi(x,y)\left[\hat{\mathbf{x}} \cos\left(\beta_x z - \omega t\right) + \hat{\mathbf{y}} \sin\left(\beta_y z - \omega t\right)\right] \quad (6.50)$$

where $\psi(x, y)$ is the transverse field distribution of the fundamental mode which is assumed to be (approximately) the same for both x- and y-polarizations. [It may be readily seen that if $\beta_x = \beta_y$ as is indeed true for circular core fibres, the beam will remain circularly polarized for all values of z.] Now, at $z = 0$,

$$\begin{aligned} E_x &= \psi(x,y)\cos\omega t \\ E_y &= -\psi(x,y)\sin\omega t \end{aligned} \quad (6.51)$$

These shows a wave which is left circularly polarized. (see fig 6.17)

$$z = z_1 = \frac{\pi}{2\left(\beta_y - \beta_x\right)} \quad (6.52)$$

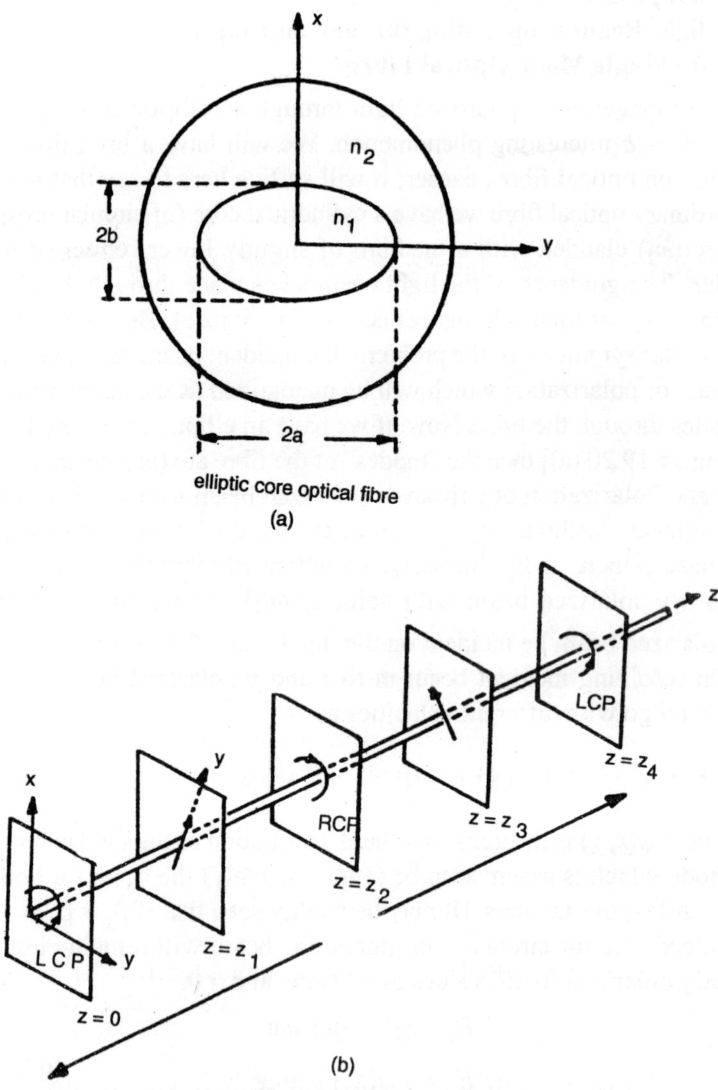

Figure (6.17) : ***(a) The transverse cross-section of an elliptic core fibre; the "modes" are (approximately)*** x***-polarized and*** y***-polarized. (b). Propagation of a left-circularly polarized beam incident on an elliptic core fibre. If we view along the*** y***'-axis then dark spots will be observed at*** $z = z_1, 5z_1, 9z_1, ...$

i.e., for $\beta_y z_1 = \beta_x z_1 + \pi/2$,

$$E_x = \psi(x,y)\cos(\phi_1 - \omega t) = +\psi(x,y)\cos(\omega t - \phi_1)$$

$$E_y = \psi(x,y)\sin\left(\phi_1 + \tfrac{\pi}{2} - \omega t\right) = +\psi(x,y)\cos(\omega t - \phi_1)$$

where $$\phi_1 = \beta_x z_1$$

which represents a linearly polarized wave [see Figure 6.17 (b)]; we assume the direction of the E-vector to be along the y' axis. Similarly, at

$$z = z_2 = \frac{\pi}{(\beta_y - \beta_x)} = 2z_1$$

$$\left.\begin{aligned} E_x &= \psi(x,y)\cos(\phi_2 - \omega t) = \psi(x,y)\cos(\omega t - \phi_2) \\ E_y &= \psi(x,y)\sin(\phi_2 + \pi - \omega t) = \psi(x,y)\sin(\omega t - \phi_2) \end{aligned}\right\} \quad (6.53)$$

where $$\phi_2 = \beta_x z_2$$

Here we get the right circularly polarized wave (See fig 6.17)

$$z = z_3 = \frac{3\pi}{2(\beta_x - \beta_y)} = 3z_1$$

we will have

$$E_x = \psi(x,y)\cos(\phi_3 - \omega t) = \psi(x,y)\cos(\omega t - \phi_3)$$

$$E_y = \psi(x,y)\sin\left(\phi_3 + \tfrac{3\pi}{2} - \omega t\right) = -\psi(x,y)\cos(\omega t - \phi_3)$$

where $$\phi_3 = \beta_x z_3$$

Again we have the same linear polarization but-the direction of oscillating electric field would be at light angle to the field at $z = z_1$. In a similar manner, we can easily continue to determine the SOP of the propagating beam. Thus at $z = 5z_1, 9z_1, 13z_1, \ldots$ the SOP will be the same as at $z = z_1$ and at $z = 7z_1, 11z_1, 15z_1$ the SOP will be the same as at z =3z1. Likc that at $z = 4z_1, 8z_1,$

$12z_1$, amd $z = 2z_1, 6z_1, 10z_1$... beam would be LCP and RCP respectively.

Thus the rotation of fibre will be in *a* way that the *y*' axis is alongs the vertical line means the x' and z axis are in the horizontal plane. Thus if we put our eyes vertically above the fibre and view vertically down then the regions $z = z_1, 5z_1, 9z_1$,... will appear dark. This is because of the fact that in these regions the electric field is oscillating in the y' direction (which is the vertical direction) and we know that if the dipole oscillates along the y' direction, there is no radiation emitted in that particular direction. Thus by measuring the distance between two consecutive black spots (= $4z_1$) one can calculate z_1 and hence $\beta_y - \beta_x$. Furthermore, by moving the eyes to the horizontal plane, i.e., viewing along the x' axis we will see the regions $z = z_1, 5z_1, 9z_1$.... appear bright and the regions $z = 3z_1, 7z_1, 11z_1$,... appear dark. Thus the experiment not only allows one to understand the changing SOP of a beam propagating through a birefringent fibre, but also helps us understand the radiation pattern of an oscillating dipole.

Let an Elliptie core fibre to give *a* numerical example—

$$2a = 2.14\ \mu\text{m}, \quad 2b = 8.85\ \mu\text{m}$$
$$n_1 = 1.535, \quad n_2 = 1.47$$

(see Figure 6.17(a)). For such a fibre operating at $\lambda_0 = 6328$ Å $k_0 \simeq 9.929 \times 10^9\ \text{cm}^{-1}$),

$$\frac{\beta_x}{k_0} \simeq 1.506845 \text{ and } \frac{\beta_y}{k_0} \simeq 1.507716$$

The quantity

$$L_b = \frac{2\pi}{\Delta\beta} = \frac{2\pi}{\beta_y - \beta_x} \simeq 0.727\ \text{mm}$$

is known as the coupling length.

Plane wave Propagation in Anisotropic Media

Here we will discuss Mazwell equation of plane wave solution for an anisotropic medium and give many assumptions. The difference between an isotropic and an anisotro r medium is in the relationship between the displacement vector **D**, and electric vector **E**; the displacement vector **D**. The direction of D and E is same in an isotropic medium i.e.—

$$\mathbf{D} = \varepsilon \mathbf{E} \tag{6.54}$$

Where ε is the dielectric permittivity of the medium. On the other hand, in an anisotropic medium **D** is not, in general, in the direction of **E** and the relation between **D** and **E** can be written in the form

$$\begin{aligned} D_x &= \varepsilon_{xx} E_x + \varepsilon_{xy} E_y + \varepsilon_{xz} E_z \\ D_y &= \varepsilon_{yx} E_x + \varepsilon_{yy} E_y + \varepsilon_{yz} E_z \\ D_z &= \varepsilon_{zx} E_x + \varepsilon_{zy} E_y + \varepsilon_{zz} E_z \end{aligned} \tag{6.55}$$

where $\varepsilon_{xx}, \varepsilon_{xy}, \ldots$ are constants. One can show that

$$\varepsilon_{xy}, \varepsilon_{yx}, \quad \varepsilon_{xz} = \varepsilon_{zx} \quad \text{and} \quad \varepsilon_{yz} = \varepsilon_{zy} \tag{6.56}$$

Thus, we can always choose the directions of x, y and z axes i.e.

$$\begin{aligned} D_x &= \varepsilon_x E_x \\ D_y &= \varepsilon_y E_y \\ \text{and} \qquad D_z &= \varepsilon_z E_z \end{aligned} \tag{6.57}$$

This coordinate system is known as the principal axis system and the quantities ε_x, ε_y and ε_x are known as the principal dielectric permitivities of the medium. If

$$\varepsilon_x \neq \varepsilon_y \neq \varepsilon_z \text{ (biaxial)} \tag{6.58}$$

we have what is known as a biaxial medium and the quantities

$$n_x = \sqrt{\varepsilon_x / \varepsilon_0}, \quad n_y = \sqrt{\varepsilon_y / \varepsilon_0}, \quad n_z = \sqrt{\varepsilon_z / \varepsilon_0} \tag{6.59}$$

are said to be the principal refractive indices of the medium; in the above equation E_0 represents the dielectric permittivity of free space (= 8.8542×10^{-12} $C^2/N\text{–}m^2$). If

$$\varepsilon_x = \varepsilon_y \neq \varepsilon_z \text{ (uniaxial)} \tag{6.60}$$

In above equation we have uniaxial medium its z axis shows the optic axis of the medium and further it would be.

$$n_o = \sqrt{\varepsilon_x / \varepsilon_0} = \sqrt{\varepsilon_y / \varepsilon_0} \text{ and } n_e = n_z = \sqrt{\varepsilon_z / \varepsilon_0} \quad (6.61)$$

are known as ordinary and extraordinary refractive indices; typical values for some uniaxial crystals are given in Table 6.1. For a uniaxial medium, since $\varepsilon_x = \varepsilon_y$ the x and y directions can be arbitrarily chosen as long as they are perpendicular to the optic axis, i.e. any two mutually perpendicular axes (which are also perpendicular to the z-axis) can be taken as the principal axes of the medium.

On the other hand, if

$$\varepsilon_x = \varepsilon_y = \varepsilon_z \text{ (isotropic)} \quad (6.62)$$

In as isotropic medium three mutually perpendicular axes can be choose as the principal axis system.

Let us consider the propagation of a plane electromagnetic wave; for such a wave the vectors **E, H, D** and **B** would be proportional to exp $[i(\text{k. r} - wt)]$. Thus

$$\left.\begin{aligned} \mathbf{E} = \mathbf{E}_0 e^{i(\mathbf{k}\cdot\mathbf{r}-\omega t)}, \quad \mathbf{H} = \mathbf{H}_0 e^{i(\mathbf{k}\cdot\mathbf{r}-\omega t)} \\ \mathbf{D} = \mathbf{D}_0 e^{i(\mathbf{k}\cdot\mathbf{r}-\omega t)}, \quad \mathbf{B} = \mathbf{B}_0 e^{i(\mathbf{k}\cdot\mathbf{r}-\omega t)} \end{aligned}\right\} \quad (6.63)$$

Table (6.1) : Ordinary and Extraordinary Refractive Indices for Some Uniaxial Crystals

Name of the crystal	Wavelength	n_o	n_e
Calcite	4046Å	1.68134	1.49694
	5890 Å	1.65835	1.48640
	7065 Å	1.65207	1.48359
Quartz	5890 Å	1.54424	1.55335
Lithium niobate	6000Å	2.2967	2.2082
KDP	6328 Å	1.50737	1.46685
ADP	6328 Å	1.52166	1.47685

where the vectors $\mathbf{E}_0$, $\mathbf{H}_0$, $\mathbf{D}_0$ and $\mathbf{B}_0$ are independent of space and time; **k** represents the propagation vector of the wave and ω the angular frequency. The relation between velocity (or phase velocity) v_w and refractive index n_w of the wave is as follows—

$$v_w = \frac{\omega}{k} = \frac{c}{n_w} \tag{6.64}$$

Thus
$$|\mathbf{k}| = k = \frac{\omega}{c} n_w$$

In this part we will discuss about the possible values i.e. n_w for *a* plane wave which is propagate through an anisotropic equation. In the absence of any currents (i.e., **J** = 0) Maxwell's curl equations become

$$\nabla \times \mathbf{E} = -\frac{\partial \mathbf{B}}{\partial t} = i\omega \mathbf{B} = i\omega\mu_0 \mathbf{H} \tag{6.65}$$

and

$$\nabla \times \mathbf{H} = \frac{\partial \mathbf{D}}{\partial t} = -i\omega \mathbf{D} \tag{6.66}$$

where we have assumed the medium to be non-magnetic (i.e., $\mathbf{B} = \mu_o \mathbf{H}$).

Now, if

$$\mathbf{E} = \mathbf{E}_0 \exp\left[i\,(\mathbf{k}.\mathbf{r} - \omega t)\right]$$

then

$$(\nabla \times \mathbf{E})_x = \frac{\partial E_z}{\partial y} - \frac{\partial E_y}{\partial z} = \left(ik_y E_{0z} - ik_z E_{0y}\right) e^{i(\mathbf{k}\cdot\mathbf{r} - \omega t)}$$
$$= i\left(k_y E_z - k_z E_y\right) = i(\mathbf{k} \times \mathbf{E})_x$$

Thus

$$\nabla \times \mathbf{E} = i(\mathbf{k} \times \mathbf{E}) = i\omega\mu_0 \mathbf{H} \Rightarrow \mathbf{H} = \frac{1}{\omega\mu_0}(\mathbf{k} \times \mathbf{E}) \tag{6.67}$$

and

$$\nabla \times \mathbf{H} = i(\mathbf{k} \times \mathbf{H}) = i\omega \mathbf{D} \Rightarrow \mathbf{D} = \frac{1}{\omega}(\mathbf{H} \times \mathbf{k}) \tag{6.68}$$

Equations (6.67) and (6.68) show that **H** is at right angles to

k, E and D; thus **k**, **E** and **D** must lie in the same plane. Further **D** is at right angles to **k**. From eq. (6.68) we get—

$$\mathbf{D} = \frac{1}{\omega^2 \mu_0}[(\mathbf{k} \times \mathbf{E}) \times \mathbf{k}]$$

$$= \frac{1}{\omega^2 \mu_0}[(\mathbf{k} \cdot \mathbf{k})\mathbf{E} - (\mathbf{k} \cdot \mathbf{E})\mathbf{k}] \tag{6.69}$$

where we have used the vector identity

$$(\mathbf{A} \times \mathbf{B}) \times \mathbf{C} = (\mathbf{A} \cdot \mathbf{C})\mathbf{B} - (\mathbf{B} \cdot \mathbf{C})\mathbf{A}$$

Thus

$$\mathbf{D} = \frac{k^2}{\omega^2 \mu_0}\left[\mathbf{E} - (\hat{\kappa} \cdot \mathbf{E})\hat{\kappa}\right]$$

$$= \frac{n_\omega^2}{c^2 \mu_0}\left[\mathbf{E} - (\hat{\kappa} \cdot \mathbf{E})\hat{\kappa}\right] \tag{6.70}$$

where

$$\hat{\kappa} = \frac{\mathbf{k}}{k}$$

represents the unit vector along **k**. Since

$$D_x = \varepsilon_x E_x = \varepsilon_0 n_x^2 E_x$$

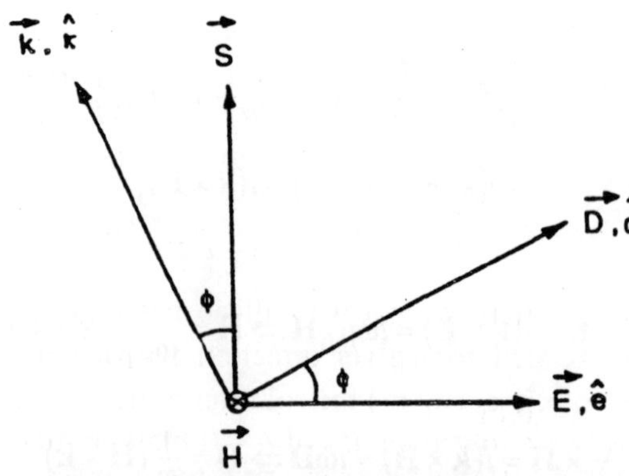

Figure 6.18 : ***In an anisotropic medium D · κ = 0, E · s – 0 and* H *is normal to* E, D, s *and* κ.**

From equation (6.70) for x-component—

$$\frac{\varepsilon_0 \mu_0 c^2 n_x^2}{n_\omega^2} E_x = E_x - \kappa_x\left(\kappa_x E_x + \kappa_y E_y + \kappa_z E_z\right)$$

Since $c^2 = 1/(\varepsilon_0 \mu_0)$, we have

$$\left(\frac{n_x^2}{n_\omega^2} - \kappa_y^2 - \kappa_z^2\right) E_x + \kappa_x \kappa_y E_y + \kappa_x \kappa_z E_z = 0 \qquad (6.71)$$

where we have used the relation $\kappa_x^2 + \kappa_y^2 + \kappa_z^2 = 1$ (since $\hat{\kappa}$ is a unit vector). Similarly,

$$\kappa_x \kappa_y E_x + \left(\frac{n_y^2}{n_\omega^2} - \kappa_x^2 - \kappa_z^2\right) E_y + \kappa_y \kappa_z E_z = 0 \qquad (6.72)$$

$$\kappa_x \kappa_z E_x + \kappa_y \kappa_z E_y + \left(\frac{n_z^2}{n_\omega^2} - \kappa_x^2 - \kappa_y^2\right) E_z = 0 \qquad (6.73)$$

Because in above equation we have a set from three homogenous equations, for non-trivial solutions, we must have

$$\begin{vmatrix} \frac{n_x^2}{n_\omega^2} - \kappa_y^2 - \kappa_z^2 & \kappa_x \kappa_y & \kappa_x \kappa_z \\ \kappa_x \kappa_y & \frac{n_y^2}{n_\omega^2} - \kappa_x^2 - \kappa_z^2 & \kappa_y \kappa_z \\ \kappa_x \kappa_z & \kappa_y \kappa_z & \frac{n_z^2}{n_\omega^2} - \kappa_x^2 - \kappa_y^2 \end{vmatrix} = 0 \qquad (6.74)$$

We should remember that we still do not know the possible values of n_w. Indeed, for a given direction of propagation (i.e., for given values of κ_x, κ_y and κ_z) the solution of the above equation) gives us the two allowed values of n_w. It may be mentioned that from Eq. (6.74) it appears as if we. Will have a cubic equation in n_w^2 which would give us three roots of n_w^2; however, the coefficient of n_w^6 will always be zero and hence there will be always two roots. A general procedure can be concluded from some example.

Propagation along a Principal Axis

Suppose we have *a* wave which is propagating along the *z* axis. Thus $\kappa_x = 0 = \kappa_y$ and $\kappa_z = 1$ and Equations (6.71)-(6.73) simplify to

$$\left.\begin{aligned} \left(\frac{n_x^2}{n_w^2} - 1\right) E_x &= 0 \\ \left(\frac{n_y^2}{n_w^2} - 1\right) E_y &= 0 \\ \frac{n_z^2}{n_w^2} E_z &= 0 \end{aligned}\right\} \tag{6.75}$$

and

$$\begin{vmatrix} \left(\frac{n_x}{n_w}\right)^2 - 1 & 0 & 0 \\ 0 & \left(\frac{n_y}{n_w}\right)^2 - 1 & 0 \\ 0 & 0 & \left(\frac{n_z}{n_w}\right)^2 \end{vmatrix} = 0 \tag{6.76}$$

which implies

$$\left(\frac{n_x^2}{n_w^2} - 1\right)\left(\frac{n_y^2}{n_w^2} - 1\right)\frac{n_z^2}{n_w^2} = 0 \tag{6.77}$$

giving us only two roots for n_w:

$$n_w = n_{w1} = n_z \text{ with } E_y = 0 = E_z \text{ (}x\text{-polarized wave)}$$

$$n_w = n_{w2} = n_y \text{ with } E_x = 0 = E_z \text{ (}y\text{-polarized wave)} \tag{6.78}$$

These are the two modes of the system. Corresponding to the solution $n_w = n_{w1} = n_x$, the **E** and **D** vectors are along the **x**-axis, and for the solution $n_w = n_{w2} = n_y$, the **E** and **D** vectors are along the y-axis $n_x = n_y = n_o$ would be applicable in case of Uniaxial crystal.

$$n_{w1} = n_{w2} = n_o \tag{6.79}$$

and the x and y polarized modes are 'degenerate', i.e. both waves propagate with the same wave velocity

$$v_w = v_{w1} = v_{w2} = \frac{c}{n_0} \tag{6.80}$$

Similarly, for a wave propagating along the x-axis, $\kappa_x = 1$, $\kappa_y = 0 = \kappa_z$ and we would have the following two solutions:

and
$$\left.\begin{aligned} n_w &= n_{w1} = n_y \quad (y\text{-polarized}) \\ n_w &= n_{w2} = n_z \quad (z\text{-polarized}) \end{aligned}\right\} \tag{6.81}$$

In case of uniaxial medium the velocities of propagation of waves would be different and $n_y = n_o$, $n_z = n_e$. As will be shown later, they-polarized wave is the ordinary wave and the z-polarized wave the extraordinary wave.

Propagation in Uniaxial Crystals

In this section, we will completely restrict ourselves to uniaxial crystals for which

$$n_x = n_y = n_o \text{ and m } n_z = n_e \tag{6.82}$$

The x and y directions can be arbitarity chosen in case of uniaxial crystal because of perpendicular to the optic axis. Now, for a wave propagating along any direction **k**, we can choose our y axis in such a way that it is at right angles to **k**. i.e the y-axis is normal to the plane defined by **k** and the z-axis; obviously, the x-axis will lie in the same plane. Thus we may write

$$\kappa_x = \sin\Psi, \quad \kappa_y = 0, \quad \text{and} \quad \kappa_z = \cos\Psi$$

where ψ is the angle that the **k** vector makes with the optic axis and Eqs (6.71-(6.73) become

$$\left(\frac{n_o^2}{n_w^2} - \cos^2\psi\right)E_x + \sin\psi\cos\psi\, E_z = 0 \tag{6.83}$$

$$\left(\frac{n_o^2}{n_w^2}-1\right)E_y = 0 \tag{6.84}$$

and

$$\sin\psi\cos\psi\, E_x + \left(\frac{n_e^2}{n_w^2}-\sin^2\psi\right)E_z = 0 \tag{6.85}$$

From Equation (6.84), one readily obtains the following as one of the solutions:

$$n_w = n_{wo} = n_o \quad \text{with} \quad E_x = 0 = E_z \ (y\text{-polarized}) \tag{6.86}$$

The corresponding wave velocity is given by

$$v_w = v_{wo} = \frac{c}{n_o} \quad \text{(ordinary wave)} \tag{6.87}$$

Because the velocity of an ordinary wave doesn't depend upon direction therefore the subscript '*O*' on n_w and v_w. From Equation (93) it is seen that the wave is *y*-polarized, i.e. the electric vector associated with the ordinary wave is perpendicular to **k** and the optic axis. In order to obtain the other solution we use Equations (6.82)-(6.84) and (91) to obtain

$$\frac{E_z}{E_x} = -\frac{\dfrac{n_o^2}{n_w^2}-\cos^2\psi}{\sin\psi\cos\psi} = -\frac{\sin\psi\cos\psi}{\dfrac{n_e^2}{n_w^2}-\sin^2\psi}$$

and obviously E_y must vanish. Simple manipulations give us

$$\frac{1}{n_w^2} = \frac{1}{n_{we}^2} = \frac{\cos^2\psi}{n_o^2} + \frac{\sin^2\psi}{n_e^2} \tag{6.88}$$

with

$$\frac{n_e^2 E_z}{n_o^2 E_x} = -\tan\psi$$

or

$$\frac{\varepsilon_z E_z}{\varepsilon_x E_x} = \frac{D_z}{D_x} = -\tan\psi \quad \text{and} \quad D_y = \varepsilon_y E_y = 0 \tag{6.89}$$

The corresponding wave velocity is given by

$$v_{we}^2 = \frac{c^2}{n_{we}^2} = \frac{c^2}{n_o^2}\cos^2\psi + \frac{c^2}{n_e^2}\sin^2\psi \tag{6.90}$$

In case of extraordinary wave the velocity is dependent on the direction therefore the subscript e. From Equations (6.68) and (6.89) it follows that the displacement vector **D** of the normal to **k** and also to they-axis implying that *the displacement vector D associated with the extraordinary wave lies in the plane containing the propagation vector* ***k*** *and the optic axis and is normal to* **k**.

Thus the small dashes on the extraordinary ray in Figs 6.12 (a) and (b) represent the directions of the **D** vector· Thus from fig 6.18 at can get the angle ϕ.

$$\cos\phi = \frac{\mathbf{E}\cdot\mathbf{D}}{|\mathbf{E}|\,|\mathbf{D}|} = \frac{E_x D_x + E_z D_z}{\left(E_x^2 + E_x^2\right)^{1/2}\left(D_x^2 + D_z^2\right)^{1/2}}$$

$$= \frac{1 + \dfrac{E_z}{E_x}\dfrac{D_z}{D_x}}{\left[1 + \left(\dfrac{E_z}{E_x}\right)^2\right]^{1/2}\left[1 + \left(\dfrac{D_z}{D_x}\right)^2\right]^{1/2}}$$

$$= \frac{1 + \dfrac{n_o^2}{n_e^2}\tan^2\psi}{\left[1 + \left(\dfrac{n_o^2\tan\psi}{n_e^2}\right)^2\right]^{1/2}\left(1 + \tan^2\psi\right)^{1/2}}$$

$$= \frac{\left(n_e^2 + n_o^2 \tan^2 \psi\right)\cos\psi}{\left(n_e^4 + n_o^4 \tan^2 \psi\right)^{1/2}} \tag{6.91}$$

As a numerical example, we consider calcite for which (at λ = 5893 Å and 18°C)

$$n_o = 1.65836, n_e = 1.48641$$

Suppose **k** has angle of 30° with optic axis so ψ = 30° and thus we get ϕ = 5.7°.

Ray Velocity and Ray Refractive Index

Direction of energy propagation or ray propagation depends on poynting vectors so—

$$\mathbf{S} = \mathbf{E} \times \mathbf{H} \tag{6.92}$$

Clearly, **S** is at right angles to **E** and **H** and will therefore be in the same plane containing the vectors **k, E** and **D** (see Figure 19.21). As can be seen from the figure, the direction of the propagation of the wave ($\hat{\kappa}$) is not along the direction of energy propagation ($\hat{\mathbf{s}}$), where ($\hat{\mathbf{s}}$) is the unit vector along **S**. v_r is the velocity of ray (or energy transmission velocity) in following equation.

$$v_r = \frac{S}{u} \tag{6.93}$$

where u is the energy density. Now,

$$u = ½(\mathbf{D}\cdot\mathbf{E} + \mathbf{B}\cdot\mathbf{H}) = 2(\mathbf{D}\cdot\mathbf{E} + \mu_0\mathbf{H}\cdot\mathbf{H}) \tag{6.94}$$

Substituting for **H** and **D** from Equations (6.67) and (.75), we obtain

$$u = \frac{1}{2\omega} [(\mathbf{H} \times \mathbf{k}) \cdot \mathbf{E} + (\mathbf{k} \times \mathbf{E}) \cdot \mathbf{H}]$$

$$= \frac{1}{2\omega} [\mathbf{k} \cdot (\mathbf{E} \times \mathbf{H}) + \mathbf{k} \cdot (\mathbf{E} \times \mathbf{H})]$$

$$= \frac{1}{2\omega} \mathbf{k} \cdot \mathbf{S} \tag{6.95}$$

Thus Equation (6.93) becomes

$$v_r = \frac{\omega S}{\mathbf{k}\cdot\mathbf{S}} = \frac{\omega}{k\cos\phi} = \frac{v_w}{\cos\phi} \tag{6.96}$$

Here ϕ represents the angle between $\hat{\kappa}$ and $\hat{s}$ which we can see in fig (6.18). The ray refractive index, n_r is defined as—

$$n_r = \frac{c}{v_r} = \frac{c}{v_w} \cos\phi = n_w \cos\phi \tag{6.97}$$

In order to express **E** in terms of **D**, we refer to Figure 19.21 and write

$$\mathbf{D} = (\mathbf{D} \cdot \hat{\mathbf{e}})\,\hat{\mathbf{e}} + (\mathbf{D} \cdot \hat{\mathbf{s}})\,\hat{\mathbf{s}}$$

where a is a unit vector along the direction of the electric field **E**. Thus

$$\mathbf{D} - (\mathbf{D} \cdot \mathbf{s})\,\hat{\mathbf{s}} = (\mathbf{D} \cdot \hat{\mathbf{e}})\,\hat{\mathbf{e}}$$

$$= (D\cos\phi)\frac{\mathbf{E}}{E} \tag{6.98}$$

Similarly,

$$\mathbf{E} = (\mathbf{E} \cdot \hat{\mathbf{d}})\,\hat{\mathbf{d}} + (\mathbf{E} \cdot \hat{\kappa})\,\hat{\kappa} \tag{6.99}$$

Here. $\hat{\mathbf{d}}$ unit vector is the displacement vector **D** which shown in fig (6.18). If we now substitute for **E** –(**E** · $\hat{\kappa}$)$\hat{\kappa}$ in Equation (77), we would get

$$\mathbf{D} = \frac{n_w^2}{\mu_0 c^2}(\mathbf{E} \cdot \hat{\mathbf{d}})\,\hat{\mathbf{d}}$$

or

$$\mathbf{D} = \frac{n_w^2}{\mu_0 c^2}\mathbf{E}\cos\phi \tag{6.100}$$

Substituting in Equation (105), we get

$$\mathbf{D} - (\mathbf{D} \cdot \hat{\mathbf{s}})\,\hat{\mathbf{s}} = \frac{n_w^2}{\mu_0 c^2}\cos^2\phi\,\mathbf{E}$$

$$= \frac{n_r^2}{\mu_0 c^2}\mathbf{E}$$

This is considered from equation (6.97). Taking the x component of the above equation (where x represents the direction of one of the principal axes), we obtain

$$D_x - \left(D_x s_x + D_y s_y + D_z s_z\right) s_x = \frac{n_r^2}{\mu_0 c^2} E_x = \frac{n_r^2}{\mu_0 c^2 \varepsilon_x} D_x$$

If we use the relations

$$n_x^2 = \frac{\varepsilon_x}{\varepsilon_0}, \quad c^2 = \frac{1}{\varepsilon_0 \mu_0} \quad \text{and} \quad s_x^2 + s_y^2 + s_z^2 = 1$$

we would get,

$$\left(\frac{n_r^2}{n_x^2} - s_y^2 - s_z^2\right) D_x + s_x s_y D_y + s_x s_z D_z = 0 \tag{6.101}$$

Similarly

$$s_x s_y D_x + \left(\frac{n_r^2}{n_y^2} - s_x^2 - s_z^2\right) D_y + s_z s_y D_z = 0 \tag{6.102}$$

$$s_x s_z D_x + s_z s_y D_y + \left(\frac{n_r^2}{n_z^2} - s_x^2 - s_y^2\right) D_z = 0 \tag{6.103}$$

As we discussed earlier we can have the above set from three homogeneous equation for non trivial solutions, then we get—

$$\begin{vmatrix} \dfrac{n_r^2}{n_x^2} & s_x s_y & s_x s_z \\ s_x s_y & \dfrac{n_r^2}{n_y^2} - s_x^2 - s_z^2 & s_z s_y \\ s_x s_z & s_z s_y & \dfrac{n_r^2}{n_z^2} - s_x^2 - s_y^2 \end{vmatrix} = 0 \tag{6.104}$$

We still do not know the possible values of n_r. Indeed for a given ray direction (i.e. for given values of s_x, s_y and s_z) the solution of the above equation gives us the two allowed values of n_r and hence two possible values of the ray velocities. This can be simply from same examples.

Ray Propagation along One of the Principal Axes

Suppose a ray is propagating along z which is the principal axes. Thus $s_x = 0 = s_y$ and $s_z = 1$ and Eqs (6.101) to (6.103) simplify to

$$\left(\frac{n_r^2}{n_x^2} - 1\right) D_x = 0$$

$$\left(\frac{n_r^2}{n_y^2} - 1\right) D_y = 0$$

$$\frac{n_r^2}{n_z^2} D_z = 0$$

Obviously, D_z will always be zero and the two possible values of the rays refractive index would be

$$\left.\begin{aligned} & n_r = n_{r1} = n_x \left[\text{with } D_x \neq 0 \text{ and } D_y = D_z = 0\right] \\ \text{and} \quad & n_r = n_{r2} = n_y \left[\text{with } D_y \neq 0 \text{ and } D_x = D_z = 0\right] \end{aligned}\right\} \quad (6.105)$$

Thus the values of v_r which are the velocities of two corresponding ray would be—

$$\left.\begin{aligned} v_r = v_{r1} = \frac{c}{n_{r1}} = \frac{c}{n_x} \quad (x\text{-polarized wave}) \\ v_r = v_{r2} = \frac{c}{n_{r2}} = \frac{c}{n_y} \quad (y\text{-polarized wave}) \end{aligned}\right\} \quad (6.106)$$

Comparing these with Equation (6.78), we find that along the principal axes the ray and wave velocities are equal and that they both propagate along the same direction.

Ray Propagation in Uniaxial Crystals

For a uniaxial crystal the optic axis is along with z direction. Thus

$$n_x = n_y = n_o \quad \text{and} \quad n_z = n_e$$

As discussed in the previous section, x and y directions can be arbitrarily chosen as long as they are perpendicular to the z-axis. F ray propagating along x–z plane and making angle θ with the z axis so we get—

$$s_x = \sin\theta, \quad s_y = 0 \quad \text{and} \quad s_z = \cos\theta$$

and Equations (6.101)-(6.103) become

$$\left(\frac{n_r^2}{n_o^2} - \cos^2\theta\right) D_x + \sin\theta\cos\theta\, D_z = 0 \tag{6.107}$$

$$\left(\frac{n_r^2}{n_o^2} - 1\right) D_y = 0 \tag{6.108}$$

$$\sin\theta\cos\theta\, D_x + \left(\frac{n_r^2}{n_e^2} - \sin^2\theta\right) D_z = 0 \tag{109}$$

Obviously, one of the roots is given by

$$n_r = n_{ro} = n_o \quad \text{with} \quad D_x = 0 = D_z \text{ (y-polarized)} \tag{110}$$

The corresponding ray velocity is given by

$$v_r = v_{ro} = \frac{c}{n_{ro}} = \frac{c}{n_o} \text{ (ordinary ray)} \tag{6.111}$$

In case of ordinary ray in which the velocity is in dependent to the direction, therefore the subscript 'O' on V_r and n_r.

From equations 6.107 and 6.109 we can get an another solution which is as follows—

$$\frac{D_z}{D_x} = \frac{\frac{n_r^2}{n_o^2} - \cos^2\theta}{\sin\theta\cos\theta} = -\frac{\sin\theta\cos\theta}{\frac{n_r^2}{n_e^2} - \sin^2\theta}$$

and obviously,

$$D_y = 0$$

Simple manipulations give us

$$n_r^2 = n_{re}^2 = n_o^2 \cos^2\theta + n_e^2 \sin^2\theta \quad \text{(extraordinary ray)} \tag{6.112}$$

with

$$\frac{D_z/n_e^2}{D_x/n_o^2} = \frac{E_z}{E_x} = -\tan\theta, \quad (D_y = 0) \tag{6.113}$$

The corresponding ray velocity is given by [cf Equation (6.28)].

$$\frac{1}{v_r^2} = \frac{1}{v_{re}^2} = \frac{n_{re}^2}{c^2} = \frac{\cos^2\theta}{c^2/n_o^2} + \frac{\sin^2\theta}{c^2/n_e^2} \tag{6.114}$$

which corresponds to the extraordinary ray and hence the subscript 'e' *on* n_r and v_r. If we use the above equation to plot the variation of v_{re} with θ we will get an ellipse and if $n_o > n_e$ (as in calcite) the major and the minor axes will be along the x and z axes. See, for example, Figure 19.17 (a) which shows that the minor axis is along the optic axis. Equation (6.114) shows the equation of ellipse in the x-z plane is as follows—

$$\frac{x^2}{a^2} + \frac{z^2}{b^2} = 1 \tag{6.115}$$

If we use polar coordinates (ρ, θ) such that

$$x = \rho\sin\theta, \qquad z = \rho\cos\theta$$

then Equation (6.115) becomes

$$\frac{1}{\rho^2} = \frac{\sin^2\theta}{a^2} + \frac{\cos^2\theta}{b^2}$$

Above equation is similar to the equation (6.114). In three dimensions, Equation (6.114) will represent an ellipsoid of revolution with the optic axis as the axis of rotation. These *ray velocity* surfaces are used in constructing Huygens' secondary wavelets while discussing: For example, in Figure 6.19 (a) we have a plane wave incident normally. The extraordinary wave also propagates t) in a direction which is normal to the surface. But in case of traveling extraordinary wave which is along *BE* and *DE'* with *EE'* represents the wave frants.

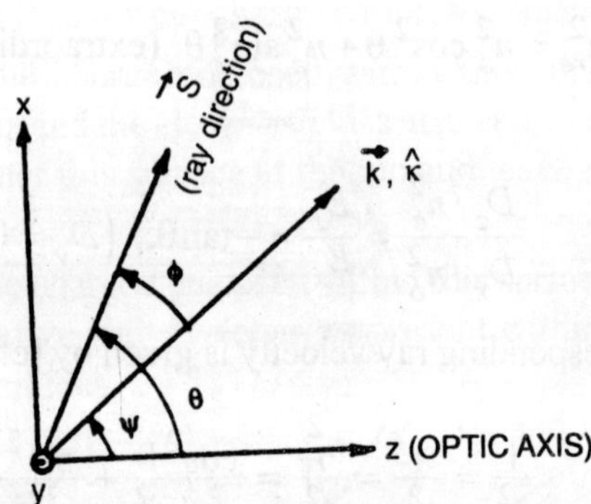

Figure 6.19 : *While considering propagation propagation in uniaxial crystals in an uniaxial crystal, the x-axis may be assumed to He in the plane defined by k and the optic axis, the latter is assumed to lie along the z-axis.*

On comparing this equation with (6.113) we get—

$$\tan\theta = -\frac{D_z / n_e^2}{D_x / n_o^2} = \frac{n_o^2}{n_e^2}\tan\psi \qquad (6.116)$$

Thus when the wave propagates along a direction which makes an angle ψ with the optic axis, then the ray will propagate along the direction

$$\theta = \tan^{-1}\left[\frac{n_o^2}{n_e^2}\tan\psi\right] \qquad (6.117)$$

As an example, for calcite

$$n_0 = 1.65836,\ n_e = 1.48641, \text{ with } \psi = 45^\circ$$

we obtain $\theta \simeq 51.2°$

In fig 6.12 we get that the direction of ray is always from the optic axis.

7

The Electric Field

In a room, the temperature at every point has a definite value. The temperature at any given point can be measured by using a thermometer. We call the resulting distribution of temperatures *a temperature field.* In much the same way, you can imagine a *pressure field* in the atmosphere: it consists of the distribution of air pressure values, one for each point in the atmosphere. As the temperature and air pressure are scalar quantities, so these are example of scalar fields.

The electric field consists of distribution of vectors one for each point around the region around a charged object, so the electric field is a vector field. In principle, we can define the electric field at some point near the charged object, such as point P in Figure 7.1 (a), by placing *a positive* charge q_0, called *a test charge,* at the point. We then measure, the electrostatic force **F** that acts on the test charge.

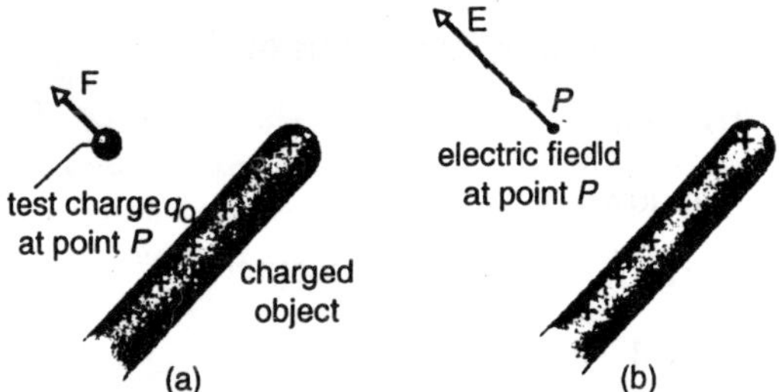

Figure: 7.1 : ***(a) A positive test charge q_0 placed at point* P *near a charged object. An electrostatic force* F *acts on the test charge.***
(b) The electric field* E *at point* P *produced by the charged object.

The electric field **E** at point *P* due to the charged object is defined as electric field

$$\mathbf{E} = \frac{\mathbf{F}}{q_0} \text{ (electric field)} \tag{7.1}$$

So the direction of electric fields *E* in same as that of the force *F* which acts on the positive test charge and the magnitude of the electric field *E* at point *P* in given by $\mathbf{E} = \frac{\mathbf{F}}{q_0}$. As shown in Figure 7.1 (b), we represent the electric field at *P* with a vector whose tail is at *P*. To define the electric field within some region, we must similarly measure it at all points in the region. The *SI* unit for the electric field is the newton per coulomb (N/C). Table 7.1 shows the electric fields that occur in a few physical situations.

The electric field of a schared particle exist independently of the test charge q_0. We put q_0 at point *P* only to determine the magnitude of electric field at *P* but the electric field existed both before and after the object was put there. (We assume that in our defining procedure, the presence of the test charge does not affect the charge distribution on the charged object, and thus does not alter the electric field we are defining.)

Table: 7.1 Some Electric Fields

Field location or Struation	Value (N/C)
At the surface of a uranium nucleus	3×10^{21}
Within a hydrogen atom, at a radius of 5.29×10^{-11} m	5×10^{11}
Electric breakdown occurs in air	3×10^{6}
Near the charged drum of a photocopier	10^{5}
Near a charged plastic comb	10^{3}
In the lower atmosphere	10^{2}
Inside the copper wire of household circuits	10^{-2}

To examine the role of an electric field in the interaction between charged objects. We have two tasks: (1) calculating the electric field produced by a given distribution of charge and (2) calculating the force that a given field exerts on charge placed in it. The calculation of electric field produced by given distribution of charge will be discussed from. By imagine a point charge and a pair of point charge we calculate the force exerted by the electric field on the charge particles. But first, we discuss a way to visualize electric fields.

Electric field Lines

Michael Faraday, who, introduced the idea of electric fields in the 19th century, thought of the space around a charged body as filled with lines of force. These lines now known as electric field lines are no longer used as reality.

There are two relation between the electric field vector *s* and the field lines—(i) the direction of **E**, at any point, can be determined by the direction of straight field line to any curved field line. (ii) the field line are drawn so that the number of lines per unit area, measured in a plane that is perpendicular to the lines, is proportional to the *magnitude* of **E**. This indicates that when the field lines are close to each other the magnitude of **E** is large similarly where the fields are far from each other **E** is small.

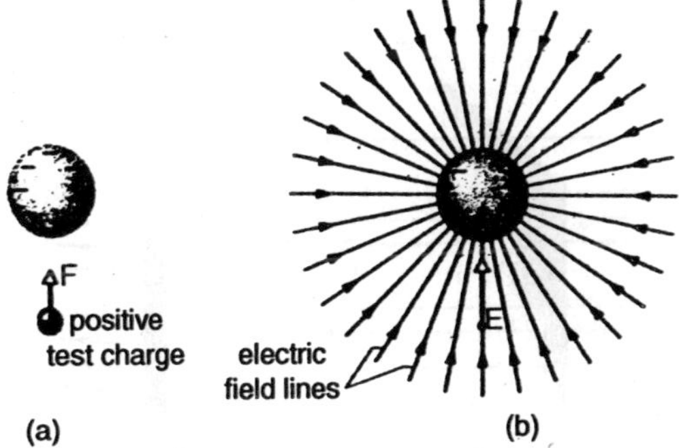

Figure 7.2 : ***(a) represents uniformly distributed negative charge sphere. When we put* a tve *charge near the sphere then, an elecrostatic force directing towards the centre of the sphere will arise.***

Figure 7.2 (a) represents uniformly distributed negative charge sphere. When we put a +ve charge near the sphere than, an electrostatic force directing towards the center of the sphere will arise. In other words, the electric field vectors at all points near the sphere are directed radially toward the sphere. This pattern of vectors is neatly displayed by the field lines in Figure 7.2 (b) which point in the same directions as the force and field vectors. Moreover, the spreading of the field lines with distance from the sphere tells us that the magnitude of the electric field decreases with distance from the sphere.

If the sphere (7.2a) consist uniformly distributed positive change, them the direction of the electric field vectors would be opposite *i.e.*, they would be directed radially away from the centre of the sphere. Thus the electric field lines would also extend radially away from the sphere. We then have the following rule:

Electric field lines extend away from positive charge and toward negative charge.

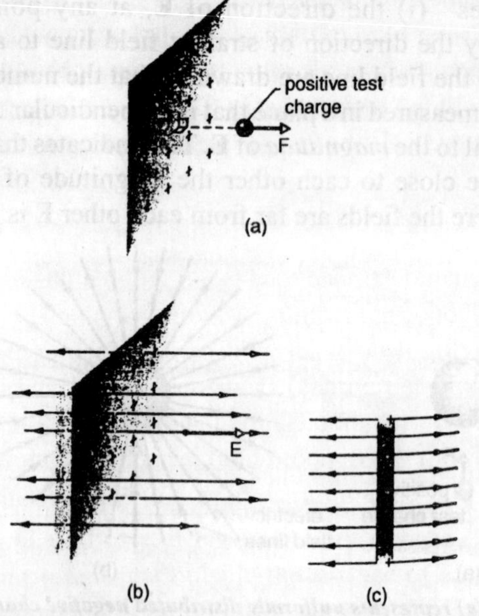

Figure 7.3 : *(a) The electrostatic force* F *on a positive test charge near a very large, nonconducting sheet with uniformly distributed positive charge on one side. (b) The electric field vector* E *at the location of the test charge, and the electric field lines in the space near the sheet. The field lines extend away from the positively charged sheet. (c) Side view of (b).*

In fig 7.3 (a) A large non conducting sheet is telcen and on a small part of the sheet positive charge is uniformly distributed on one side.

If we were to place a positive test charge at any point near the sheet of Figure 7.3 (a) the net electrostatic force acting on the test charge would be perpendicular to the sheet, because forces acting in all other directions would cancel one another as a result of the symmetry. The net force exerted on the charge will be perpendicular to the sheet. So the electric field vector would also be perpendicular to the sheet and away from the sheet. Since the charge is uniformly distributed along sheet, all the field vectors have the same magnitude. The electric field is uniform as the magnitude and direction is same at any given point of the field.

Of course, no real nonconducting sheet (such as a flat expanse of plastic) is infinitely large, but if we consider a region that is near the middle of a real sheet and not near its edges, the field lines through that region are arranged as in Figure 7.3 (b).

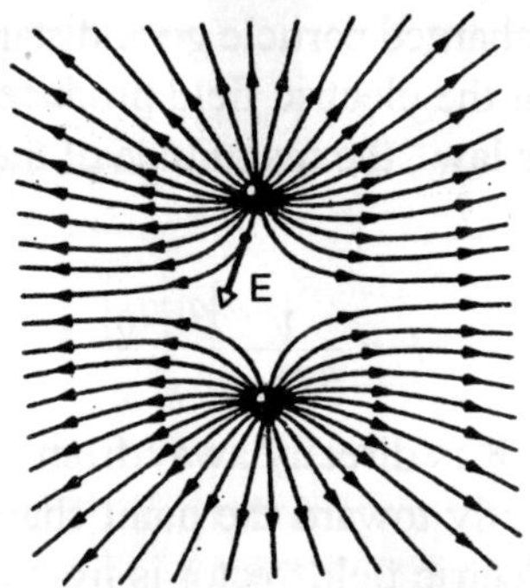

Figure 7.4 : ***Field lines for two equal positive point charges. The charges repel each other. (The lines terminate on distant negative charges.) To "see" the actual three-dimensional pattern of field lines, mentally rotate the pattern shown here about an axis passing through both charges in the plane of the page. The three-dimensional pattern and the electric field it represents are said to have*** **rotational symmetry** ***about that axis. The electric field vector at one point is shown; note that it is tangent to the field line through that point.***

When two equal positive charges will be the field lines obtained will be as fig 7.4. But if we will fake two charges that are equal in magnitude but one is +ve and other is *-ve*, them the case of electric dipole will seen. Although we do not often use field lines quantitatively, they are very useful to visualize what is going on. Can you not almost "see" the charges being pushed apart in Figure 7.4 and pulled together in Figure 7.5

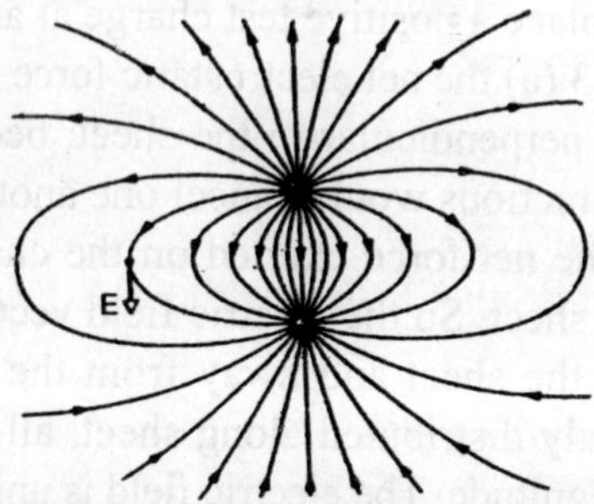

Figure 7.5 : ***Figure shows field lines of the case of electric dipole. The charges attract each other. The pattern of field lines and the electric field it represents have rotational symmetry about an axis passing through both charges. The electric field vector at one point is shown the vector, is tangent to the field line through the point.***

The Electric Field Due to a Point Charge

We will place a charged particle *go* at distance *r* from the point charge *q* to find out the electric field produced by point charged *q*. From Coulomb's law , the magnitude of the electrostatic force acting on q_0 is

$$F = \frac{1}{4\pi\varepsilon_0} \frac{|q|\,|q_0|}{r^2} \tag{7.2}$$

The direction of **F** is directly away from the point charge f *q is* positive and directly toward the point charge if negative. The magnitude of the electric field vector is from eq. 7.1.

$$E = \frac{F}{q_0} = \frac{1}{4\pi\varepsilon_0} \frac{|q|}{r^2} \text{ (point charge)} \tag{7.3}$$

The direction of electric field **E** depends upon the nature of the point charge. If the point charge is *+ve*, the direction of **E** will be directly away from it and if point charge is *–ve*, **E** will be towards the point charge. We find the electric field in the space around a point charge by moving the test charge around in that space. The field for a positive point charge is shown in Figure 7.6 in vector form (not as field lines).

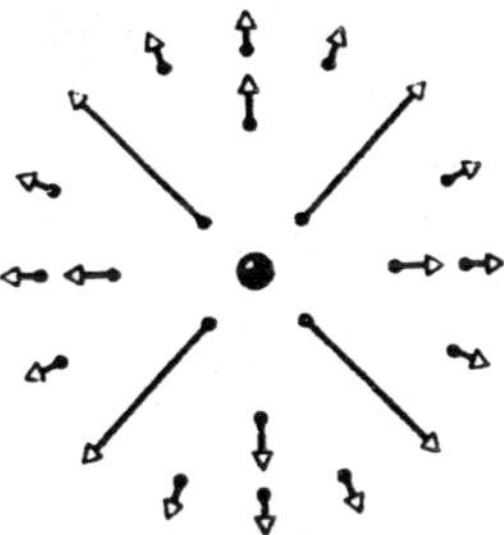

Figure 7.6: ***The electric field vectors at several points around a positive point charge.***

If more than one point charge is present in an electric field then the net electric field can be determined by the use of principle of super-position. If we place a positive test charge q_o near n point charges $q_1, q_2, \ldots, q_n$ then, from Equation 7.7 the net force $\mathbf{F}_0$ from the *n* point charges acting on the test charge is

$$\mathbf{F}_0 = \mathbf{F}_{01} + \mathbf{F}_{02} + \cdots + \mathbf{F}_{0n}.$$

So, from equation 7.1, the net electric field at the position of the test charge is

$$\mathbf{E} = \frac{\mathbf{F}_0}{q_0} = \frac{\mathbf{F}_{01}}{q_0} + \frac{\mathbf{F}_{02}}{q_0} + \cdots + \frac{\mathbf{F}_{0n}}{q_0}$$
$$= \mathbf{E}_1 + \mathbf{E}_2 + \cdots + \mathbf{E}_n.$$

Here $\mathbf{E}_i$ is the electric field that would be set up by point charge *i* acting alone. With the above equation, we now know that principle of superposition **b**, applies to electric field as well as electrostaric forces.

The Electric Field due to an Electric Dipole

Now we will discuss the case of electric field produced by electric dipole i.e, when two equal but opposite charges are separated by distance-**d.** Let us find the electric field due to the dipole of figure 7.7 *a* at a point *P*, a distance *z* from the midpoint of the dipole and on its central axis, which is called the *dipole axis*.

All the three electric field i.e (i) the electric field **E** at point *P*,

(ii) the electric field E due to $+ve$ point charge and (iii) the electric field $E_{(-)}$ must lie along the dipole axis. This axis will be z-axis. Applying the superposition principle for electric field, we find that the magnitude E of the electric field at P is

$$E = E_{(+)} - E_{(-)}$$

$$= \frac{1}{4\pi\varepsilon_0} \frac{q}{r_{(+)}^2} - \frac{1}{4\pi\varepsilon_0} \frac{q}{r_{(-)}^2}$$

$$= \frac{q}{4\pi\varepsilon_0\left(z - \frac{1}{2}d\right)^2} - \frac{q}{4\pi\varepsilon_0\left(z + \frac{1}{2}d\right)^2} \tag{7.5}$$

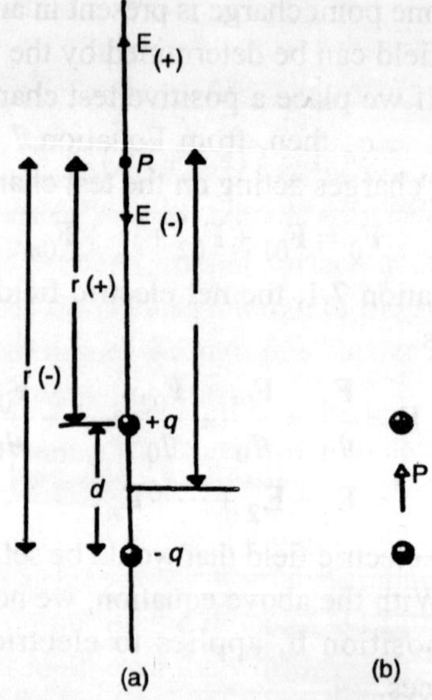

Figure 7.7 **(a)** ***An electric dipole. The electric field vectors*** **E** ***and*** **E** ***at point*** **P** ***on the dipole axis resulting from the two charges are shown.*** **P** ***is at distances*** $r_{(+)}$ ***and*** $r_{(-)}$ ***from the individual charges that make up the dipole.*** **(b)** ***The dipole moment*** **p** ***of the dipole points from the negative charge to the positive charge.***

We can write this equation in algebric form as—

$$E = \frac{q}{r\pi\varepsilon_0 z^2}\left[\left(1 - \frac{d}{2z}\right)^{-2} - \left(1 + \frac{d}{2z}\right)^{-2}\right] \tag{7.6}$$

Generally, we consider electric field of a electric dipole only at large distances that are compared with the dimension of the dipole. At such large distances, we have $d / 2z \ll 1$ in Equation 7.6. By expanding these quantities using binomial theorem, we get the equation as—

$$\left[\left(1+\frac{2d}{2z(1!)}+\cdots\right)-\left(1-\frac{2d}{2z(1!)}+\cdots\right)\right]$$

So,

$$E=\frac{q}{4\pi\varepsilon_0 z^2}\left[\left(1+\frac{d}{z}+\cdots\right)-\left(1-\frac{d}{z}+\cdots\right)\right] \tag{7.7}$$

The unwritten terms in the two expansions in Equation 7.7 involve d/z raised to progressively higher powers. Since $d / z \ll$, the contributions of those terms are progressively less, and to approximate E at large distances, we can neglect them. Then, in our approximation, we can rewrite Equation

$$E=\frac{q}{4\pi\varepsilon_0 z^2}\frac{2d}{z}-\frac{1}{2\pi\varepsilon_0}\frac{qd}{z^3} \tag{7.8}$$

The product qd, is the magnitude p of a vector quantity electric dipole moment of dipole. Now the equation become—

$$E=\frac{1}{2\pi\varepsilon_0}\frac{p}{z^3} \text{ (electric dipole).} \tag{7.9}$$

The direction of **p** is taken to be from the negative to the positive end of the dipole, as indicated in Figure 7.8 (b) **P** can be used to specify the orientation of a dipole.

From the above equation it is clear that when we measure electric field of *a* dipole at distance point then we will **P** find product of q and d instead of q and d separately. The field at distant points would be unchanged if, for example, q were doubled and d simultaneously halved.

The basic property of dipole is the dipole moment. Equation 7.9 is true for all the distance point on the dipole axis. But now it is found that **E** for a dipole differ by $1/r^3$ for all distance points, where r is the distance between the centre of dipole axis and the distance point taken. When we give a serious look on the figure of field lines and electric dipole, it becomes clear that the direction of **E** for distant points on the dipole axis is same as that of the vector of diple moment p, regardless of the position of p. If the distance between the point and the dipole center i.e., 'r' is increased 2 times men, the electric field at the point is decreased by 8 times.

If you double the distance from a single point charge, however the electric field drops only by a factor of 4. Thus the electric field of a dipole decreases more rapidly with distance than does the, electric field of a single charge. The physical reason for this rapid decrease in electric field for a dipole is that from distant points a dipole looks like two equal but opposite charges that almost-but not quite-coincide.

When we give a serious look on the figures of field lines and electric dipole, it became clear that the direction of E for distant points on the dipole axis is same as that of the vector of dipole moment p, regardless of the position of P. If the distance between the point and the diple centre i.e. 'r' is increased two times decrease by eight times. So, the electric field of a dipole (one *+ve* and other –ve will almost cancel each other.

The Electric Field Due to a Line of charge

We have discussed electric field produced by one or *a* few point charges. We now consider charge distributions that consist of a great many closely spaced point charges (perhaps billions) that are spread along a line, over a surface, or within a volume. Such distributions are said to be continuous rather than discrete. These continuous distribution consist of large no of point charges, so to find the electric field used by them we have to use calculus method. In this section we discuss the electric field caused by a line of charge. We consider a charged surface in the next section. We will discuss electric field inside the sphere in the next chapter.

In the case of continuous charge distribution, the charge on an object is expressed as charge density instead of total charge. For a line of charge, for example, we would report the linear charge density (or charge per length) λ, whose **SI** unit is the coulomb per meter. Some other charge density are shown in the table 7.2.

Let us take a thin ring with uniform linear charges λ around its circumference, the radius of ring is *R*. We may imagine the ring to be made of plastic or some other insulator, so that the charges can be regarded as fixed in place. Now let us take a point *p* along the central axis of the ring and distance of *P* from the centre is *Z*. Now we have to find the *E* at point *P*.

Table: 7.2 Some Measures of Electric Charge

Name	Symbol	SI Unit
Charge	q	C
Linear charge density	λ	C/m
Surface charge density	σ	C/m^2
Volume charge density	p	C/m^3

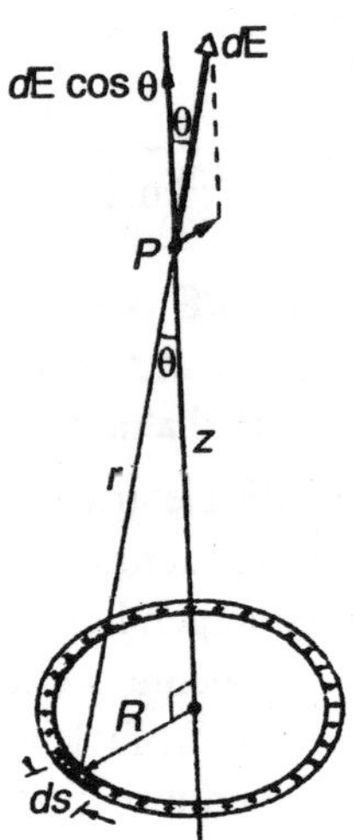

Figure: 7.8. A ring of uniform positive charge. A differential element of charge occupies a length* ds *(greatly exaggerated for clarity). This element sets up an electric field* dE *at point* P. *The component of dE along the central axis of the ring is dE cos θ.

To find *E* at P_1 we cannot use equation for point charge, as it is a case of ring which is not point charge. However, we can mentally divide the ring into differential elements of charge that are so small that they are like point charges, and then we can apply Equation 7.3 to each of them. Then, we will add all the electric field produced at *P* by differential elements and the sum will be the result.

Let *ds* be the (arc) length of any differential element of the ring. Since λ is the charge per unit length, the element has a charge of magnitude

$$dq = \lambda \; ds \qquad (7.10)$$

Each of these differential element will produced a differential electric field at point *P*. If the distance of the differential element *d* from the centre is *r* then the equation can be rewrite by using point charge equation as 7.11.

$$dE = \frac{1}{4\pi\varepsilon_0}\frac{dq}{r^2} = \frac{1}{4\pi\varepsilon_0}\frac{\lambda \, ds}{r^2} \qquad (7.11)$$

From Figure, we can rewrite Equation 7.11 as

$$dE = \frac{1}{4\pi\varepsilon_0}\frac{\lambda \, ds}{\left(z^2 + R^2\right)} \qquad (7.12)$$

Figure 7.8 shows us that $d\mathbf{E}$ is at an angle θ to the central axis (which we have taken to be a *z*-axis) and has components perpendicular to and parallel to that axis.

All the $d\mathbf{E}$ vectors, of $d\mathbf{E}$ electrical field sets up by differential element charge, have same magnitude and direction. All these $d\mathbf{E}$ vectors have components perpendicular to the centre axis as well; these perpendicular components are identical in magnitude but point in different directions. In fact, for any perpendicular component that points in a given direction, there is another one that points in the opposite direction. As the two components are oppositely directed, so they will cancel each other and sum is zero.

The components which are perpendicular to each other will also cancel each other. This leaves the parallel component: they are all in the same direction, so the net electric field at *P* is their sum.

The component paralleled to $d\mathbf{E}$ has the magnitude $d\mathbf{E}$ cos we can also write that—

$$\cos\theta = \frac{z}{r} = \frac{z}{\left(z^2 + R^2\right)^{1/2}} \tag{7.13}$$

Then Equations 7.13 and 7.12 give us, for the parallel component of $d\mathbf{E}$,

$$dE\cos\theta = \frac{z\lambda}{4\pi\varepsilon_0\left(z^2 + R^2\right)^{3/2}}\,ds \tag{7.14}$$

We have to integrate the equation 7.14, to find the sum of the parallelled component of $d\mathbf{E}$ cos θ produced by all element around the circumference of the ring. Since the only quantity in Equation 7.14 that varies during the integration is s, the other quantities can be moved outside the integral sign. We integrate from $s = 0$ to $s = 2\pi r$ and the final result we will obtain is the equation 7.15.

$$E = \int dE\cos\theta = \frac{z\lambda}{4\pi\varepsilon_0\left(z^2 + R^2\right)^{3/2}}\int_0^{2\pi R} ds$$

$$= \frac{z\lambda(2\pi R)}{4\pi\varepsilon_0\left(z^2 + R^2\right)^{3/2}} \tag{7.15}$$

Since λ is the charge per length of the ring, the term $\lambda\,(2\pi R)$ in Equation 7.15 is q, the total charge on the ring. We then can rewrite Equation 7.15 as

$$E = \frac{qz}{4\pi\varepsilon_0\left(z^2 + R^2\right)^{3/2}} \text{ (charged-ring)} \tag{7.16}$$

The magnitude of electric field at P will remain same independent to the –*ve* or +*ve* charge of ring. But the direction of the vector P will be opposite if we change the charge form +*ve* to –*ve*. In –*ve*, the direction of P will points toward the ring.

Now if we will rewrite the equation 7.16 for *a* point on the central axis not away from the axis. For this point the term $Z^2 + R^2$ will become Z^2 as the point is on central axis.

$$E = \frac{1}{4\pi\varepsilon_0} \frac{q}{z^2} \text{ (charged ring at large distance)} \qquad (7.17)$$

From large distance the ring looks like a point charge, so the result shown in above equation is correct. If we replace *z* by *r* in equation 7.17, we indeed do have equation 7.3, the electric field due to a point charge.

If we have to prove the equation 7.16 for *a* point which is at the centre of the axis, so we will take Z = 0. At that point, Equation 7.16 tells us , that *E* = 0. This is a reasonable result, because if we were to place a test charge at the center of the ring, there would be no net electrostatic force acting on it: the force due to any element of the ring would be canceled by the force due to the element on the opposite side of the ring. In this case, if *F*, force is zero then the electric field at centre will also be zero.

The Electric Field Due to a Charged Disk

Let us consider a circular plastic dise which has a *+ve* surface charge of uniform density σ. Now we will find electric field at point P_1 which is at *a* distance *Z*, from the central axis of the disc.

To find electric field *E* at P_1 we will first divide the disc into many rings with *a* common centre. Now we will add up the contribution from all the rings to find *E* at *P* up (that is, by integrating) Figure. 7.9 shows one such ring, with radius *r* and radial width *dr*. As we know that charge per unit area is σ, so the net charge on the ring is—

$$dq = \sigma\, dA = \sigma\,(2\pi\, r\, dr), \qquad (7.18)$$

where *dA* is the differential area of the ring.

We can obtain an expression for the electric field *d*E at P due to that ring by substituting *dq* for *q* in equation 7.16 and by replacing *R* in equation 7.16 with *r* as follow—

$$dE = \frac{z\sigma 2\pi r\, dr}{r\pi\varepsilon_0 \left(z^2 + r^2\right)^{3/2}}$$

which we may write as

$$dE = \frac{\sigma z}{4\varepsilon_0} \frac{2r\,dr}{\left(z^2 + r^2\right)^{3/2}}$$

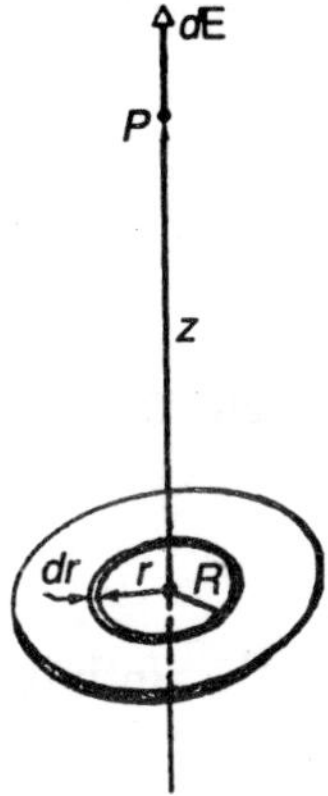

Figure 7.9 : ***A disk of radius R and inform positive charge. The ring shown has radius r and raial width dr. It sets up a differential electric field dE at point P on its central axis.***

The value of E can be found by integrating the equation by variable r, from $r = o$ to $r = R$ while the Z remains constant.

$$E = \int dE = \frac{\sigma z}{4\varepsilon_0} \int_0^R \left(z^2 + r^2\right)^{-3/2} (2r)\,dr \qquad (7.19)$$

We can charge the form of integral as $\int X^m dX$ by using $x = (z^2 + r^2)$, $m = \frac{3}{2}$ and $dX = (2r)\,dr$. By changing the form we have the equation as—

$$\int X^m dX = \frac{X^{m+1}}{m+1}$$

so Equation 7.23 becomes

$$E = \frac{\sigma z}{4\varepsilon_0} \left[\frac{\left(z^2 + r^2\right)^{-1/2}}{-\frac{1}{2}} \right]_0^R$$

Taking the limits and rearranging, we find

$$E = \frac{\sigma}{2\varepsilon_0}\left(1 - \frac{z}{\sqrt{z^2 + R^2}}\right) \text{ (charged disk)} \qquad (7.20)$$

as the magnitude of the electric field produced by a flat, circular, charged disk on its central axis. (In carrying out the integration, we assumed that $z \geq 0$.)

If this equation is written for a disc such has $R \to \infty$ than the term $\left(b\frac{z}{\sqrt{Z^2 + R^2}}\right)$ will tens to zero and the equation becomes as (7.25).

$$E = \frac{\sigma}{2\varepsilon_0} \text{ (infinite sheet)} \qquad (7.21)$$

This E is produced by non conductor sheet as plastic one of which side has informaly distributed charge. The electric field lines for such a situation are shown in Figure 7.3.

The value of $E = \frac{\sigma}{2\varepsilon_0}$ can also be obtained if Z approaches to 0 and R remains finite. This shows that at points very close to the disk, the electric field set up by the disk is the same as if the disk were infinite in extent.

A Point Charge in an Electric Field

Till now, we have discussed as how to find the electric field if charge distribution is given to us. Now, we will determine, that if *a* charge particle placed in an electric field, what happens to the particle.

When a charge particle is placed in a electric field an electro static force will act on the particle. The electro static force acting on particle is a vector quantity and given by —

$$\mathbf{F} = q\mathbf{E}, \qquad (7.22)$$

in which q is the charge of the particle (including its sign) and **E** is the electric field that other charges have produced at the location of the particle. (The field is not the field set up by the particle itself; to distinguish the two fields, the acting on the particle in Equation 7.22 is often called the external field. The particle will not affected by the electrical field produced by it. Equation 7.22 tells us:

The electrostatic force **F** acting on a charged particle in an external electric field **E** points in the direction of **E** if the charge q of the particle is positive the opposite direction if q is negative.

Measuring the Elementary Charge

American physicist Robert A. Millikan took the help of equation $\mathbf{F} = q\mathbf{E}$ to measure the elementary charge e. Figure 7.10 a representation of his apparatus. When tiny oil drops are sprayed into chamber A, some of them become charged, either positively or negatively, in the process. Let us assume that a drop which has a negative charge q, shifts down ward through a small hole in plate P_1 into chamber C.

If the switch S will open as shown in Fig 7.10 then there will be no effect on the chamber C. If the switch is closed (the connection between chamber C and the positive terminal of the battery is then complete), the battery causes an excess positive charge on

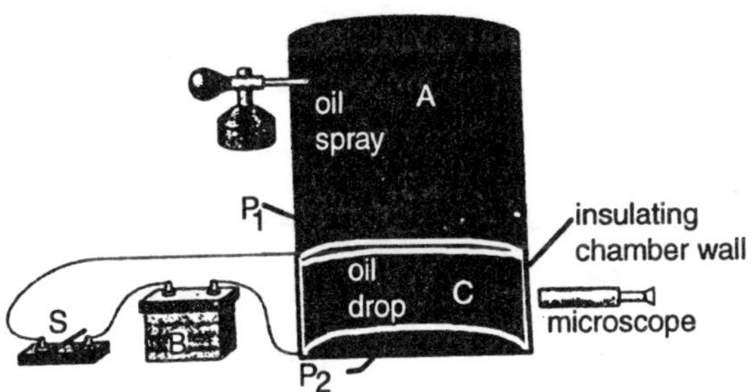

Figure 7.10: ***The Millikan oil-drop apparatus for measuring the elementary charge* e. *When a charged oil drop drifted into chamber C through the hole in plate P_1 its motion could be controlled by closing and opening switch* S *and thereby setting up or eliminating an electric field in chamber C. The microscope was used to view the drop, to permit timing of its motion.***

conducting plate P1 and an excess negative charge on conducting plate P_2. The charged plates set up a downward-pointing electric field **E** in chamber C. As suggested by equation 7.22 this field will exert an electrostatic force on any charged drop present in side the chamber. Our drop will experience a upward drift.

By adjusting the motion of oil drops with switch opened and closed Millikan determined the effect of charge q. Millikan discovered that the values of q were always given by—

$$q = ne, \quad n = 0, \pm 1, \pm 2, \pm 3, \cdots \tag{7.23}$$

in which e turned out to be the fundamental constant we call the *elementary charge,* 1.60×10^{-19} C. Millikan's experiment is convincing proof that charge is quantized, and he earned the 1923 Nobel Prize in physics in part for this work. Now *a* days the measurement of elementary charge depends more on many interlocking experiments.

Ink Jet Printing

To get high quality and high speed printing attempts were made to search an alternative for typewriter. Building up letters by squirting tiny drops of ink at the paper is one such alternative.

A charged drop moving between two conducting plates, between which *a* uniform downward pointing electric field E has been set up, is shown in figure 7.11.

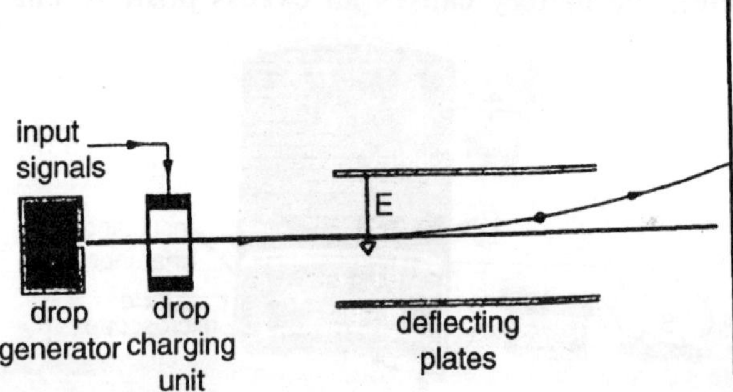

Figure 7.11 : ***The essential features of an prints An input signal from a computer controls the charge given to each drop and thus the position on the paper at which the drop add. About 100 tiny drops are needed to form a single character***

The drop is deflected upward according to Equation 7.22 and strikes the paper at a position that is determined by the magnitudes of **E** and the charge q of the drop.

Practically, E is kept constant and the position of the drop is determined by the charge q, which is delivered to the drop in the charging unit. The charging unit is activated by electronic signals which encoded the printing material.

A Dipole in an Electric Field

As we know, the electric dipole moment P of an electric dipole is defined as *a* vector that points from –ve to the +*ve* end of the dipole. As you will see, the behaviour of a dipole in a uniform external electric field **E** can be described completely in terms of the two vectors **E** and **P** with no need of any details about the dipole's structure.

We know that the molecule of water in an electric dipole. Figure 7.12 shows why. There the black dots represent the oxygen nucleus (having eight protons) and the two hydrogen nuclei (having one proton each). The orbits in which electron revolve around nucleus, are shown by coloured enclosed areas.

The two hydrogen atom and one oxygen atom an water molecule do not lie in a straight line, rather they form an angle of about 105°.

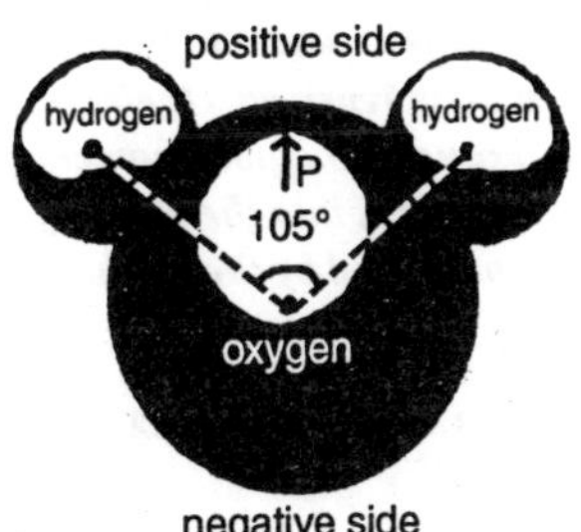

Figure 7.12: ***A molecule of H_2O, showing the three nuclei (represented by dots) and the, regions in which the electrons orbit the nuclei. The electric dipole moment* p *points from the (negative) Oxygen side to the (positive) hydrogen side of the molecule.***

As a result, the molecule has a definite "oxygen side" and "hydrogen side. All the electrons of water molecule remain closer to the oxygen molecule so it is slightly –*ve*. This makes the oxygen side of the molecule slightly more negative than the hydrogen side and creates an, electric dipole moment **p** that points along the symmetry axis of the molecule as shown. When the water molecule is introduced in to an electric field, the molecule will behave to show more abstract electric dipole.

Now, we will consider *a* case of abstract dipole in an uniform electric field **E**. We assume, that the dipole is a rigid structure (due to internal electrostatic: forces) that consists of two centers

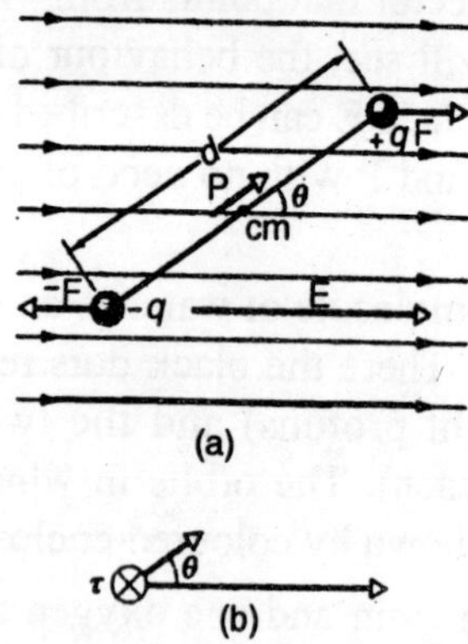

Figure 7.13 : (a) An electric dipole in a uniform electric field* E. *Two centers of equal but opposite charge are separated by distance* d. *Their center of mass cm is assumed to be midway between them. The bar between them represents their rigid connection. (b) Field* E *causes a torque* r *on dipole moment* p. *The direction of the torque vector* r *is into the plane of the page, as represented by the symbol ⊗.

of opposite charge, each of magnitude *q* separated by a distant *q*. The electric field **E** has a dipole moment *P* which makes an angle θ with it.

The forces acting on the two ends of the dipole is *F* and –*F*. Both these force have same magnitude **F** = *q***E** and opposite in direction. Thus the net force exerted on the dipole by the field is zero,. However, these forces exert a net torque τ on the dipole about its center of mass, which we can take to be midway along the line connecting the charged ends. By putting $r = \frac{d}{2}$ we can write the equation for the torque acting on the dipole as—

$$\tau = F\frac{d}{2}\sin\theta + F\frac{d}{2}\sin\theta = Fd\sin\theta \qquad (7.24)$$

Another form for magnitude of torgue can be obtained by using terms of magnitude of E and magnitude of dipole moment qd. To do so, 've substitute qE for F and p/q for d in Eq. 7.24, finding that the magnitude of τ is

$$\tau = pE\sin\theta \quad (7.25)$$

We can generalize this equation to vector form as

$$\tau = \mathbf{p} \times \mathbf{E} \quad \text{(torque on a dipole)}$$

As shown in fig 7.17(b) torque acting on P tends to rotate P in the direction of E and hence, it reduced the angle θ between P and E. As we have discussed earlier, we can represent a froque which gives rise to a rotation by including a minus sing with the magnitude of torque. By using this statement, we can write—

$$\tau = -pE\sin\theta \quad (7.26)$$

Potential Energy of an Electric Dipole

There is as unique relation between orientation of an electric dipole and potential energy. The dipole has its least potential energy when it is in its equilibrium orientation, which is when its moment **p** is lined up with the field **E** (then $\tau = \mathbf{p} \times \mathbf{E} = 0$). It has greater potential energy in all other orientations. Thus the dipole is like a pendulum which has its least gravitational potential energy in its equilibrium orientation—at its lowest point. An external force is required to change the direction or orientation of dipole.

In any case of potential energy, only differences in potential energy have physical meaning, so we are free to define zero-potential energy configuration. It turns out that the expression for the potential energy of an electric dipole in as external electric field is simplest if we choose the potential energy to be zero when the angle θ in Figure 7.13 is 90°. We then can find the potential energy U of the dipole at any other value of θ ($\Delta U = -W$) by calculating the work W done by the field on the dipole when the dipole is roated to that value of θ from 90°. The potential energy U at any angle θ can be measured by the help of equation 7.35

and the relation $W = \int r d\theta$ as —

$$U = -W = -\int_{90°}^{\theta} \tau \, d\theta$$

$$= \int_{90°}^{\theta} pE \sin\theta \, d\theta \qquad (7.27)$$

Evaluating the integral leads to

$$U = -pE\cos\theta \qquad (7.28)$$

The general form of the equation is as follow—

$$U = -\mathbf{p}\cdot\mathbf{E} \quad \text{(potential energy of a dipole)} \qquad (7.29)$$

From the above equations, it is clear that potential energy U is minimum when P and E are in the same direction ($U = -pE$) and $\theta = 0$. And the value of U is most when P and E are in opposite direction i.e. at $\theta = 180°$, $U = PE$.

Microwave Cooking

The water molecules are relatively free so the electric field produced by each dipolar molecule of water affects the surrounding dipole electric field. As a result, the molecules bond together in groups of two or three, because the negative (oxygen) end of one dipole and a positive (hydrogen) end of another dipole attract each other. When groups of 2 or 3 water molecules are formed the potential energy is transferred to surrounding medium. And each time collisions among the molecules break up a group, the transfer is reversed. The net transfer of energy is zero, so the temperature of water close not change.

While the case is different in microwave oven when the oven is started, the electric field produced by even, oscillates back and forth in direction. If there is water in the oven, the oscillating field exerts oscillating torques on the water molecules, continually rotating them back and forth to align their dipole moments with the field direction. The molecules which are in pair will remain stable as they can rotate around their bond, but in the case of three or more molecules, there must be breakage of one of the molecules.

The energy produced of microwave electric field breaks the

bond between molecule. Then molecules that have broken away from groups can form new groups, transferring the energy they

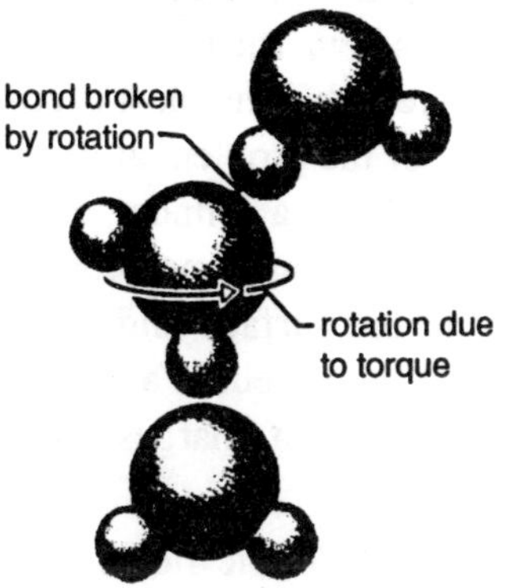

Figure 7.14: ***A group of three water molecules. A torque due to an oscillating electric field in a microwave oven breaks one of the bonds between the molecules and thus breaks up the group.***

just gained into thermal energy. When new groups are formed thermal energy is supplied by the microwave electric field, so the thermal power of water increases. Food that contain water can be cooked in a microwave oven because of the heating of that water. So, the microwave operates on the basis of the dipolar nature of water molecule.

A New Look at Coulomb's Law

The centre of mass of a potato can be find by hard calculation involving the numerical evaluation of *a* triple integral. However, if the potato happens to be a uniform ellipsoid, you know from an symmetry exactly where the center of mass is without calculation. Such are the advantages of symmetry. In physics, we have to deal with situation of symmetry at every step.

In electrostatic, Coulombios law in very helpful but in case of symmetrical situation, this law is not much helpful. In this chapter we introduce a new formulation of Coulomb's law, derived by German mathematician and physicist Carl Friedrich Gauss (1777-1855). This law, called **Gauss' law,** can be used to take advantage of special symmetry situations. Gauss's law is another from of coulomb's law & which of these, we choose, depends only on situation.

The "Gausssian Surface" hypothetical closed surface, is present at central to the Gauss's law. The Gaussian surface can be of any shape you wish to make it, but the most useful surface is one that mimics the symmetry of the problem at hand. Thus the Gaussian surface will often be a sphere, a cylinder, or some other symmetrical form. To make clear distinctions between the points present on the surface, inside the surface and outside the surface, the Gaussian surface must be closed. If a gaussian surface is established around *a* distribution of charges, The surface will give *a* relation as—

Gauss' law relates the electric fields at points on a (closed) Gaussian surface and the net charge enclosed by that surface.

Let us consider *a* gaussian surface which is sphere and there is an electric field at every point at the surface and all the electric field have same magnitude and direction outward from the centre. Without knowing anything about Gauss' law, you can guess that some net positive charge must be inside the Gaussian surface. If you *do* know Gauss' law, you can calculate just how much net positive charge is inside the surface. The quantity of electric field intercepted by the surface depends upon the flux of the electric field, through the surface

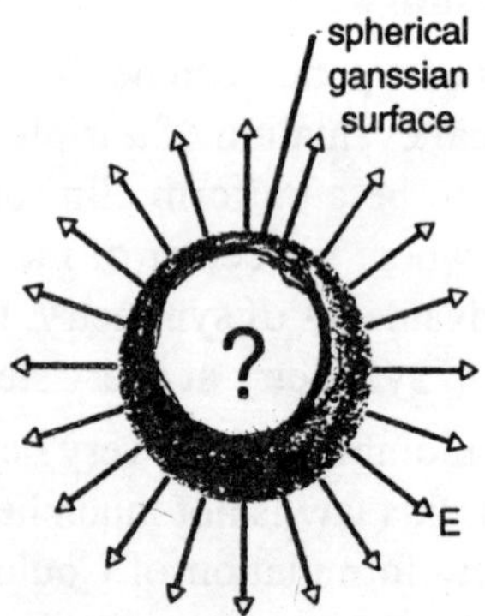

Figure 7.15 : ***A spherical Gaussian surface. If the electric field vectors are of uniform magnitude and point radially outward at all surface points, you can Conclude that a net positive distribution of charge must lie within the surface and have spherical symmetry.***

Flux

Now we will consider *a* thin small square loop of area. A within *a* wide airstream of uniform velocity v. Let Φ represent the volume flow rate (volume per unit time) at which air flows through the loop. The angle between velocity of air v and the plane of loop will determine the rate of air flow. The rate of flow of air is vA if v is perpendicular to the loop. The rate of flow ϕ will be zero, if the velocity of air v is parallel to the plane of loop as air moves through the loop. For an intermediate angle θ, the rate Φ depends on the component of v that is normal to the plane Fig: 7.16 (b). The component of v, which is normal to the plane is $v \cos \theta$, so the rate of volume flow will be—

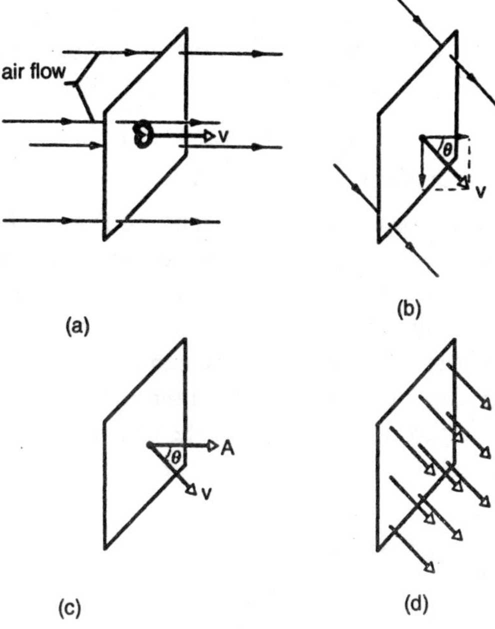

Figure: 7.16: (a) A uniform airstream of velocity* v *is perpendicular to the plane of a square loop of area* A. *(b) The component of* v *perpendicular to the plane of the loop is* v *cos θ, where θ is the angle between* v *and a normal to the plane. (c) The area vector* A *is perpendicular to the plane of the loop and makes an angle θ with* v. *(d) The velocity field intercepted by the area of the loop.

$$\Phi = (v \cos \theta) A \qquad (7.30)$$

As the volume of air flows through an area, so this is an example of volume flux. We can write above equation in the form of vectors. Now we will determine an area vector A. The magnitude of A is equal to the area of the loop and its direction should be normal to the plane of the loop. We then rewrite Equation 7.30 as the scalar (or dot) product of the velocity vector v of the airstream and the area vector **A** of the loop:

$$\Phi = vA\cos\theta = \mathbf{v}\cdot\mathbf{A}, \tag{7.31}$$

where θ is the angle between **v** and **A**.

" Flux" is a latin word, which means "to flow". That meaning makes sense if we talk about the flow of air-volume through the loop. However, Equation 7.31. can be regarded in a more abstract way. At each point of air stream passing through the loop a velocity vector can be assigned. So we can interpret Equation 7.31 as giving the flux of the velocity field through the loop. With this interpretation, flux no longer means the actual flow of something through an area. Now the Flux means product of area and the field of that area.

Flux of an Electric Field

We will take the help of an asymmetrical Gausian surface immersed in *a* non-uniform electric field, to define the term "flux of electric field. Let us divide the surface into small squares of area ΔA, each square being small enough to permit us to neglect any curvature and consider the individual square to be flat. We represent each such element of area with an area vector ΔA, whose magnitude is the area ΔA. All the area vector ΔA will be at 90° angle to the Gaussian surface i.e. perpendicular to the surface and going away from it.

As the squares are assumed to be very small, so the electric field E can be assumed to be constant at any given field. The vectors $\Delta\mathbf{A}$ and **E** for each square then make some angle θ with each other. Three squares from the Gaussian surface are shown in figure 7.17 the angle θ between the electric field and the area vector $\Delta\mathbf{A}$. Provisional, definition for-the-flux of the electric field for the Gattssian surface of Figure 7.17

$$\Phi = \sum \mathbf{E} \cdot \Delta \mathbf{A} \tag{7.32}$$

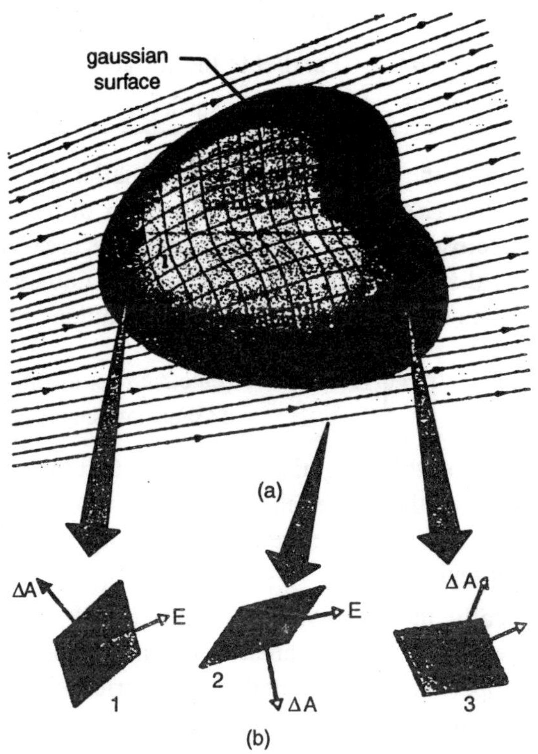

Figures 7.17: ***(a) A Gaussian surface of arbitrary shape immersed in an electric field. The surface is divided into small squares of area Δ A. (b) The electric field vectors* E *and tie area vectors* ΔA *for three representative squares, masked 1, 2, and 3.***

By above equation, it is clear that we have to determine the scalar product E. ΔA for each square and add them all. The sign resulting from each scalar product determines whether the flux through any given square is positive, negative, or zero. As: Table (7.3) shows, squares like 1. in which **E** points inward make a negative contribution to the sum of 7.32. Squares like 2, in which **E** lies in the surface, make zero contribution. The squares which have E directed out ward will give *a +ve* value.

Table 7.3: Three squares on the Gausslan Surface of Figure 7.17

Square		Direction of **E**	Sign of $\mathbf{E} \cdot \Delta\mathbf{A}$
1	$> 90°$	Into the surface	Negative
2	$= 90°$	Parallel to the surface	Zero
3	$< 90°$	Out of the surface	Positive

By making the area of the squares smaller and smaller approaching limit *d*A, the definition of the flux of the electric field through *a* closed surface can be made. The area vectors then approach a differential limit *d***A**. The sum of Equation 7.33 then becomes an integral and we have, for the definition of electric flux,

$$\Phi = \oint \mathbf{E} \cdot d\mathbf{A} \qquad \text{(electric flux through a Gaussian surface).} \qquad (7.34)$$

The circle on the integral sign indicates that the integration is to be taker over the entire (closed) surface. The *SI* unit for electric field is newton-square metre per Coulomb i.e. ($N.m^2/c$) and hence its is scalar quantity.

Specifically, the magnitude *E* is proportional to the number of electric field lines per unit area. Thus, the dot product **E**.*d***A** in 7.34: is proportional to the number of electric field lines passing through area *d***A**. From the above equation two interpritation can be made

(i) The density of electric field lines passing through an area can be used to measure the electric field *E* at that area.

(ii) As the above equation is made on a closed Gaussian surface, so the electric flux and through a Gaussian surface is proportional to the number of electric field lines passing through that surface.

Gauss' Law

Gauss' law gives the relation between the net flux of an electric field of Gaussian surface with the net charge of Gaussian surface. It tells us that

$$\varepsilon_0 \Phi = q_{\text{enc}} \text{ (Gauss' law).} \tag{7.35}$$

By substituting Equation 7.34 the definition of flux, we can also write Gauss' law as

$$\varepsilon_0 \oint \mathbf{E} \cdot d\mathbf{A} = q_{\text{enc}} \quad \text{(Gauss' law).} \tag{7.36}$$

The equation for Gauss' law is applicable only the net charge is situated or present in a vacuum in air. We will discuss the modification of Gauss law when metal or glass material is used later in the chapter.

The result of Gauss law is the sum of the electric charge present in the vacuum, so it may be *–ve*, *+ve* xero depending upon the charges present.

The sign of the charge determine the flux, as when the sign is *+ve*, then the flux is outward but if the sign is *–ve*, the flux is inward.

The Gauss' law doesn't include the charges outside the surface in q_{enc}. The exact form or location of the charges inside the Gaussian surface is also of no concern the only things that matter, on the right side of Equation 7.7 are the magnitude and sign of the net enclosed charge The E on the left side of Equation 7.36, however, is the electric field resulting from all charges, both those inside and those outside the Gaussian surface. The charge present outside the surface, exert or producer zero net flux on through the surface because the number of field lines entering the surface due to that outside charge is equal to the no of field lines clearing the surface.

Now, if these ideas are applied to two charges that are equal in magnitudes but opposite in sign and the field lines produced by them. Four Gaussian surfaces are also shown in cross section. We will discuss each Gaussian surface in turn.

Surface 1. All the electric field of this surface are out wardly directed, So the flux of the electric field through this surface is *+ve*. According to Gauss' law, the net charge within the surface is also *+ve*, as the flux is +ve.

Surface 2. On this surface, all the electric field for every point is directed inwardly. So according to Gauss' law the net flux through the surface as well as the net charge within the surface are *–ve.*

Surface 3. The charge enclosed within this surface is zero,. Hence the net flux and the electric field is also zero. This is because the field lines enter this surface at the top and leave it through the bottom.

Surface 4. The charged enclosed within this surface is also zero because the negative and *+ve* charges present within this surface have equal magnitudes.

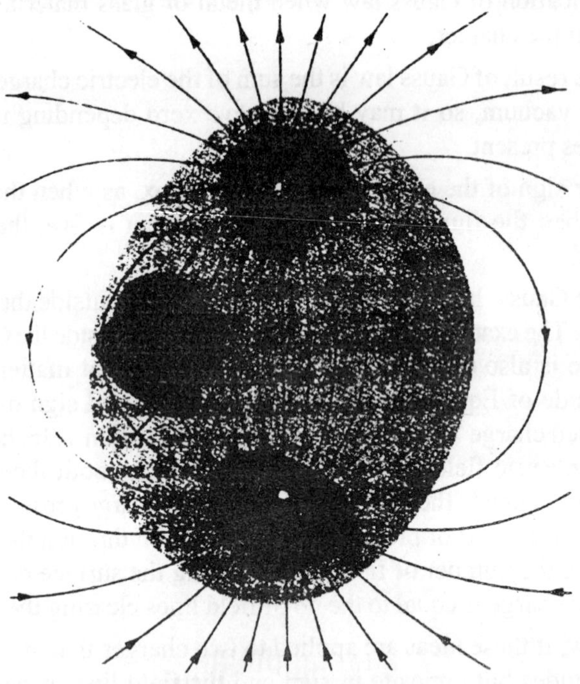

Figure 7.18: ***Two point charges, equal in magnitude but opposite in sign, and the field lines that represent their net electric field. Four Gaussian surfaces are shown in cross section. Surface*** S_1 ***encloses the positive charge. Surface*** S_2 ***encloses the negative charge: Surface*** S_3 ***encloses no charge. And surface*** S_4 ***encloses both charges, and thus no net charge.***

Gauss' law requires that the net flux of the electric field through this surface be zero. The fields lines entering and leaving this surface are also same in number.

Imagine, that we have to bring an enormous charge Q up close to the surface 4. The pattern of the field lines would certainly change, but the net flux for the four Gaussian surfaces would not change. We can understand this because the field lines associated with the added Q would pass entirely through each of the four Gaussian surfaces, making no contribution to the net flux through any of them. Hence, it is clear that the electric field as wall to as field lines produced by Q will not enter the four Gaussian surface, because Q is outside from the all four Gaussian surface.

Gauss' law and Coulomb's law

We should be able to derive the Gauss' law from Coulomb's law and vice-versa, if both laws are equivalent. We will try to drive Coulomb's law from Gauss law.

Suppose, a *+ve* point charge q is present at the centre of a concentric Gaussian surface of radius r. Imagine dividing this surface into differential areas $d\mathbf{A}$. By definition, the area vector $d\mathbf{A}$ at any point is perpendicular to the surface and directed outward from the interior. From the symmetry of the situation, we know that at any point the electric field **E** is also perpendicular to the surface and directed outward from the interior. As the direction of electric field E and the area rector $d\mathbf{A}$ is same i.e. away from the surface, so angle between them θ is also zero, thus Gauss' law can be written as—

$$\varepsilon_0 \oint \mathbf{E} \cdot d\mathbf{A} = \varepsilon_0 \oint E \cdot dA = q_{\text{enc}} \tag{7.37}$$

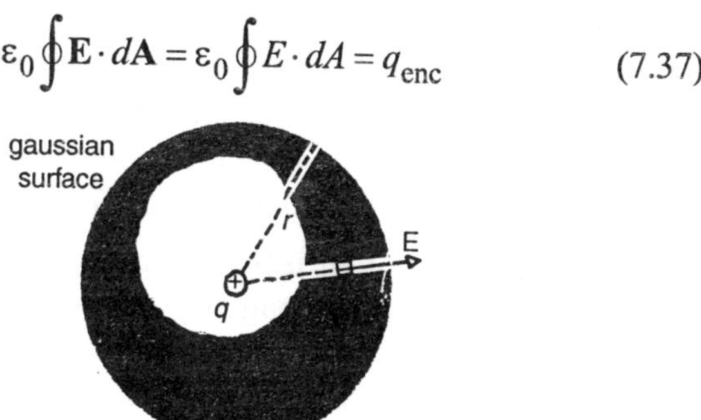

Figure 7.19: ***A spherical Gaussian surface centered on a point charge* q.**

So, wet got that $q = q$ (enclosed). E varies radically with the distance from q, but the value of E remains same every where on the surface. Since the integral in Equation 7.37 is taken over that Surface, E is a constant in the integration and can be brought out in front of the integral sign. And the equation changes as—

$$\varepsilon_0 E \oint dA = q \tag{7.38}$$

The integral represents the surface area i.e., $4\pi r^2$ as it is the sum of all differential area of the surface, so the equation becomes—

$$\varepsilon_0 E\left(4\pi r^2\right) = q$$

or

$$E = \frac{1}{4\pi\varepsilon_0}\frac{q}{r^2} \tag{7.39}$$

The equation gives the electric field E produced by *a* point charge, which is obtained by using Coulomb's law. Hence Coulomb's law and Gauss' law are equivalent.

A Charged Isolated Conductor

By using Gauss' law we can prove an important theorem regarding isolated conductors.

If an excess charge is placed on an isolated conductor, that amount of charge will move entirely to the surface of the conductor. None of the excess charge will be found within the body of the conductor.

This is possible because charges with similar sign repel each other. We might imagine that, by moving to the surface, the added charges are getting as far away from each other as they can. This theorem can be verified with the help of Gauss' law.

The cross section of an isolated lump of copper is shown in figure 7.20. The copper is hanging from an insulating thread. A Gaussian surface is inserted just inside the surface of conductor. The conductor is having an excess charge q.

The conductor must have zero electric field within it. If this were not so, the field would exert forces on the conduction (free) electrons, which are always present in the conductor, and thus current would always exist within the conductor. As the isolated conductor is always free of internal current, so the electric field inside the conductor must be zero.

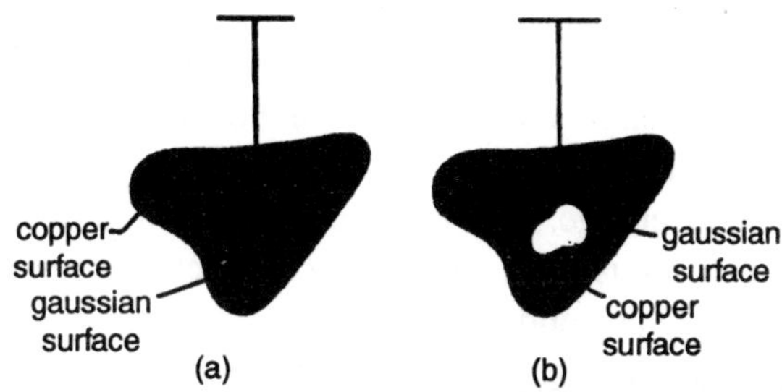

Figure 7.20: (a) A lump of copper with a charge q hangs from an insulating thread. A Gaussian surface is drawn within the metal, just inside the actual surface. (b) The lump of copper now has a cavity within it. A Gaussian surface lies within the metal, close to the cavity surface.

When the conductor is charged, an internal electric field appears. However, the added charge quickly distributes itself in such a way that the net internal electric field-the vector sum of the electric fields due to all the charges-is zero. The forces acting on each charge become zero as the movement of charges stop so an electrostatic equilibrium of charges is established.

As the isolated conductor has zero electric field, so the Gaussian surface should also have zero electric fields as the Gaussian surface is inside the surface of conductor. This means that the flux through the Gaussian surface must be zero. Gauss' law then tells us that the net charge inside the Gaussian surface must also be zero.

So as the electric field is zero for the Gaussian surface, the flux through this surface must be zero. Hence the charge is not present in the Gaussian surface. As the charges is not present in the Gaussian surface, then it must be outside *it i.e.* on the surface of the isolated conductor.

An Isolated Conductor with a Cavity

Now, a cavity is made within the same conductor we have used earlier Figure 7.20 (b). It is perhaps reasonable to suppose that when we scoop out the electrically neutral material to form the cavity, we should not change the distribution of charge or the pattern of the electric field that exists in Figure 8.9a. The Gauss' law will be again used in this case also for verification.

Imagine a Gaussian surface which surrounds the cavity and which is close to the cavity but inside the conductor. Because **E** = 0 inside the conductor, there can be no flux through this new Gaussian surface. Therefore, from Gauss' law that surface can enclose no net charge. It is again proved that the net charge on the wall of cavity is zero, so all the excess charge remain on the outer surface of the conductor.

The Conductor Removed

Imagin that the charges are enclosed in a thin plastic bag and fined on the conductor, then the conductor is removed completely. This is equivalent to enlarging the cavity of Figure 7.20 (b). Until it consumes the entire conductor, leaving only the charges. The electric field would not change at all; it would remain zero inside the thin shell of charge and would remain unchanged for all external points. This shows us that the electric field is set up by the charges and not by the doctor. The conductor helps the charges to take their respective position.

The External Electric Field

It has been studied that the excess charges remains moving on the surface of conductor. However, unless the conductor is spherical, the charge does not distribute itself uniformly. Put another way,. the surface charge density σ (charge per unit area) varies over the surface of any nonspherical conductor. It becomes difficult to determine the electric field produced by the charges on the conductor surfaces due to density variation.

By using Gauss' law, it becomes very easy to determine the electric field present just outside the surface of a conductor. To do this, we consider a section of the surface that is small enough to permit us to neglect any curvature and take the section to be flat. Then let us consider that a tiny cylindrical Gaussian surface is inserted in the section. One end of the cylindrical surface is fully inside the conductor while the other end is fully outside and cylinder is perpendicular to the conductor surface.

As the cylinder is perpendicular to the surface so the electric field **E** at and just outside the conductor surface should also be perpendicular. If it were not, then it would have a component along the conductor's surface that would exert forces on the surface charges, causing them to move. But such motion would violate our implicit assumption that we are dealing with electrostatic equilibrium. So, the electric field E is at 90° to the conductor surface i.e., Perpendicular to it.

Now, all the flux through Gaussian surface should be added. At the internal cap, the flux is zero, as the electric field is zero. There is no flux through the curved surface of the cylinder, because internally (in the conductor) there is no electric field and externally the electric field is parallel to the curved surface. The only flux through the Gaussian surface is that through the external end cap, where E is perpendicular to the plane of the cap. We assume that the cap area A is small enough that the field magnitude

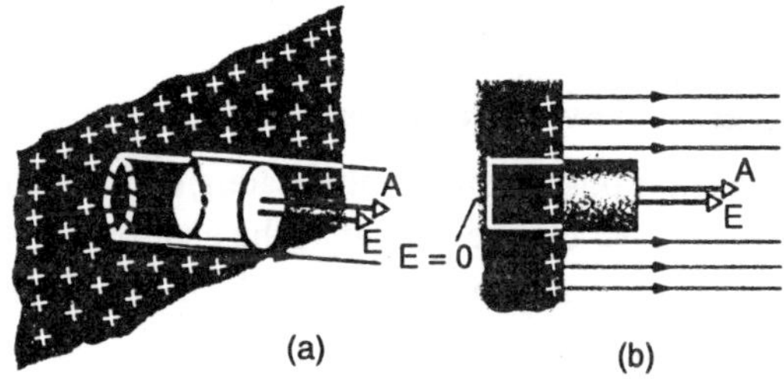

Figure 7.21: ***(a). Perspective view* (a) *and side view* (b) *of a tiny portion of a large, isolated conductor with excess positive charge on its surface. A (closed) cylindrical Gaussian surface, embedded perpendicularly in the conductor, encloses some of the charge. Electric field lines pierce the external end cap of the cylinder, but not the internal end cap. The external end cap has area* A *and area vector* A.**

E is constant over the cap. So the external end cap is the only area on the as Gaussian surface where flux is present. The net flux Φ at the external end cap is $E\mathbf{A}$.

As the flux is present only on external end cap which has area **A**, the charge enclosed in Gaussian surface will be also present in area A. If σ is the charge per unit area, then q_{enc} is equal to

σA. If we will replace σA for q_{enc} and EA for Φ the Gauss' law becomes—

$$\varepsilon_0 EA = \sigma A,$$

from which we find

$$E = \frac{\sigma}{\varepsilon_0} \text{(conducting surface).} \tag{7.40}$$

Thus from the above equation, it is clear that the electric field present just outside the conductor surface is proportional to the surface charge density at that point.

The electric field directed away from the body if the charges are *+ve* but if the charges are *–ve*, the electric field is directed inwardly.

If negative charges come in the way of field lines of figure (7.21), (b). The lines will be deflected from their direction. If we bring those charges near the conductor, the charge density at any given location changes and so does the magnitude of the electric field. But the relation between E and σ will remain same as in equation 7.11.

Applying Gauss' law: Cylindrical Symmetry

Imagine a short portion of infinitely long cylindrical plastic rod with a uniform linear charge density λ the charge is *+ve*. Our work is to determine the value of electric field E at a distance r from the axis of the rod.

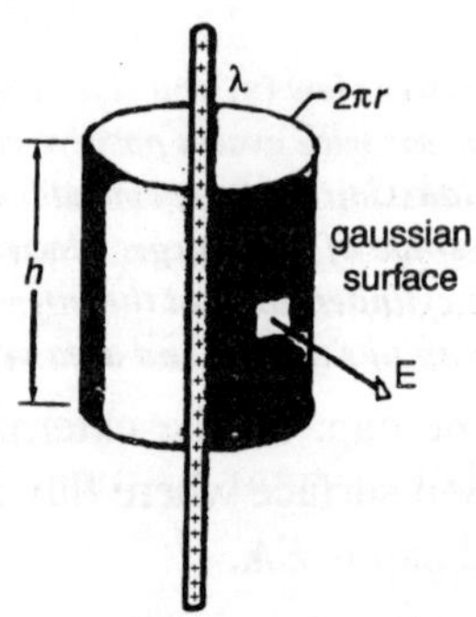

Figure 7.22: ***A Gaussian surface in the form of a closed cylinder surrounds a section of a very long, uniformly charged, cylindrical plastic rod.***

The Gaussian surface, chosen by us, must be in relation with the symmetry of the cylindrical rod. We choose a circular cylinder of radius *r* and length *h,* coaxial, with the rod. Both the ends of the selected Gaussian surface should be closed, as we can use the ends as Gaussian surface.

Suppose, some one rotating the cylindrical rod around its longitudinal axis, when we are not looking at the rod. When we turn to the rod again. When we look again at the rod, we will not be able to detect any change. We conclude from this symmetry that the only uniquely specified direction in this problem is along a radial line. From the above observation, it becomes quite clear that **E** must have same magnitude at every point of the Gaussian surface and it must be directed radially outward from the surface. As the height of the cylinder is *h* and the circumference is $2\pi r$, so the area of the cylinder is $2\pi rh$ Now, the net flux of *E* through this cylindrical surface is given by—

$$\Phi = EA\cos\theta = E(2\pi rh)$$

As the direction of *E* is radially outward, so *it* becomes paralleled to the surface of end caps, hence flux is zero at end caps.

The charge enclosed by the surface is λh so that; Gauss' law,

$$\varepsilon_0 \Phi = q_{enc},$$

changes to $$\varepsilon_0 E(2\pi rh) = \lambda h,$$

giving $$E = \frac{\lambda}{2\pi\varepsilon_0 r} \text{ (line of charge)} \tag{7.41}$$

So, equation 7.41 gives us the value of *E* which is due to infinitely longest line of charge at a point which is at a distance *r* from the line of charge, If the charge is +*ve* *E* is directed radially outward while if the line of charge is –*ve*, then *E* is directed radially inward.

Applying Gauss' law: Planar Symmetry

Nonconducting Sheet

A portion of infinite long non conducting sheet which has a uniform surface +*ve* charge density σ, is shown in the figure 7.23.

A sheet of thin plastic wrap, uniformly charged on one side, can serve as a simple model. We will find electric field E at a distance r from the sheet.

The Gaussian surface is perpendicular to the sheet in its long axis and both ends of the cylindrical Gaussian surface are closed with area $d\mathbf{A}$. From symmetry, **E** must be perpendicular to the sheet, hence to the end caps. Furthermore, since the charge is positive, **E** must point away from the sheet, and thus the electric field lines pierce the two Gaussian and caps in an outward direction. Because the field lines do not pierce the cylinder walls, there is no flux through this portion of the Gaussian surface. Hence, the net flux is present only at the two end caps of the cylindrical Gaussian surface, given by $E.d\mathbf{A}$, so then by Gauss' law—

$$\varepsilon_0 \oint \mathbf{E} \cdot d\mathbf{A} = q_{\text{enc}}$$

becomes $$\varepsilon_0 (EA + EA) = \sigma A,$$

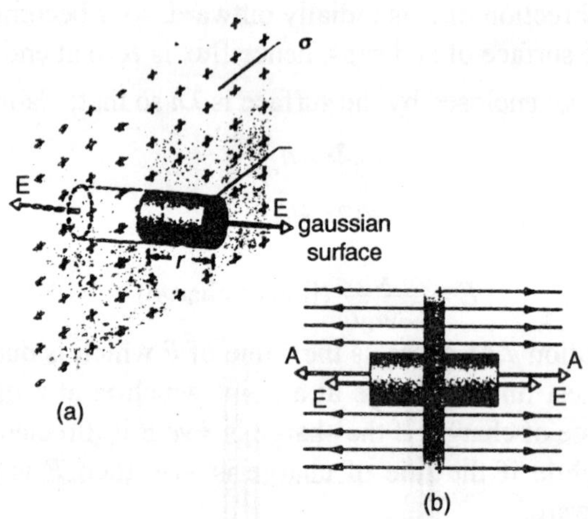

***Figure 7.23:** Perspective view (a) and side view (b) of a potion of a very large, thin plastic sheet, uniformly charged on one We to surface charge density σ. A closed cylindrical Gaussia surface passes through the sheet and is perpendicular to it.*

Where σA is the charge enclosed by the Gaussian surface. Now we get—

$$E = \frac{\sigma}{2\varepsilon_0} \text{ (sheet of charge).} \tag{7.42}$$

These equations are useful for any point at a finite distance from the sheet, as we are assuming an infinite sheet with uniform charge density. Equation 7.13 agrees with Equation 7.21, which we found by integration of the electric field components that are produced by individual charges. (Look back to that time-consuming and challenging integration, and note how much more easily we obtain the result with Gauss' law. For symmetric arrangements of charge. The results can be obtained very easily with the Gauss' law without going through the long processes of integration.

Two Conducting Plates

Suppose, a thin infinite conducting plate with excess *+ve* charge is present as in figure 7.24 (a) As we know that the excess charge is present on the surface, so assume that the charge is distributed on the two false of the large infinite conducting plate.

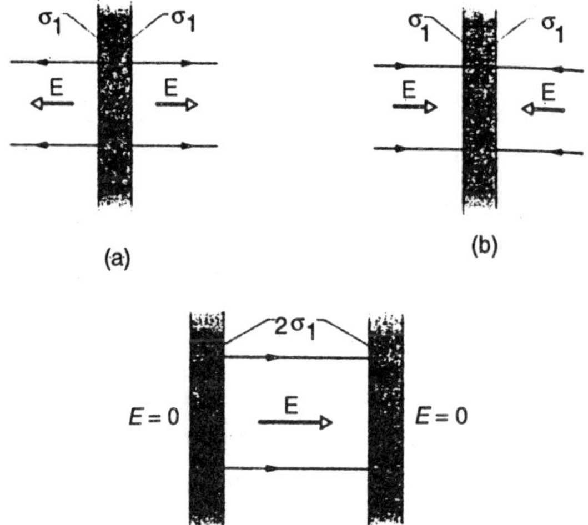

Figure 7.24 : ***(a) A thin, very large conducting plate with excess positive charge. (b) An identical plate with excess negative charge. (c) The two plates arranged to be parallel and close.***

In the absence of external electric field, the charge will distribute itself on the two faces of the conductor with a uniform surface charge density σ, From Equation 7.40 we know that just out-side' the plate this charge sets up an electric field of magnitude $E = \sigma_1 / \varepsilon_0$. The electric field points away from the surface as the excess charge density is *+ve.*

In the second case that is shown in figure 7.24: (b), the magnitude of the charge distributed on the conductor sheet is same but the excess charge is *–ve* and have the electric field will point toward the surface.

Imagine that is bring both the conductor sheet of 7.24 (a) and 7.24 (b) close to each other paralleled. Since the plates are conductors, when we bring them into this arrangement, the excess charge on one plate attracts the excess charge on the other plate, and all the excess charge moves onto the inner faces of the plates as in Figure 7.24 (c). With twice as much charge now on each inner face is new surface charge density (call it σ) on each inner face is twice σ_1. So the magnitude of electric field E at any point between the two plates would be as—

$$E = \frac{2\sigma_1}{\varepsilon_0} = \frac{\sigma}{\varepsilon_0} \tag{7.43}$$

As the electric field will be directed towards the negative charged plate and away from the +vely charged plate. So on the outer sides of the plates the magnitude of electric field will be zero.

The charges on the two plates will move while the two plates will bring closer to each other, Figure 7.24 (c) is not the superposition of Figures 7.24 (a) and *b*; that is, the charge distribution of the two-plate system is not merely the sum of the charge distributions of the two-plate system is not merely the sum of the charge distribution of the individual plates

It feels quite strange sometime to discuss some unrealistic situations, such as electric fields produced by infinite sheet of charge, or infinite line of charge. It is not enough to say that we do so because it is simple to analyse such situations with Gauss' law although that is indeed true. The proper answer is that analyses for "infinite" situations yield good approximations to many real-world problems. Thus 7.13 holds well for a finite nonconducting sheet as long as you are close to the sheet and not too near its edges. Similarly equation 7.14 can be applied very well to the pair of finite conducting plates till the point in question is not too close to the edges of the conducting plates.

We are avoiding the points very close to the edges of the conducting plates because, at the edges or near the edges the field lines are curved, so it become difficult to use planar symmetry to find the electric field, And it becomes difficult to express the electric field algebrically.

Applying Gauss' law: Spherical Symmetry

The two shell theorems discussed earlier, will be now proved by the use of Gauss' law.

A shell of uniform charge attracts or repels a change particle that is outside the shell as if all the shell's charge were concentrated at the centre of the shell.

A shell of uniform charge exerts no electrostantic force on a charged particle that is located inside the shell.

A charged spherical shell of total charge q and radius R is shown in figure 7.25. Tow Gaussian surfaces which are concentric are also present around the spherical shell. By applying Gauss' law to the second Gaussian surface S_2 we will get—

$$E = \frac{1}{4\pi\varepsilon_0}\frac{q}{r^2} \text{ (spherical shell, field at } r \geq R) \quad (7.44)$$

The determined value of E is same as that would be produced by *a* point charge q at the centre of the shell. Thus the magnitude of the force exerted by the shell on a charged particle placed outside the shell is the same as if the shell were replaced with a point charge q at the center of the shell. Hence, the first shell theorem is proved.

If we apply Gauss' law to second Gaussian surface we will find that—

$$\boldsymbol{E} = \mathbf{O} \text{ (spherical shell, field at } r < R), \quad (7.45)$$

because this Gaussian surface encloses no charge. Thus if a

charged particle were enclosed by the shell, the shell would exert no net electrostatic force on it. So, by Gauss' law application second shell theorem is also proved.

If the charge distribution is in symmetrical spherical shape then it can be devided into number of concentric spherical shalls. For purposes of applying the two shell theorems, the volume charge density ρ should have a uniform value for each shell but need not be the same from shell to shell, That is, for the charge

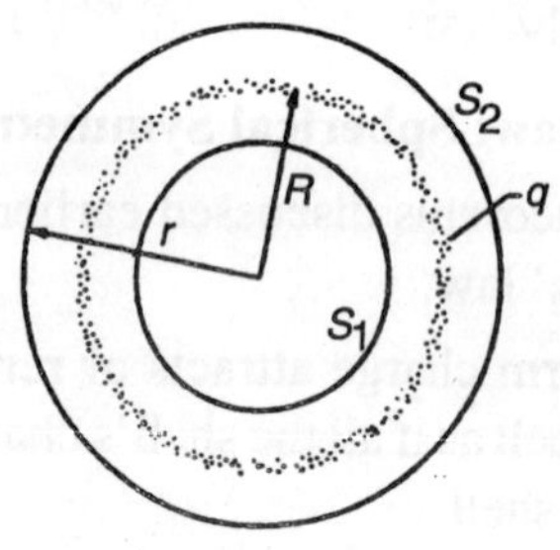

Figure 7.25: ***A thin, uniformly charged, spherical shell with total charge q, in cross section. Two Gaussian surfaces*** S_1 ***and*** S_2 ***are also shown in cross section. Surface*** S_2 ***encloses the shell, and*** S_1 ***encloses only the empty interior of the shell.***

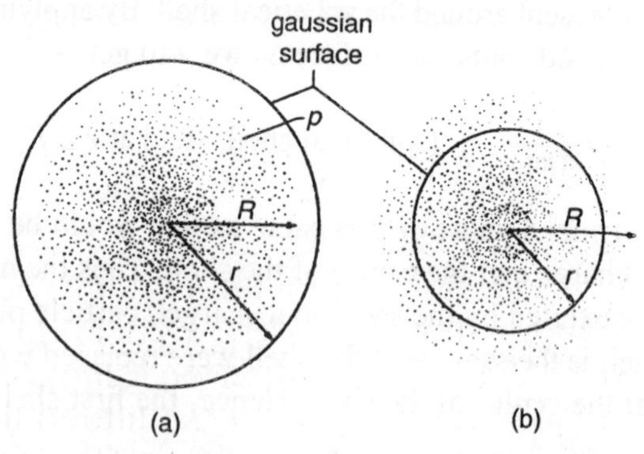

Figure 7.26: ***The dots represent a spherically symmetric distribution of charge of radius R, whose volume charge density p is a function only of distance from the center. The charged object is not a conductor, and the charge is assumed to be fixed in position.***

(a) A concentric spherical Gaussian surface with r > R is included.

(b) A similar Gaussian surface with r < R is included.

distribution as a whole, ρ can vary, but only with r, the radial distance from the center. Then the electric field produced by such charge can be discussed as shell by shell,

According to figure 7.26 (a) the charge is distributed within the spherical Gaussian surface, and the radius of Gaussian surface $r > R$, radius of charged spherical shell. So the charge will produce *a* electric field at the Gaussian surface a point charge so the equation 7.44 hold for this case.

Another spherical shell is shown in figure 7.26 (b), where the radius of Gaussian surface $r < R$, radius of shell. We can imagine two sets of charged shell to find electric field produced at the Gaussian surface. Equation 7.45 says that the charge lying outside the Gaussian surface does not set up an electric field on the Gaussian surface. And Eq. 7.44 says that the charge enclosed by the surface sets UP an electric field as if that enclosed charge were concentrated at the center. By assuming that the charge enclosed within in q the equation 7.44 becomes—

$$E = \frac{1}{4\pi\varepsilon_0}\frac{q'}{r^2} \quad \text{(spherical distribution, field at } r \le R\text{)}. \tag{7.46}$$

Electric Potential Energy

The general features of the gravitational force can be applied to the electrostatic force because the Newton's law for the gravitational force and Coulomb's law for electrostatics are mathematically identical.

It is widely known that the electrostatic force is *a* conservative force. Thus when that force acts between two or more charged particles within a system of particles, we can assign an electric potential energy U to the system. Moreover, if the system changes its configuration from an initial state i to a different final state f, the electrostatic force does work W on the particles. So the change in the potential energy of the system i.e. ΔU is given by—

$$\Delta U = U_f - U_i = -W \tag{7.47}$$

The work done by electrostatic force are path independent like all other conservative forces. Suppose a charged particle within the system moves from point i to point f while an electrostatic force between it and the rest of the system acts on it. If the system remains at rest, the work done by force is same for any path between the points $i < f$.

Imagin a reference configuration of *a* system of changed particles to be that in which the particles are away from each other. And we usually set the corresponding reference potential energy to be zero. Suppose that several charged particles come together from initially infinite separations (state i) to form a system of nearby particles (state f). Let the initial potential energy U_i be zero, and let W_∞ represent the work done-by the electrostatic forces between the particles during the move in from infinity. Thus the final potential energy of the system becomes—

$$U = -W_\infty \qquad (7.48)$$

The electrical potential energy, like other potential energy, is considered to be mechanical energy, that if only conservative forces act within a (closed) system, the mechanical energy of the system is conserved. This fact will be used in this chapter more than once.

Electric Potential

We know, that the potential energy of *a* charged particle in an electric field depends on the magnitude of charge. But the potential energy per unit charge is unique in distribution.

If we place a test particle of positive charge 1.60×10^{-9} C in an electric field where the particle has an electric potential energy of 2.40×10^{-7} J, then the potential energy per unit charge on that test particle is—

$$\frac{2.40 \times 10^{-17}\,\text{J}}{1.60 \times 10^{-19}\,\text{C}} = 150\ \text{J}/\text{C}$$

Now, if the first particle is replaced with another test particle which has double charge than the first one *i.e.*, 3.20×10^{-19} *C*, then the electric potential energy of the 2nd particle would also be double than the first particle *i.e.* 4.80×10^{-17} *J*. But the potential energy per unit charge will remain same as that of the first particle *i.e.*, 150J/C

So, we conclude that potential energy per unit charge is independent of the magnitude of the charge of the particle rather it is characteristic of only the electric field. The potential energy per unit charge at a point in an electric field is called the electric potential V (or simply the potential) at that point.

So $$V = \frac{U}{q} \tag{7.49}$$

Note that electric potential is a scalar, not a vector.

If we look at two points i and f on the surface then the electric potential difference between the two points is equal to difference in their electric potential energy per unit charge.

$$\Delta V = V_f - V_i = \frac{U_f}{q} - \frac{U_i}{q} = \frac{\Delta U}{q} \tag{7.50}$$

Using Equation 7.47 to substitute $-W$ for ΔU in Equation 7.50 can define the potential difference between points i and f as

$$\Delta V = V_f - V_i = -\frac{W}{q} \text{ (potential difference defined).} \tag{7.51}$$

The potential difference between two points is thus the negative of the work done by the electrostatic force per unit charge that moves from one point to the other. According to magnitudes of work done W and charge q and sign of charge q, the potential energy difference may be $+ve$, $-ve$ or zero.

If this initial potential energy V_i of a point is zero, than the electric potential will alos be zero, we can define the electric potential V at any point f in an electric field to be

$$V = -\frac{W_\infty}{q} \text{ (potential defined),} \tag{7.52}$$

where W_∞ is the work done by the electric field on a changed particle as that particle moves in from infinity to point f. Depending on q and W, the potential energy may be $+ve$ or 0.

In S_1 system, the unit for potential is joule coulomb. This combination occurs so often that a special unit, the volt (abbreviated V) is used so represent it. That is,

$$1 \text{ volt} = 1 \text{ joule per coulomb.} \tag{7.53}$$

This new unit allows us to adopt a more conventional unit for the electric field **E**, which we have measured up to now in newtons per coulomb. We will obtain another unit for E with two unit conversions.

$$\begin{aligned} \text{N/C} &= \left(1\frac{\text{N}}{\text{C}}\right)\left(\frac{1\text{V}\cdot\text{C}}{1\text{ J}}\right)\left(\frac{1\text{ J}}{1\text{ N}\cdot\text{m}}\right) \\ &= 1\text{ V/m} \end{aligned} \tag{7.54}$$

The conversion factor in the second set of parecheses comes from Equation 7.53; that in the third set of parentheses is derived from the definition of the joule. So, we got a new unit for electric field E *i.e.* volt/meter. From now we will use this unit in our further discussions.

Finally, we are now in a position to define the electron volt, the energy unit that was introduced in Section 7.1 as a convenient one for energy measurements in the atomic and sutatomic domain. One *electron-volt* (eV) is the energy equal to the work required to move a single elementary enrage e, such as that of the electron or the proton, through a potential difference of exactly one volt. From equation 7.51, the magnitude of work done is equal to $q\,\Delta\,V$, so —

$$\begin{aligned} \text{e V} &= e\,(1\text{ V}) \\ &= \left(1.60 \times 10^{-19}\text{C}\right)(1\text{ J/C}) = 1.60 \times 10^{-19}\text{ J} \end{aligned}$$

Work Done by an Applied Force

Imagin that we move a charged particle q by using force from initial position i to final position f. During the move, our applied force does work W_{app} on the charge while the electric' field does work W on it. By the work kinetic energy theorem the change ΔK in the kinetic energy of the particle is

$$\Delta K = K_f - K_i = W_{\text{app}} + W \tag{7.55}$$

Now suppose the particle is stationary before and after the

move. Thus initial kinetic energy K_i and final kinetic energy K_f, *both* becomes zero, so the above equation changes —

$$W_{\text{app}} = -W \tag{7.56}$$

From the above equation, the work done W_{app} by the force, applied to move the charged particle is the negative of the work done by the electric field.

If we substitute the equation 7.47 to 7.56 we get *a* relation between the work done by the applied force and the potential energy difference of the charged particle. We find

$$\Delta U = U_f - U_i = W_{\text{app}} \tag{7.57}$$

By similarly substituting Equation 7.56 into Equation 7.51 we can relate our work W_{app} to the electric potential difference ΔV between the initial and final points of the particle. We find

$$W_{\text{app}} = q\,\Delta V. \tag{7.58}$$

W_{app} can be positive, negative, or zero depending on the signs and magnitudes of q and ΔV. So, to avoid change in kinetic energy of the charged particle, it is necessary to do work in moving the charged particle through *a* potential diffenence ΔV.

Equipotential Surfaces

The points which have same electric field and are adjacent to each other, form *an* together *a* equipotential surface. No net work W is done on a charged particle by an electric field when the particle moves between two points i and f on the same equipotential surface. This follows from which tells us that W must be zero if $V_f = V_i$. As the work done (W) is path independent, so $W = 0$ for any path connecting points i and f whether the path is entirely within equipotential surface.

Many points join to form the equipotential surfaces, which are shown in figure 7.27 along with electric fields. The work done by the electric field on a charged particle as the particle moves from one end to the other of paths I and II is zero because each of these path begins and ends on the same equipotential surface. The work done as the charged particle moves from one end to the other of paths III and IV is not zero but has the same value

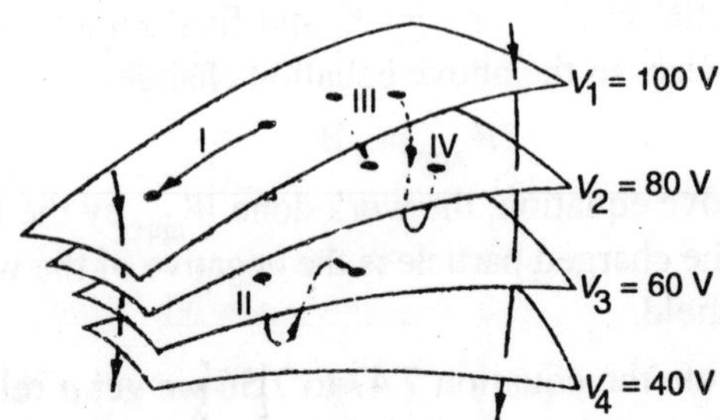

Figure 7.27: ***Portions of four equipotential surfaces paths along which a test charge may move are also shown. Two electric field lines are indicated.***

for both these paths because the initial and final potentials are identical for the two paths. It indicates that the paths III and IV joins the same pair of equipotential surfaces

By using law of symmetry all the equipotential surfaces formed by point charge or spherically symmetrical charge distribution, found to be concentric spheres. For a uniform field, the surfaces

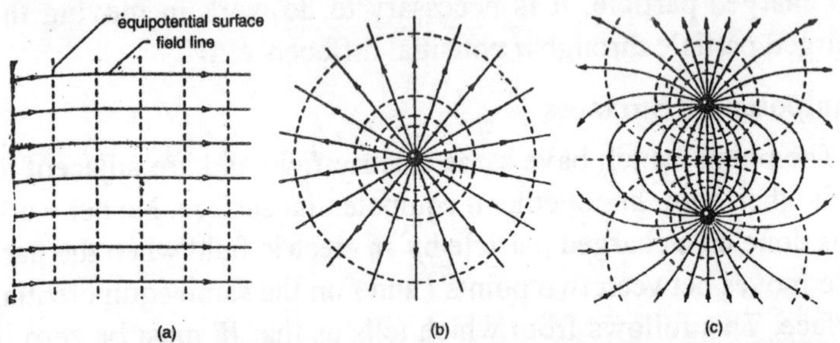

Figure 7.28: ***Electric field lines (purple) and cross sections of equipotential surfaces (gold) for (a) a uniform field, (b) the field of a point charge, and (c) the field of an electric dipole.***

are a family of planes perpendicular to the field lines. In fact, equipotential surfaces are always 'perpendicular to electric field lines and thus to **E**, which is always tangent to these lines. If **E** were *not* perpendicular to an equipotential surface, it would have a component lying along that surface. This component would then do work on a charged particle as it moved along the surface. But by Equation 7.51 work cannot be done if the surface is truly equipotentional surface; the only possible conclusion is that **E** must be everywhere perpendicular to the surface. The electric field lines and cross section of the equipotential surfaces are shown in figure 7.28 associated with a uniform electric field (b) point charge and (c) electric dipole.

Calculating the Potential from the Field

If the field vector E at all position along any path connecting the points i and f are known, then the potential difference between the two points can be calculated. We determine the work done by the field on *a* positive test charge to calculate the potential difference between the two positions.

An electric field with its field lines is shown in figure 7.29 and also *a +ve* charge q_0 that moves along the path connecting i to f. At any point on the path, an electrostatic force q_{oE} acts on the charge as it moves through a differential displacement $d\mathbf{s}$. From previous chapter, we know that the differential work dW done on a particle by a force **F** during a displacement $d\mathbf{s}$ is—

$$dW = \mathbf{F} \cdot d\mathbf{s}. \tag{7.59}$$

For the situation of Figure 7.29 $\mathbf{f} = q_0\mathbf{E}$ and Equation 7.59 becomes

$$dW = q_0\mathbf{E} \cdot d\mathbf{s}.$$

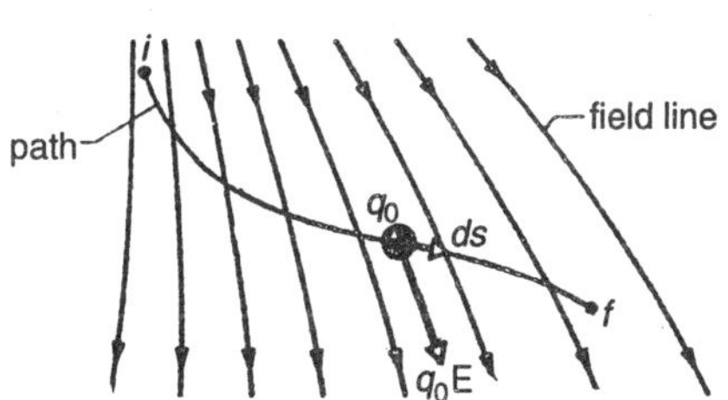

Figure 7.29: ***A test charge q_0 moves from point i to point f along the path, shown in: a non-uniform electric field. During a displacement ds, an electrostatic force q_0E acts on the test charge. This force points in the direction: of the field line at the location of the test charge.***

To calculate the work done by the field on the particle during its from point i to f, we will add all the differential work done on the charge for all the differential displacement ds along the path form i to f.

$$W = q_0 \int_i^f \mathbf{E} \cdot d\mathbf{s} \tag{7.61}$$

If we substitute the total, work W from Equation 7.61 into Equation 7.7 we find

$$V_f - V_i = -\int_i^f \mathbf{E} \cdot d\mathbf{s} \tag{7.62}$$

Thus the potential difference $V_f - V_i$ between any two points i and f in an electric field is equal to the negative of the *line* integral (meaning the integral along the path) of **E**. *d***s** from i to f. The result obtained in the above equation is independent of the charge q_0, used in the example.

If the electric field is known for *a* region, that the potential difference between any two points in the said field can calculated because, as electrostatic force is conservative, so same result would be obtained in any path chosen.

Imagine that initial potential V_i is zero at point i then the equation becomes—

$$V = -\int_i^f \mathbf{E} \cdot d\mathbf{s} \tag{7.63}$$

in which we have dropped the subscript f on V_f. Equation 7.63 gives us the potential V at any point f in the electric field relative to the zero potential at point i. Suppose point i is in infinity then equation 7.63 gives th potential V at f relative to the zero potential at infinity.

Potential due to a Point Charge

We well derive on equation for the electric potential v in the space around a point charge relative to zero potential at infinity by the help of equation 7.63

Suppose, a point P is situated at a distance r from a fixed point charge q. To use Equation 7.63 imagine that a positive test charge moves from in infinity to point P. The charge moves along a line that extends from infinity to P along a radius from the point charge q.

The next step is to evaluate the dot product; **E**.d**s** in equation 7.63 along the line taken by the charge. In 7.30 the test particle is at some intermediate point, at distance r from the point charge. The electric field E at the location of the test particle is directed radially outward. The differential displacement d**s** of the test charge as it moves toward point P is radially inward. Thus the angle between **E** and d**s**

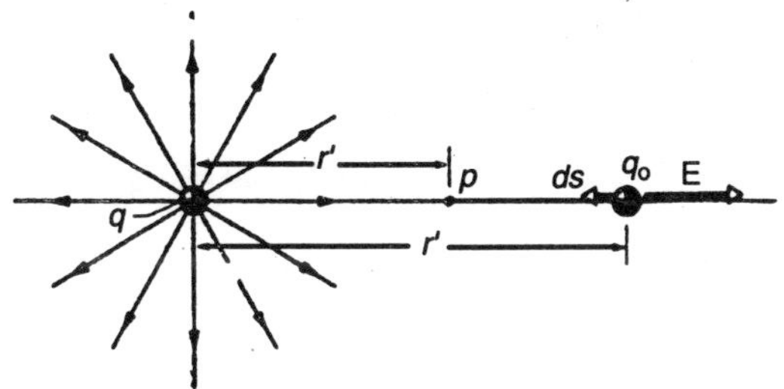

Figure 7.30: ***The positive point charge* q *produces an elect c field* E *and an electric potential Vat point P. We find the potential by moving a test charge q_0 to P from infinity. The tea charge is shown at distance r' from the point charge undergoing differential displacement ds.***

is 180°. By substituting the displacement d**s** by dr and using the angle 180° we get the value of **E**.d**s** as—

$$\mathbf{E}\cdot d\mathbf{s} = |E|(dr')(\cos 180°) = -|E|\,dr' \qquad (7.64)$$

where IEI is the absolute value of **E**. (We use the absolute value sign to show explicitly that we want only the magnitude of **E** here.) Substituting Equation 7.64 and the limits of our integration into Equation 7.63 gives us

$$V = -\int_i^f \mathbf{E}\cdot d\mathbf{s} = \int_\infty^r |E|\,dr' \qquad (7.65)$$

The magnitude of the electric field at the site of the test charge is given by

$$E = \frac{1}{4\pi\varepsilon_0}\frac{q}{r'^2} \qquad (7.66)$$

Substituting this result into Equation 7.65 and integrating lead to

$$V = \frac{q}{4\pi\varepsilon_0}\int_\infty^r \left|\frac{1}{r'^2}\right| dr' = \frac{q}{4\pi\varepsilon_0}\left[\frac{1}{r'}\right]_\infty^r \qquad (7.67)$$

or

$$V = \frac{1}{4\pi\varepsilon_0}\frac{q}{r} \text{ (point charge } +q) \tag{7.68}$$

So, if the charge is positive then the *v, around* the charge will be positive relative to the zero potential at infinity.

As we know that if the charge is *–ve,* then electric field *E* produced by the charge will a be directed towards the charge and the angle between *ds* and *E* is zero. The integration in Equation 7.67 now yields

$$V = -\frac{1}{4\pi\varepsilon_0}\frac{q}{r} \text{ (point charge } -q) \tag{7.69}$$

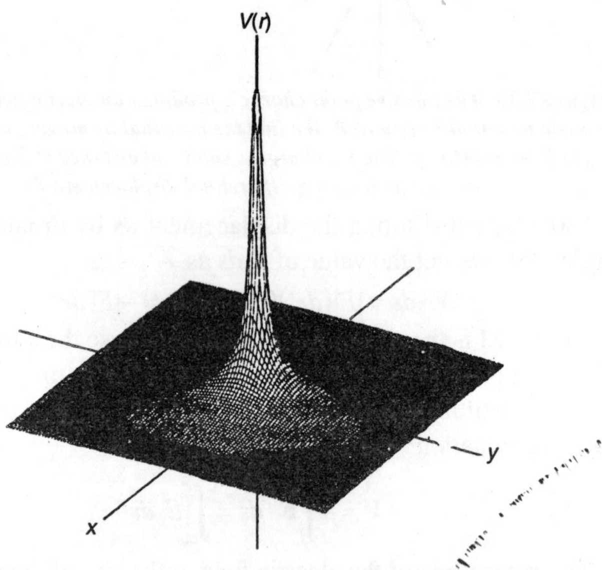

Figure 7.31: ***A computer-generated plot of electric potential* V(r) *due to a positive point charge, which is located at the origin of an* xy *plane. The potentials at points in that plane are plotted vertically. (Curved lines have been added to help you visualize the plot.) The infinite value of* V *predicted by Eq. 7.70 for* r *= 0 is not plotted.***

So, if the charge is negative then the *v* around the charge will be *–ve,* relative to the zero potential at infinity.

So if we assume that *q* is representing just the magnitude of the charge and it may be *–ve* or *+ve*, then the equation can be generalized as—

$$V = \frac{1}{4\pi\varepsilon_0}\frac{q}{r} \text{ (positive or negative point charge } q). \quad (7.70)$$

Now the sign of V is the same as the sign of q. Figure 7.31 shows a computer-generated plot of Equation 7.69 for a positive point, charge. So according to equation, v for a point charge is infinite at $r = 0$.

This equation is also valid for the electric potential outside or on the surface of spherically symmetrical charge distribution. We can prove this by using one of the shell theorems to replace the actual charge with an equal charge concentrated at the center of the spherical distribution. Provided we don't consider *a* point charge within actual distribution, the derivation for shell theorem equation follows.

Potential Due to a Group of Point Charges

The net potential produced by a ϕ group of point charges can be measured with the help of the superposition principle. We calculate the potential resulting from each charge at the given point separately, using Equation 7.70 with the sign of the charge included. Then we sum the potentials. For n charges, the net potential is

$$V = \sum_{i=1}^{n} V_i = \frac{1}{4\pi\varepsilon_0}\sum_{i=1}^{n}\frac{q_i}{r_i} \text{ (} n \text{ point charges).} \quad (7.71)$$

Here q_i is the value of the ith charge, and r_i is the radial distance of the given point from the ith charge. The sum in Equation 7.71 is an *algebraic sum*, not a vector sum like the sum that would be used to calculate the electric field resulting from a group of point charges. This equation has an important computational advantage of potential over electric field, as it involves sum of scalar quantities which is quite easier than sum of vector quantities.

Potential Due to an Electric Dipole

w will, now, apply the equation 7.71 to an electric dipole case.

At P, the positive point charge (at distance $r_{(+)}$) sets up potential $V_{(+)}$ *and* the negative point charge (at distance $r_{(-)}$)sets potential

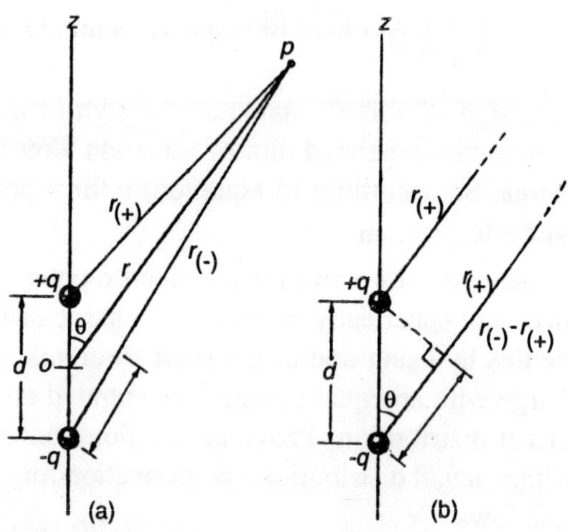

Figure 7.32: ***(a) Point P is a distance r from the midpoint of a dipole. The line OP makes an angle θ with the dipole axis. If P is far from the dipole, the lines of lengths $r_{(+)}$ and $r_{(-)}$ approximately parallel to the line of length r, and the dashed b line is approximately perpendicular to the line of length $r_{(-)}$.***

$V_{(-)}$, both potentials as given by Equation 7.70. Thus, at P_i total potential is given by—

$$V = \sum_{i=1}^{2} V_i = V_{(+)} + V_{(-)} = \frac{1}{4\pi\varepsilon_0}\left(\frac{q}{r_{(+)}} + \frac{-q}{r_{(-)}}\right)$$

$$= \frac{q}{4\pi\varepsilon_0}\frac{r_{(-)} - r_{(+)}}{r_{(-)}\, r_{(+)}} \tag{7.72}$$

As, the naturally occurring dipoles are very small, so we prefer case where point is far away from the dipole. Under these conditions, the approximations that follow from Figure 7.32 (b) are

$$r_{(-)} - r_{(+)} \approx d\cos\theta \text{ and } r_{(-)}r_{(+)} \approx r^2$$

If, we substitute these quantities, into Equation 7.72 we can approximate V to be

$$V = \frac{q}{4\pi\varepsilon_0}\frac{d\cos\theta}{r^2},$$

Here θ is measured, from the dipole axis as shown in Figure We can now write V as

$$V = \frac{1}{4\pi\varepsilon_0} \frac{p\cos\theta}{r^2} \quad \text{(electric dipole).} \tag{7.73}$$

which p (= qd) is the magnitude of the electric dipole moment p defined earlier in the chapter. The vector P directed from the –*ve* to the +*ve* charge, is along the dipole axis.

Induced Dipole Moment

Many naturally occuring molecules possess permanent electric dipole moments e.g. water molecular.

In other molecules (*nonpolar molecules*) and in every atom, the centers of the positive and negative charges coincide Figure 7.33 (a) and thus no dipole moment is set up. However, if we place an atom or a nonpolar molecule in an external electric field, the field distorts the electron orbits and separates the centers of positive and negative charge Figure 7.33 (b). Because the electrons are negatively charged, they tend to be shifted in a direction opposite the field. This shift sets up a dipole moment **p** that points in the direction of the field. This dipole-moment is said to be induced by the field, and the atom or molecule is then said to be polarized by the field (it has a positive side and a negative side). However, the induced dipole moment and the polarization of the molecule will disappear if the field is removed.

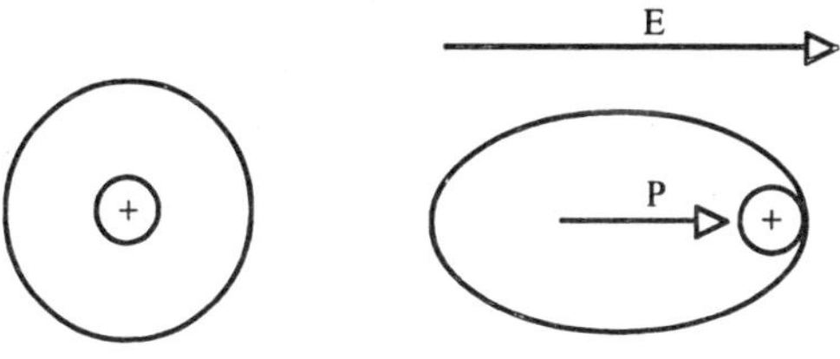

Figure 7.33: ***(a) An atom, showing the positively charged nucleus (green) and the negatively charged electrons (gold shading). The centers of positive and negative charge coincide. (b) If the atom is placed in an external electric field, the electron orbits are distorted so that the centers of positive and negative charge no longer coincide. An induced dipole moment appears. The distortion is greatly exaggerated here.***

Potential Due to a Continuous Charge Distribution

If the charge distribution is continuous, then the equation 7.70 cannot be used to find v at P. Rather we should consider *a* differential element of charge dq and after determination of dv at P due to dq we should integrate it over the charge distribution.

If we imagine zero potential to be at infinity, if we consider dq as point charge, then dv at P due to dq is given by—

$$dV = \frac{1}{4\pi\varepsilon_0}\frac{dq}{r} \text{ (positive or negative } dq) \qquad (7.74)$$

Here r is the distance between P and dq. To find the total potential V *at* P, we integrate to sum the potentials due to all the charge elements:

$$V = \int dV = \frac{1}{4\pi\varepsilon_0}\int \frac{dq}{r} \qquad (7.75)$$

The integral is to be taken over the entire charge distribution. Next we will discuss two cases of continuous charge distribution first *a* line of charge and second is a charged dise.

Line of Charge

We will take the case of a non-conducting thin rod of length L the rod has a $+ve$ charge distribution of charge density λ. Let us determine the electric potential V due to the rod at point P a perpendicular distance d from the left end of the rod.

Imagin, a short portion of the rod, differential element dx of the rod as shown in figure 7.34 (b). This (or any other) element of the rod has a differential charge of

$$dq = \lambda\, dx \qquad (7.76)$$

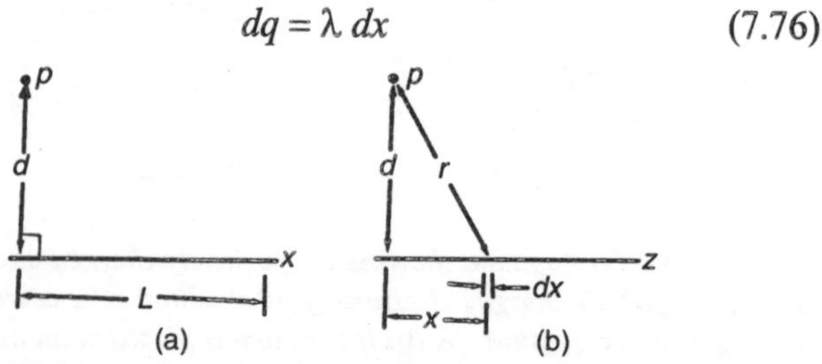

Figure 7.34: ***(a) A thin, uniformly charged rod produce an electric potential* V *at point P. (b) An element of charge produces a differential potential* dV *at P.***

This element produces a potential dV at point P, which is a distance $r = (x^2 + d^2)^{1/2}$ from the element. Treating the element as a point charge, we can use Equation 7.74 to write the potential dV as

$$dV = \frac{1}{4\pi\varepsilon_0}\frac{dq}{r} = \frac{1}{4\pi\varepsilon_0}\frac{\lambda\, dx}{\left(x^2 + d^2\right)^{1/2}} \tag{7.77}$$

As we know that the charge on rod is $+ve$ and $v = 0$ at infinity, then dv in above equation (7.77) must be positive.

Total potential v at point P can be measured by integrating equation 7.77 along the length of the rod, taking value $n = 0$ to $n = L$ and we get—

$$V = \int dV = \int_0^L \frac{1}{r\pi\varepsilon_0}\frac{\lambda}{\left(x^2 + d^2\right)^{1/2}}\, dx$$

$$= \frac{\lambda}{4\pi\varepsilon_0}\int_0^L \frac{dx}{\left(x^2 + d^2\right)^{1/2}}$$

$$= \frac{\lambda}{4\pi\varepsilon_0}\left[\ln\left(x + \left(x^2 + d^2\right)^{1/2}\right)\right]_0^L$$

$$= \frac{\lambda}{4\pi\varepsilon_0}\left[\ln\left[L + \left(L^2 + d^2\right)^{1/2}\right] - \ln d\right]$$

We can simplify this result by using the general relation $\ln A - \ln B = \ln (A/B)$. We then find

$$V = \frac{\lambda}{4\pi\varepsilon_0}\ln\left[\frac{L + \left(L^2 + d^2\right)^{1/2}}{d}\right] \tag{7.78}$$

Because V is the sum of positive values of dV, it should be positive. But does Equation 7.78 give a positive V? As the logarithm involves more than one argument, so the logarithm is $+ve$ and hence the V $+ve$.

Charged Disk

In the previous section, we have discussed the electric field of points on the central axis of the charged disc or rod that has uniform charge density σ on one surface. Now we will find an expression for the electric field by any points on the central axis.

Imagine a differential element which is a flat ring, has a radius R and radial dR'. The charge on the differential element is—

$$dq = \sigma(2\pi R')(dR')$$

in which $(2\pi R')(dR')$ is the upper surface area of the ring. All parts of this charged element are the same distance r from point P on the disk's axis. With the aid of Figure 7.35 we can now use Equation 7.74 to write the contribution of this ring to the electric field at P as

$$dV = \frac{1}{4\pi\varepsilon_0}\frac{dq}{r} = \frac{1}{4\pi\varepsilon_0}\frac{\sigma(2\pi R')(dR')}{\sqrt{z^2 + R'^2}} \tag{7.79}$$

We find the net potential at P by adding (via integration) the contributions of all the strips from $R' = 0$ to $R = R^2$ note that the

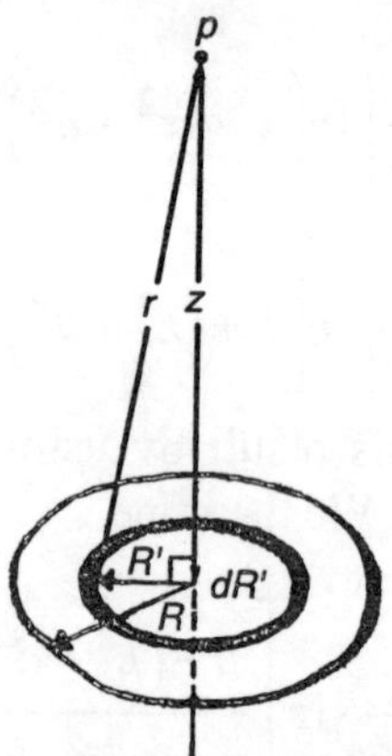

Figure 7.35: ***A plastic disk of radius R is charged on its top surface to a uniform surface charge density σ. We width to find the potential V at point P on the central axis of the disk.***

$$V = \int dV = \frac{\sigma}{2\varepsilon_0}\int_0^R \frac{R'\,dR'}{\sqrt{z^2 + R'^2}}$$

$$= \frac{\sigma}{2\varepsilon_0}\left(\sqrt{z^2 + R^2} - z\right) \tag{7.80}$$

variable in the second integral of Equation 7.80 is find not z, which remains constant while the integration the surface of the disk is carried out. (Note also that, evaluating the integral, we have assumed that $z \geq 0$).

Calculating the Field from the Potential

In the previous section we have studied how to find the potential at a point f if the electric field along a path from a reference point to point f is already known. In this section, we propose to go the other way, that is, to-find, the electric field when we know the potential, solving this problem graphically is easy; if we know the potential V at all points near an assembly of charges, we can draw in a family of equipotential surfaces. The electric field lines sketched perpendicular to those surfaces; reveal the variation of **E**. Here we are seeking the mathematical equivalent of this graphical procedure.

Figure 7.15 shows cross sections of a family of closely spaced equipotential surfaces, the potential difference between each pair of adjacent surfaces being dV. As the figure suggests the field **E** at any point P is perpendicular to the equipotential surface through P.

Let us assume a positive test charge q_0 moving through a displacement ds from one equipotential surfae to the adjacent surface.

From Equation 7.51 we see that the work that the electric field does on the test charge during the move is $-q_0\, dV$. From Equation 7.60 and Figure 7.61, we see that the work done by the electric field may also be written as $(q_0\mathbf{E}) \cdot d\mathbf{s}$, or $q_0\mathbf{E}\,(\cos\theta)\, ds$. Equating these two expressions for the work yields

$$-q_0 dV = q_0 E(\cos\theta)\, ds,$$

or

$$E\cos\theta = -\frac{dV}{ds} \tag{7.81}$$

Since $E \cos \theta$ is the component of **E** in the direction of *ds*. Equation 7.81 becomes

$$E_s = -\frac{\partial V}{\partial s} \tag{7.82}$$

We have added a subscript to *E* and switched to the partial derivative symbols to emphasize that Equation 7.82 involves only, the variation of *V* along a specified axis (here called the *s* axis) and only the component of **E** along that; axis in words.

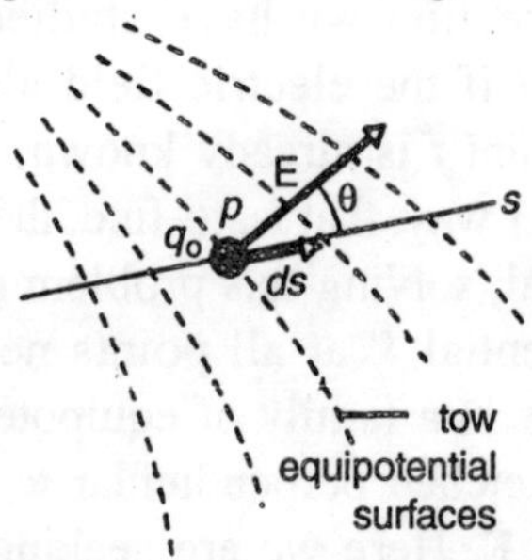

Figure 7.36: ***A test charge*** q_0 ***moves a distance*** **ds** ***from see equipotential surface to another. (The separation between the faces has been exaggerated for clearity.) The displacement makes an angle*** θ ***with the direction of the electric field*** **E**.

Equation 7.82 (which is essentially the inverse of Equation 7.62 states:

The component of, **E** in any direction, is the negative of the rate of change of the electric potential with distance in that direction.

If we take the *s* axis to be, in turn, the *x*, *y*, and *z* axes, we find that the *x*, *y*, and z components of **E** at any point are

$$E_x = -\frac{\partial V}{\partial x};\ E_y = -\frac{\partial V}{\partial y};\ E_z = -\frac{\partial V}{\partial z} \tag{7.83}$$

So, if *v* for all points in the region around *a* charge distribution, then the component of *E* can be found and thus we can find *E* at any point by taking partial derivatives.

If the electric field *E* is uniform then equation 7.82 becomes—

$$E = \frac{\Delta V}{\Delta s} \tag{7.84}$$

The '*s*' in the above equation is perpendicular to equipotential surface. If the electric field is tangent to equipotential surface then magnitude of E is zero in any direction.

Electric Potential Energy of a System of Point Charges

In the previous section, we observed the electric potential energy of a charged particle as a function of its position in an external electric field. In the section, we assumed that the charges that produced the field were fixed in place, so that the field could not be influenced by the presence of the test charge. In the current section, we will find the electric potential energy of a system of charges due to electric field produced by these charges.

If two bodies having same sign of charge, are pushed together, then the work done during pushing them is stored as electric potential energy in the two bodies. If you later release the charges you can recover this stored energy, in whole or in part as kinetic energy of the charged bodies as they rush away from each other.

The definition, of electric potential energy of *a* system of point charges, that are held in position by some unknown forces, is as follow—

The electric potential energy of a system of fixed point charges is equal to the work that must be done by an external agent to assemble the system bringing each charge in from an infinite distance.

We assume that the charges are stationary both in their initial infinitely distant positions and in their final assent bled configuration.

Two point charges q_1 and q_2 separated by distance r is shown in figure 7.37.

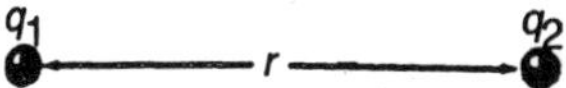

Figure 7.37: Two charges held a fixed distance r apart. What the electric potential energy of the configuration?

To find the electric potential energy of this, two-charge *system,* we mentally build the system starting with both charges infinitely far away; and at rest. When we bring q_1 in from infinity and put it in place, we do so work, because no electrostatic force acts on q_1. *When* q_2 is brought from infinity then q_1 exerts an electrostatic force on q_2 so some work must be done in to move the charge q_2.

The work done on q_1 can be calculated by removing the minus sing and substituting q_2 for q in the equation 7.52 Our work is then equal to q_2V, where V is potential that has been set up by q_1, at the point where we put q_2. From Equation 7.70 that potential is

$$V = \frac{1}{4\pi\varepsilon_0} \frac{q_1}{r}$$

Thus, from our definition, the electric potential energy of the pair of point charges of Figure 7.37 is

$$U = W = \frac{1}{4\pi\varepsilon_0} \frac{q_1 q_2}{r} \qquad (7.85)$$

If the charges have the same sign, we have to do positive work to push them together against their mutual repulsion: Hence, as Equation 7.85 shows, the potential energy of the system is then positive. If the charges have opposite signs, have to do negative work against their mutual attraction to bring them together, so that they are stationary. In that case, potential energy of the charges will be negative.

Potential of a Charged Isolated Conductor

Earlier in the chapter, we have observed that for all the points inside an isolated conductors, $E = 0$. We then used Gauss' law to prove that an excess charge placed on an isolated conductor lies entirely on its surface. (This is true even if the conductor has an empty internal cavity.) Here we use the, fact that $\mathbf{E} = 0$ for all points inside an isolated conductor to prove another fact about such conductors:

An excess charge placed on an isolated conductor will distribute itself on the surface of that conductor so that all points of the conductor—whether on the surface or inside—come to the same potential. This is true regardless of whether the conductor has an internal cavity.

Our proof follows directly from Equation 7.18, which is

$$V_f - V_i = -\int_i^f \mathbf{E} \cdot d\mathbf{s}$$

In an isolated conductor $E = 0$, for all points within the conductor so for any pair of points i and f taken within the conductor $V_f = V_i$.

If we draw a graph between potential V_s radial distance r from the centre of the spherically symmetrical shell, we will find a plot as shown in figure provided that the shell is an isolated conductor and radius of the shell is 1.0 m. For points outside the shell, we can calculate $V(r)$ from Equation 7.70 because the charge q behaves for such external points as if, it were concentrated at the center of the shell. That equation holds right up to the surface of the shell. Now let push a small test charge through the shell-assuming a small hole exists-to its center. No extra work is needed, do this because no net electric force acts on the test charge once it is inside the shell. Thus, the points inside the shell & the points on its surface, all have potential of same magnitudes.

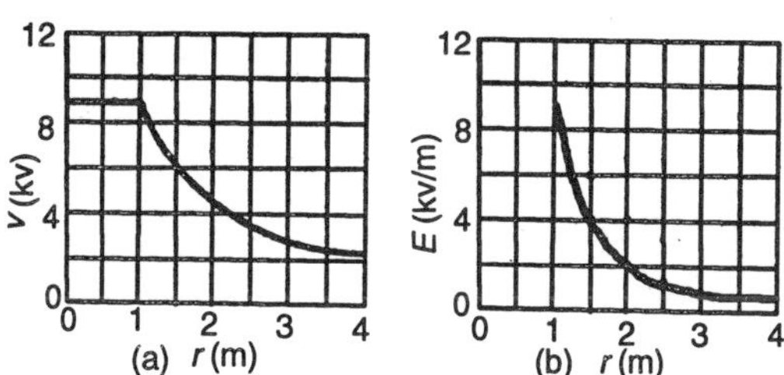

Figure 7.38: ***(a) A plot of V(r) for a charged spherical shell (b) A plot of E(r) for the same shell:***

In the fig 7.38 (b) variation of electric field with radial distance r for the same cell is. Note that $E = 0$ everywhere inside the shell. The curves of Figure 7.19 a by differentiating with respect to r, using Eq 7.82: (the derivative of a constant, recall, is zero). By integrating the curves of fig 7.38: (b) with respect to r, the curves of fig 7.38: (a) will be obtained.

The spherical charge does not distribute itself uniformly over surface of nonspherical conductors. At sharp points or edges, the surface charge density and thus the external electric field, which is proportional to it may reach very high values. The air around such sharp points may become ionized, producing the corona discharge that golfers and mountaineers see on the ups of bushes, golf clubs, and rock hammers when thunderstorms threaten. Such corona discharges, like hair that stands on end, are often the precursors of lightning strikes. So, in these cases, it is suggested to hide oneself in such *a* conducting shell where electric field would be zero.

The free conduction electrons distribute themselves on the surface in such a way that the electric field they produce at interior points cancels the external electric field that would otherwise be there. Furthermore, the electron distribution causes the net electric field at all points on the surface to be perpendicular to the surface. Imagine that the conductor is removed leaving the surface charge frozen, then also the pattern of the electric field will remain unchanged for any points, of the charge.

8

Electromagnetism

Magnets

Some stones becames magnetic naturally and those were the first known magnets, called as lodestones. When the ancient Greeks and ancient Chinese discovered these rare stones, they were amused by the stones ability to attract metal over a short distance, as if by magic. Lodestones were used in compasses to determine direction, much later.

Magnets and magnetic material are now very common. We find them in VCRs, audio cassettes, ATM and credit cards, audio headsets, and even in the inks for paper money. In fact, some breakfast cereals that are "iron fortified" contain small bits of magnetic materials (you can collect them from a slurry of cereal and water with a magnet). The modern electronics industry rely heavily on magnets and magnetic material.

The atoms and electrons of magnetic material also have magnetic properties. As we have observed, iron filings sprinkled around such a magnet tend to align with the magnetic field of the magnet, and their pattern reveals the magnetic field lines. The clustering of the lines at the ends of the magnet suggests that one end is a *source* of the lines (the field diverges from it) and the other end is a sink the lines (the field converges toward it). The end which is source of these lines is called north pole and the other end is known as south pole. With its two poles end, the magnet is also an exaple of magnetic dipoles.

Figure 8.1: ***If you break a magnet, each fragment becomes a separate magnet, with its own north and south poles.***

Imagine that *a* bar magnet is broken into pieces as *a* chalk piece is done. We should, it seems, be able to isolate a single pole, or *monopole.* But surprisingly we cannot, not even if we break the magnet down to its individual atoms and then to its electrons and nuclei. Each piece will have *a* south pole and *a* north pole, so it can be concluded that

The simplest magnetic structure that can exist is a magnetic. dipole Magnetic monopoles do not exist (as far as we know)

Gauss' Law for Magnetic Fields

According to gauss law, For magnetic fields, the magnetic monopoles do not exist. The law asserts that the net magnetic flux Φ_B through any closed Gaussian surface is zero:

$$\Phi_E = \oint \mathbf{B} \cdot d\mathbf{A} = 0$$

Contrast this with Gauss' law for electric fields,

$$\Phi_E = \oint \mathbf{B} \cdot d\mathbf{A} = \frac{1}{\varepsilon_0} \qquad \text{(Gauss' law for electric fields).}$$

Both the above equations are integrated over a closed Gaussian surface. Gauss law for electric fields says that this integral (the net electric flux through the surface) it proportional to the net electric charge *q* enclosed by the surface. Gauss' law for magnetic fields says that there can be no net magnetic flux through the surface because there can be no net "magnetic charge" (individual magnetic peics) enclosed by the surface. The simplest magnetic structure that can exist and thus be enclosed by a Gaussian surfaces is a dipole, which consists of both a source and a sink for the field lines. So, the magnetic flux going out must be equal to the total magnetic flux coming in, So as to make net magnetic flux zero.

So it has seen that Gauss' Law for magnetic fiels holds more complicated structure than *a* magnetic dipole, even if the Gaussion surface does not enclose the whole surface.

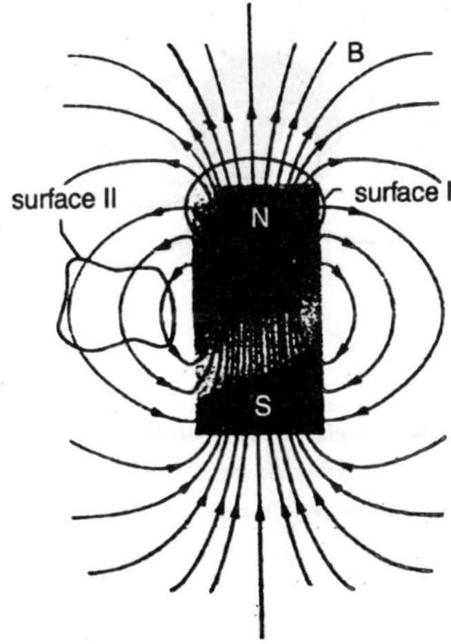

Figure 8.2 : ***The field lines for the magnetic field* B *of a short bar magnet. The red curves represent cross sections of closed, three-dimensional Gaussian surfaces.***

Gaussian surface **II** near the bar magnet of Figure 8.3 encloses no poles, and we can easily conclude that the net magnetic flux through it is zero. But Gaussian surface **I** is more difficult. As the Gaussian surface encloses only label *N* and not label 'S', so it seem to enclose only the north pole of magnet. But a south pole must be associated with the lower boundary of the surface, because magnetic field lines enter the surface there. (The enclosed section is like one piece of the broken bar magnet). So, this Gaussian surface encloses a magnetic dipole and the no total flux through this Gaussian surface is zero.

The Magnetism of Earth

Our earth acts like *a* magnetic dipole. For the points near its surface its magnetic field can be represented as a huge bar magnet. Figure 8.3 is an idealized symmetric depiction of the dipole field, without the distortion caused by passing charged particles from the Sun.

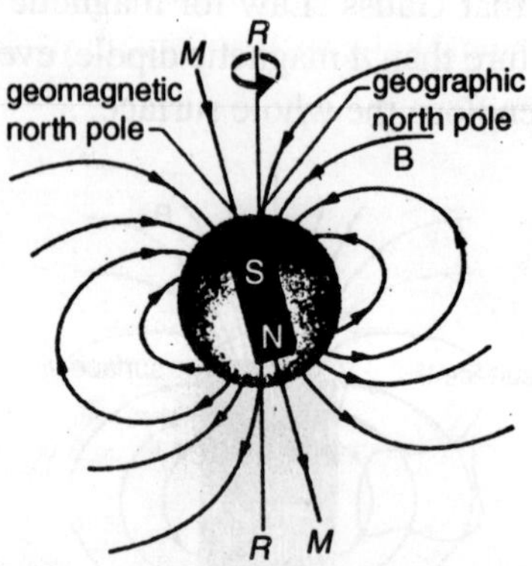

Figure 8.3 : *Earth's magnetic field represented as a dipole 'field. The dipole axis MM makes an angle of 11.5° with rotational axis RR. The south pole of the dipole is in Earth's northern hemisphere.*

A magnetic dipole moment *N* is always remain associated with the earths magnetic field due to its dipole nature. For the idealized field of Figure 8.3, the magnitude of μ is 8.0×10^{22} J/T and the direction of μ makes an angle of 11.5° with the rotation axis (*RR*) of Earth. The dipole axis lies along μ and intersects Earth surface at the *geomagnetic north pole* in northwest Green-land and the *geomagnetic south pole* in Antarctica.The lines of the magnetic field **B** generally emerge in the southern hemisphere and reenter Earth in the northem hemisphere. So, in fact, the magnetic pole present at northern hemisphere of earth and called as north pole, is actually the south pole of the earth's magnetic diple.

On earth surface, the direction of magnetic field at any location is generally specified in terms of two angle. The **field declination** is the angle (left or right) between geographic north (which is toward 90° latitude) and the horizontal component of the field. The angle between the horizontal plane and the field's direction is the field inclination.

Magnetometer measure these angles and determine the field with much precision. However, you can do reasonably well with just a *compass* and a *dip meter.* A compass is simply a needle-shaped magnet that is mounted so that it can rotate freely about a vertical axis. If *a* compass is held horizontally, the north pole end of the needle points towards geomatric north pole generally. The angle between the needle and geographic north is the field declination. A dip meter is a similar magnet that can rotate freely a horizontal axis. When the direction of the compass is aligned with the vertical plane iof rotation of the meter's needle then the angle between meter's needle and horizontal is the field inclination.

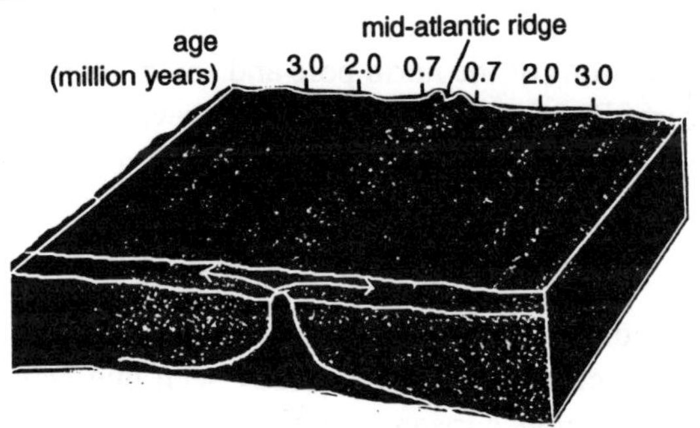

Figure 8.4: ***A magnetic profile of the seafloor on either side of the mid-Atlantic Ridge. The seafloor, extruded through the ridge and spreading out as part of the tectonic drift system, displays a record of the past magnetic history of Earth's core. The direction of the magnetic field produced by the core reverses about every million years.***

However, the measured magnetic field, at any point on earth's surface, may differ widely in both magnitude and direction. In fact, the point where the field is actually vertical to Earth's surface and inward is not located at the geomagnetic north pole in Greenland as we would expect; instead this so called *dip north pole* is located in the Queen Elizabeth Islands in northern Canada, far from Greenland.

Earth's magnetic fiels at any point, also varies with time by measurable amount in few years and the difference become much great in 'Say' 100 years. For example, between 1580 and 1820 the direction indicated by compass needles in London Changed by 35°.

However the rate of difference in dipole field is rather slow over such short period of time. Variations over longer periods can be studied by measuring the weak magnetism of the ocean floor on either side of the Mid-Atlantic Ridge. This floor has been formed by molten magma that oozed up through the ridge from Earth's interior, solidified, and was pulled away from the' ridge (by the drift of tectonic plates) at the rate of a few centimeters per year. At the time of its solidification, the magma become magnetise and the direction of magnetic field of magma is same as that of the earth's magnetic field. Study of this solidified magma across the ocean floor reveals that Earth's field has reversed its *polarity* (directions of the north pole and south pole) about every million years. But the reason begind this reversal is not known the mechanism for earth's magnetic field is also, very poorly known.

Magnetism and Electrons

All the magnetic materials have their magnetic properties only due to their electrons. We have already seen one way in which electrons can generate a magnetic field: send them through a wire as an electric current, and their motion produces a magnetic field around the wire. Electorons can produces magnetic field in two more ways, each involving a magnetic dipole moment. However, their explanation requires quantum physics that is beyond the physics presented in this book. We will give only *a* look only on outline of the results.

Spin Magnetic Dipole Moment

The intoinsic spin angular momentum (S) of electron has a spin magnetic dipole moment v_s associated with it and their relation is given by **S** and $\boldsymbol{\mu}_s$ are related by in which *e* is the elementry.

$$\boldsymbol{\mu}_s = -\frac{e}{m}\mathbf{S}, \tag{8.2}$$

charge (1.60×10^{-19}C) and m is the mass of an electron (9.11×10^{-31} kg). The negative sign in the above equation indicates that both S and N_s are oppoite to each other in direction.

The spin angular momentum S is different from the angular momenta (S) we have studied earlier in two ways

1. **S** itself cannot be measured. Instead, only its component along an axis can be measured.

2. A measured component of **S** is quantized (restricted to certain values); in fact, it always has the same magnitude (no matter which axis is chosen).

Imagine that the component of spin angular momentum '*S*' is measured along the *z*-axis of a three dimensional system. Then the measured component S_z can have only the two values given by

$$S_z = m_s \frac{h}{2\pi}, \quad \text{for} \quad m_s = \pm\frac{1}{2}, \tag{8.3}$$

where m_s is the *spin magnetic quantum number* and h (= 6.63 × 10^{-34} **J** · s) is the Planck constant, the ubiquitous constant of quantum physics. The signs given in Equation 8.3 have to do with the direction of S_z along the *z* axis. If S_z is parallel to *Z* axis and M_s is + Y_2 then electron is said to be spin up. Similarly when S_z is antiparallel to *Z* axis and M_s is $-Y_2$ then the electron is said to be spin down.

It is not possible to measure the spin magnetic dipole moment N_s rather its component is measured which has always same magnitude. We can relate the component $\mu_{s,\,z}$ measured on the *z* axis to S_z by rewriting Equation 8.2 in component form for the *z* axis as

$$\mu_{s,z} = -\frac{e}{m} S_z$$

Substituting for S_z from Equation 8.3 then gives us

$$\mu_{s,z} = \pm\frac{eh}{4\pi m} \tag{8.4}$$

where the plus and minus signs correspond to $\mu_{s,z}$ being parallel and antiparallel to the *z* axis, respectively.

In the above equation, the quantity present on the right side is known as bohr magnetone N_B.

$$\mu_B = \frac{eh}{4\pi m} = 9.27 \times 10^{-24} \text{ J/T} \quad \text{(Bohr magneton)} \tag{8.5}$$

Spin magnetic dipole moments of electrons and other elementary particles can be expressed in terms of μ_B. For the electron, the magnitude of the measured component of $\boldsymbol{\mu}_s$ is

$$\mu_{s,z} = 1\mu_B \tag{8.6}$$

(The quantum physics of the electron, called *quantum electrodynamics,* or QED, reveals that $\mu_{s,z}$ is actually slightly greater than $1\mu_B$, but we shall neglect that fact.). If we place *a* electron in an external magnetic field, then its (*U*) potential energy can be associated with the orientation of the spin magnetic dipole moment N_s of the electron.

$$U = -\boldsymbol{\mu}_s \cdot \mathbf{B}_{ext} = -\mu_{s,z} B \tag{8.7}$$

where the *z* axis is taken to be in the direction of $\mathbf{B}_{ext}$.

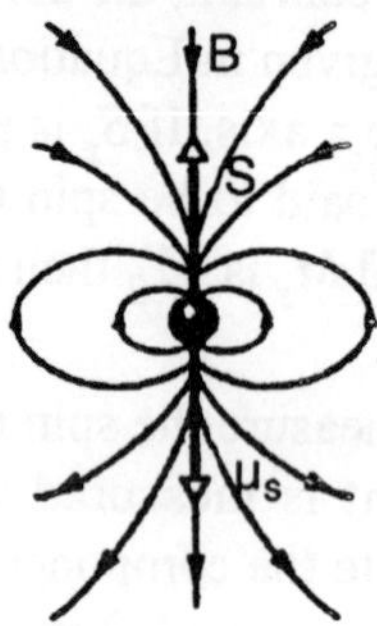

Figure 8.5 : ***The spin* S, *spin magnetic dipole moment μ_s and magnetic field* B *of an electron represented as a microscope sphere.***

Let us consider the electron to be microscopic sphere, then the representation of its spin angular momentum *S*, its magnetic dipole moment N_s will be as shown is figure 8.6. Although we use the word "spin" here, electrons do not spin like tops. How, then can something have angular momentum without actually rotating? Again, quantum physics provides the answers.

Even the other particels *i.e.* proton and neutron have the intrinsic angular momentum called spin and associated intrinsic spin magnetic dipole moment. For a proton those two vectors have the same direction, and for a neutron they have opposite directions. But their contribution to the magnetic fields of atoms is not considered as they are too small as compared to electron.

Orbital Magnetic Dipole Moment

An electron has an extra angular momentum called its orbital angular momentum when the electron is in a atom. Associated with $\mathbf{L}_{orb}$ is an **orbital magnetic dipole moment** $\boldsymbol{\mu}_{orb}$; the two are related by

$$\boldsymbol{\mu}_{orb} = -\frac{e}{2m}\mathbf{L}_{orb} \tag{8.8}$$

The *–ve* sign indicates that N_{orb} and L_{orb} are in opposite directions.

We cannot measure the L_{orb} rather *a* component along any of the axis can measured and which is quantized. The components along, say, a *z* axis can have only the values given by

$$L_{orb,z} = m_l \frac{h}{2\pi}, \quad \text{for } m_l = 0, \pm 1, \pm 2, \cdots, \pm (\text{limit}) \tag{8.9}$$

in which M_l is the *orbital magnetic quantum number* and "limit" refers to some largest allowed integer value for M_l. The ± are concerned with the direction of the L_{orb} on *z* axis.

Similar with the L_{orb}, the orbital magnetic dipole moment N_{orb} also, cannot be measured itsell but its component along any axis can be meassred which is also quantized. By writing Eq. 8.8 in component form for the *z* axis and then substituting for $L_{orb \cdot z}$ from Equation 8.9, we can write the *z* component $\mu_{orb,\,z}$ of the orbital magnetic dipole moment as

$$\mu_{orb.z} = -m_l \frac{eh}{4\pi m} \tag{8.10}$$

and, in terms of the Bohr magneton, as

$$\mu_{orb.z} = -m_l \mu_B \tag{8.11}$$

If an atom is present in an external magnetic field, then with orientation of magnetic dipole moment of each electron, a potential energy *U* can be associated which is given by

$$U = -\mu_{orb} \cdot \mathbf{B}_{ext} = -\mu_{orb.z} B_{ext}, \tag{8.12}$$

where the *z* axis is taken in the direction of $\mathbf{B}_{ext.}$

Although we have used the words "orbit" and "orbital" here,

electrons do not orbit the nucleus of an atom like planets orbiting the Sun. To know the reason behing the question that how an electron have orbital abgular momentum without orbiting, we need the help of quantum physics.

Loop Model for Electron Orbits

The relation between N_{orb} and L_{orb}, as shown in equation 8.8 can be found with the non quantum derivation that follows in which it is considered that the electron is moving in *a* circular path with *a* radius which is much larger than atomic radius but this is not applied to the atomic electron.

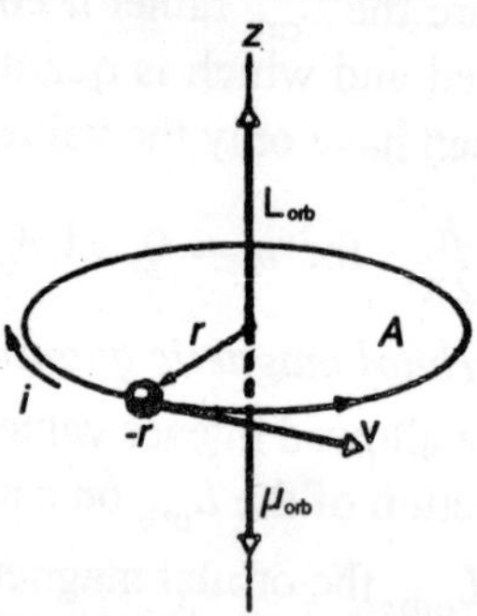

Figure 8.6 : ***An electron moving at constant speed v in a circular path of radius r that encloses an area A. The etectron has an orbital angular momentum L_{orb} and an associated orbit magnetic dipole moment μ_{orb} A clockwise current i (of positive charge) is equivalent to the counterclockwise circulation of the negatively charged electron.***

Let us consider an electron moving in a circular path of radius *r* and *a* constant speed *v* in a clockwise direction. The motion of the negative charge of the electron is equivalent to a conventional current *i* (of positive charge) that is clockwise, as also shown in Figure 8.7.

The magnitude of the orbital magnetic dipole moment of such a current loop is obtained with $N = I$.

$$\mu_{orb} = iA, \tag{8.13}$$

where *A* is the area enclosed by the loop. So, according to right hand rule the direction of the magnetic dipole moment N_s is downward.

We require current *i* to evaluate equation 8.13. Current is known as the rate at which charge flows at any point in the circuit. Here, the charge of mgnitude e takes a time $T = 2\pi r / v$ to circle from any point back through that point so

$$i = \frac{\text{charge}}{\text{time}} = \frac{e}{2\pi r / v}. \qquad (8.14)$$

Replacing this value and area $A = \pi r^2$ in the equation 8.12, we get

$$\mu_{\text{orb}} = \frac{e}{2\pi r / v}\pi r^2 = \frac{evr}{2}. \qquad (8.15)$$

We can get an expression for the orbital angular momentam L_{orb} of the electron by using $e = m\ (r \times v)$. Because **r** and **v** are perpendicular, $\mathbf{L}_{\text{orb}}$ has the magnitude

$$L_{\text{orb}} = mrv \sin 90° = mrv. \qquad (8.16)$$

$\mathbf{L}_{\text{orb}}$ is directed upward in Figure (8.7). Combining Equations 8.15 and 8.16, generalizing to a vector for mulation, and indicating the opposite directions of the vectors with a minus sign yield

$$\mu_{\text{orb}} = -\frac{e}{2m}\mathbf{L}_{\text{orb}}$$

which is Equation 8.8 : So the result of classical physics is same as that of the quantum physics as the magnitude and direction both remain same. You might wonder, since this derivation gives the correct result for an electron within an atom, why the derivation is invalid for that situation. This can be concluded that the result obtained by this line of reasoning, contradicted by the experiments.

Loop Model in a Nonuniform Field

Till now, we condider the orbit of an electron as *a* current loop. Now, however, we draw the loop in a nonumifrom magnetic field $\mathbf{B}_{\text{ext}}$ as shown in Figure 8.7 (a) (This field is the diverging field near the north pole of the magnet in Figure 8.2) : We make this change to prepare for the next several sections, in which we shall discuss the forces that act on magnetic materials when the materials are placed in a nonuniform magnetic field. To discuss about these forces we imagin that the orbits of electron inside the material are like small current loop.

Let us consider that all the magnetic field vectors around the

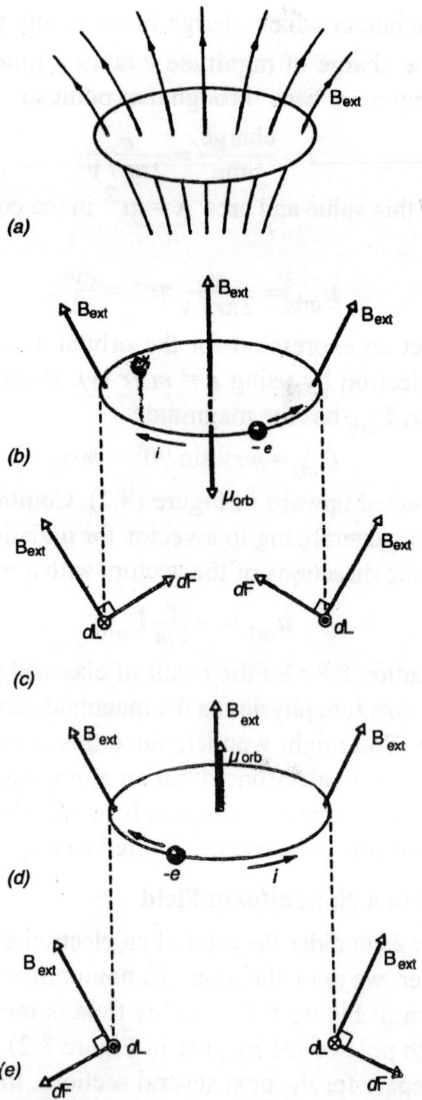

Figure 8.7: *(a) A loop model for an electron orbiting in as atom while in a nonuniform magnetic field B_{ext} (b). Charge moves counterclockwise; the associated conventional clockwise. (c) The magnetic forces dF on the left and right sides of the loop, as seen from the plane of the loop. The net force on the loop is upward. (d) Charge -e moves clockwise. (e) The net force on the loop is now downward.*

electrons circular path have the same magnitude and they make the same angle with vertical as in Fig 8.8 (b). We also assume that all the electrons in an atom move either, counterclockwise or clockwise. The orbital magnetic dipole moment N_{orb} and the associated current i around the current loop, for each of these directions, are given

Diametrically opposite views of an element dL of the loop with the same direction as i, are shown on figures 8.8 (c) and 8.8 (e), as seen from the plane of the orbit. Also shown are the field $\mathbf{B}_{ext}$ and the resulting magnetic force $d\mathbf{F}$ on $d\mathbf{L}$. Recall that a current along an element $d\mathbf{L}$ in a magnetic field $\mathbf{B}_{ext}$ experiences a magnetic force $d\mathbf{F}$ as given in previous chaper

$$d\mathbf{F} = i\, d\mathbf{L} \times \mathbf{B}_{ext.} \tag{8.17}$$

According to equation 8.17, on the left side of figure 8.8 (c) force $d\mathbf{F}$ is directed upward and rightward. On the right the force $d\mathbf{F}$ is just as large and is directed upward and leftward. Because their angles are the same, the horizontal components of these two forces cancel and the vertical components add. The same is true at any other two symmetric points on the loop. So, the net force on the current loop of Figure 8.8 (b) must be upward. The same reasoning deads to a downward net force on the loop in Figure 8.8 (d). The requairement of these two results will be done when we examin behaviour of magnetic material in uniform magnetic field.

Magnetic Materials

The orbital magnetic dipole moment and spin magnetic dipole moment of an electron are added vectorially. The resultant of these two vectors combines vectorially with similar resultants for all other electrons in the atom. And the resultants for each atom combines with those for all the other atoms in a sample of a material. If the combination of all these magnetic dipole moments produces a magnetic fiels, then the material is magnetic. We know three kinds of magnetism generally.paramagnetism, diamagnetism and ferromagnetism

Diamagnetison is characteristic of all kinds of materials, but due to its weakness it will seem to be absent if either of the other two magnetism is present. In diamagnetism, weak magnetic dipole moments are produced in the atoms of the material when the material is placed in an external magnetic field $\mathbf{B}_{ext}$; the combination of all those induced dipole moments gives the material as a whole only a feeble net magnetic field. The dipole moments and thus their net field disappear when $\mathbf{B}_{ext}$ is removed. Only those material that exhibit diamagnetism are called diamagnetic material.

The phenomenon of paramagnetism is exhibited by materials having transition elements rare earth metals and actiride elements. Each atom of such a material has a permanent resultasnt magnetic dipole moment, but the moments are randomly oriented in the material and the material as a whole lacks a net magnetic field. However, an external magnetic field $\mathbf{B}_{ext}$ can partially align the atomic magnetic dipole moments to give the material a net magnetic dipole moments to give the material a net magnetic field. The alignment and thus its field disappear when $\mathbf{B}_{ext}$ is removed. The materials, which show paramagnetism primarily, are callled paramagnetic.

Ferromagnetism is a characteristic of iron, mickel and few other elements. Some of the electrons in these materials align their resultant magnetic dipole moments to produce regions with strong magnetic dipole moments. An external field $\mathbf{B}_{ext}$ can then align the magnetic moments of these regions, producing a strong magnetic field for the material as a whole; the field partially persists when $\mathbf{B}_{ext}$ is removed. The term ferromagnetic material is used for those material which shown ferromagnetism primarially.

Maxwell's Equations

It can be said that all the electromagnetic phenomena follwo from Naxwell's wquation. These equations are based on experimental observations and are given by the following equations:

$$\frac{\partial D_x}{\partial x}+\frac{\partial D_y}{\partial y}+\frac{\partial D_z}{\partial z}=\rho \qquad (8.18)$$

$$\frac{\partial B_x}{\partial x}+\frac{\partial B_y}{\partial y}+\frac{\partial B_z}{\partial z}=0 \qquad (8.19)$$

$$\left.\begin{aligned}\frac{\partial E_z}{\partial y}-\frac{\partial E_y}{\partial z}&=-\frac{\partial B_x}{\partial t}\\\frac{\partial E_x}{\partial z}-\frac{\partial E_z}{\partial x}&=-\frac{\partial B_y}{\partial t}\\\frac{\partial E_y}{\partial x}-\frac{\partial E_x}{\partial y}&=-\frac{\partial B_z}{\partial t}\end{aligned}\right\}\qquad(8.20)$$

and

$$\left.\begin{aligned}\frac{\partial H_z}{\partial y}-\frac{\partial H_y}{\partial z}&=J_x+\frac{\partial D_x}{\partial t}\\\frac{\partial H_x}{\partial z}-\frac{\partial H_z}{\partial x}&=J_y+\frac{\partial D_y}{\partial t}\\\frac{\partial H_y}{\partial x}-\frac{\partial H_x}{\partial y}&=J_z+\frac{\partial B_z}{\partial t}\end{aligned}\right\}\qquad(8.21)$$

where ρ represents the charge density and **J** the current density; **E, D, B** and **H** represent the electric field, electric displacement, magnetic induction and magnetic field respectively. The physical implications of all the above equation will be discussed later. By using vector terms, the above equation can be rewritten as

$$\text{div }\mathbf{D}=\rho \qquad (8.22)$$

$$\text{div }\mathbf{B}=0 \qquad (8.23)$$

$$\text{curl }\mathbf{E}=-\frac{\partial \mathbf{B}}{\partial t} \qquad (8.24)$$

$$\text{curl }\mathbf{H}=\mathbf{J}+\frac{\partial \mathbf{D}}{\partial t} \qquad (8.25)$$

The above equations can be solved only if the 'constitutive relations' are known which relate **D** to **E, B** to **H** and **J** to **E**. The "constitutive relations" for a linear, isotropic and homogeneous medium are given as follow

$$\mathbf{D}=\varepsilon\mathbf{E} \qquad (8.26)$$

$$\mathbf{B} = \mu\mathbf{H} \tag{8.27}$$

and

$$\mathbf{J} = \sigma\mathbf{E} \tag{8.28}$$

In the above three equation the notations ε, μ and σ are used for dielectric primitivity, magnetic permeability and conductivity of the medium respectively.

Plane Waves in a Dielectric

We know that if *a* dielectric is non-charged and current free then

$$\rho = 0 \tag{8.29}$$

and

$$\mathbf{J} = 0 \tag{8.30}$$

Physical significance of Maxwell's Equations

Imagine the equation to be

$$\text{div}\,\mathbf{D} = \rho \tag{8.31}$$

In free-space

$$\mathbf{D} = \varepsilon_0\mathbf{E} \tag{8.32}$$

and Equation (8.31) becomes

$$\text{Div}\,\mathbf{E} = \rho / \varepsilon_0 \tag{8.33}$$

If we integrate the above equation over a volume *V*, we obtain

$$\int \text{div}\,E\,dV = \frac{1}{\varepsilon_0}\int \rho\,dV$$

Applying the divergence theoren, we get

$$\oint \mathbf{E}\cdot d\mathbf{a} = \frac{1}{\varepsilon_0}Q \tag{8.34}$$

The above equation is the simple form of Gauss'law that is the electric flux through *a* clossed Gaussian surface is the total amount of charge present inside the body divided by ε_o.

In a similar manner, the equation

$$\text{div}\,\mathbf{B} = 0 \tag{8.35}$$

gives

$$\oint \mathbf{B} \cdot d\mathbf{a} = 0 \tag{8.36}$$

i.e. the magnetic flux through a closed surface is always zero; this implies the absence of magnetic monopoles.

Now we will concertrate on the equation given below

$$\text{curl } \mathbf{E} = -\frac{\partial \mathbf{B}}{\partial t} \tag{8.37}$$

This gives a relation between time and space dependent electric field with a varying magnetic field. Now, Stokes' theorem tells us that

$$\oint_{\Gamma} \mathbf{E} \cdot d\mathbf{l} = \int_{S} \text{curl } \mathbf{E} \cdot d\mathbf{a} \tag{8.38}$$

where the LHS represents a line integral over a closed path Γ and the **RHS** represents a surface integral over any surface bounding the path Γ. So

$$\oint_{\Gamma} \mathbf{E} \cdot d\mathbf{l} = \int_{S} \text{curl } \mathbf{E} \cdot d\mathbf{a} = \int_{S} \frac{\partial \mathbf{B}}{\partial t} \cdot d\mathbf{a} \tag{8.39}$$

or

$$\oint_{\Gamma} \mathbf{E} \cdot d\mathbf{l} = -\frac{d}{dt} \int_{S} \mathbf{B} \cdot d\mathbf{a} \tag{8.40}$$

The LHS of the above equation represents the induced emf in a closed circuit which is equal to the negative of the rate of change of the magnetic flux through the circuit. This represent's the famous Law of indection given by Faraday's. It is worthwhile to mention that although this law was discovered by Faraday, it was put in the differential form by Maxwell.

Now, we have arrived at the last equation of Maxwell.

$$\text{curl } \mathbf{H} = \mathbf{J} + \frac{\partial \mathbf{D}}{\partial t} \tag{8.41}$$

However, Ampere's law (which was known before Maxwell),

when expressed as a differential equation, was of the form ‡

$$\text{curl } \mathbf{H} = \mathbf{J} \tag{8.42}$$

This indicates that only current can produce a magnetic field when long wire is charged *i.e* it has current flowing in it a magnetic field would be produced.

Since the divergence of the curl of any vector is zero, one obtains

$$\text{div } \mathbf{J} = 0 \tag{8.43}$$

This equation can be compared with the equation of continuty

$$\text{div } \mathbf{J} + \frac{\partial \rho}{\partial t} = 0 \tag{8.44}$$

So the differential form of Ampere's law holds only when $\frac{\partial \rho}{\partial t} = 0$. Thus, for the Ampere's law to be consistent with the equation of continuity, Maxwell argued that there must be an additional term on the RHS of Equation (8.42)

$$\text{curl } \mathbf{H} = \mathbf{J} + \frac{\partial \mathbf{D}}{\partial t}$$

The introduction of the displacement current, $\frac{\partial D}{\partial t}$, brought revolution to the physics. Physically it implies that not only a current produces a magnetic field but a changing electric field also produces a magnetic field (as it indeed happens during the charging and discharging of a condenser). It may be mentioned that it is the presence of the term $\frac{\partial \mathbf{D}}{\partial t}$ which leads to the wave equation and, therefore, the prediction of electromagnetic waves. One can thus argue on physical grounds that. a changing electric field produces a magnetic field which varies in space and time and this changing magnetic field produces an electric field varying in space and time, and so on. Due to this meltual generation, the electric and magnetic field produces electromagnetic waves.

Reflection at an Interface of Two Dielectrics

Imagine the incidence of a plane polarized electromagnetic wave on an interface of two media, consider the plane $x = 0$ to represent the interface. Let (ε_1, μ_1) and (ε_2, μ_2) represent the dielectric permittivity and magnetic permeability of the media below and above the plane $x = 0$; we will assume both media to be lossless dielectrics. Let $\mathbf{E}_1$ $\mathbf{E}_2$ and $\mathbf{E}_3$ denote the electric fields associated with the incident wave, refracted wave and reflected wave respectively. The form of these fields for an incident wave plane will be

$$\left.\begin{aligned} \mathbf{E}_1 &= \mathbf{E}_{10}\exp\left[i\left(\mathbf{k}_1\cdot\mathbf{r}-\omega_1 t\right)\right] \\ \mathbf{E}_2 &= \mathbf{E}_{20}\exp\left[i\left(\mathbf{k}_2\cdot\mathbf{r}-\omega_2 t\right)\right] \\ \mathbf{E}_3 &= \mathbf{E}_{30}\exp\left[i\left(\mathbf{k}_3\cdot\mathbf{r}-\omega_3 t\right)\right] \end{aligned}\right\} \tag{8.45}$$

where $\mathbf{E}_{10}$, $\mathbf{E}_{20}$ and $\mathbf{E}_{30}$ are independent of space and time but may, in general, be complex. The vectors $\mathbf{k}_1$, $\mathbf{k}_2$ and $\mathbf{k}_3$ represent the propagation vectors associated with the incident, refracted and reflected waves respectively. As it is necessary that these fields must satisfy Maxwell equation we must have—

$$\left.\begin{aligned} \mathbf{k}_1^2 &= \omega^2\varepsilon_1\mu_1 \\ \mathbf{k}_2^2 &= \omega_2^2\varepsilon_2\mu_2 \\ \mathbf{k}_3^2 &= \omega_3^2\varepsilon_1\mu_1 \end{aligned}\right\} \tag{8.46}$$

We have studied earlier that the fields have to satisfy certain boundary conditions at the interface e.g. the component of the electric field, which is tangetial to it, must be continuous across electric field. These conditions are to be satisfied at *all* space points on the interface $x = 0$ and at all times. Consequently, the coefficients of y, z and t in the exponents appearing in Equation 8.45 must be equal. Thus

$$\omega = \omega_2 = \omega_3 \tag{8.47}$$

showing that all the waves have the same frequency. Hence Eqs simplify to

$$k_1^2 = \omega^2\varepsilon_1\mu_1 = k_3^2 \tag{8.48}$$

$$k_2^2 = \omega^2\varepsilon_2\mu_2 \tag{8.49}$$

Now it +*ve* *x*, *y*, and *z* component of K_1 are given by K_{1x}, K_{1y} and K_{1z}, then for K_2 and K_3, we have

$$k_{1y} = k_{2y} = k_{3y} \tag{8.50}$$

and

$$k_{1z} = k_{2z} = k_{3z} \tag{8.51}$$

Without any loss of generality we may choose the *y*-axis such that

$$k_{1y} = 0$$

(*i.e.* $\mathbf{k}_1$ is assumed to lie in the *x-z* plane—see Figure 21.1). Consequently,

$$k_{2y} = k_{3y} = 0 \tag{8.52}$$

From the above equation, it is concluded that vectros $\mathbf{K}_1$ $\mathbf{K}_2$ and $\mathbf{K}_3$ will lie in the same plane. Further, from Equation (8.51), we get

$$k_1 \sin\theta_1 = k_2 \sin\theta_2 = k_3 \sin\theta_3 \tag{8.53}$$

According to equation (8.48), $\mathbf{K}_1 = \mathbf{K}_3$, So it is must that $\theta_1 = \theta_3$ *i.e.* the angle of incidence is equal to the angle of reflection. So

$$\frac{\sin\theta_1}{\sin\theta_2} = \left(\frac{\varepsilon_2\mu_2}{\varepsilon_1\mu_1}\right)^{1/2} \tag{8.54}$$

If the speeds of the propagation wave in medium is given by $V_1 = \left(-\frac{1}{\sqrt{\varepsilon_1\mu_1}}\right)$ and in medium *z*, it is given by

$V_2 = \left(-\frac{1}{\sqrt{\varepsilon_2\mu_2}}\right)$ then

$$\frac{\sin\theta_1}{\sin\theta_2} = \frac{v_1}{v_2} = \frac{n_2}{n_1} \tag{8.54}$$

where $$n_1\left(=\frac{c}{v_1}c\sqrt{\varepsilon_1\mu_1}\right) \text{ and } n_2\left(=\frac{c}{v_2}=c\sqrt{\varepsilon_2\mu_2}\right) \tag{8.55}$$

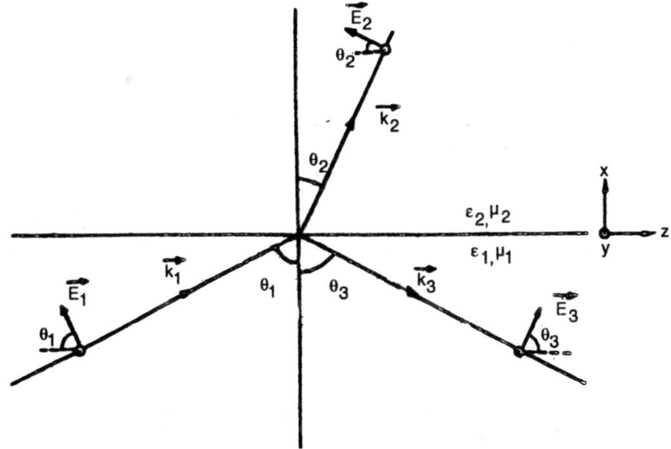

Figure 8.8 : ***The reflection of a plane wave with its electric vector parallel to the plane of incidence.***

represent the refractive indices of media 1 and 2 respectively. The above equation (Equation 8.55) is also known as Snell's law.

Let us find an equation for the reflection and transmission coefficients when *a* plane polarised wave is incident on an interface of two dielectrics. We will first consider the case when the electric vector lies in the plane of incidence which will be followed by the case when the electric vector is at right angles to the plane of incidence.

CASE 1. E *Parallel to the Plane of Incidence*

Imagine that the electric vector is lied in the plane of incidence. The magnetic vectors are along the *y*-axis. Clearly, the *z*-component of the electric field represents a tangential component which should be continuous across the surface. Thus

$$E_{1z}+E_{3z}=E_{2z}$$

or $$-E_1\cos\theta_1+E_3\cos\theta_1=-E\cos\theta_2 \tag{8.55}$$

Thus

$$\left[-E_{10}\exp\{i(\mathbf{k}_1\cdot\mathbf{r}-\omega t)\}+E_{30}\exp\{i(\mathbf{k}_3\cdot\mathbf{r}-\omega t)\}\right]_{x=0}\cos\theta_1$$

$$=\left[-E_{20}\exp\{i(\mathbf{k}_2\cdot\mathbf{r}-\omega t)\}\right]_{x=0}\cos\theta_2 \qquad (8.56)$$

As this wquation must be satisfied at all space points in the plane $x = 0$ at times, So the exponents must be identically equal which leads back to the equatons 8.47, 8.50 and 8.51.

$$\left[E_{10}-E_{30}\right]\cos\theta_1 = E_{20}\cos\theta_2 \qquad (8.57)$$

Further, the normal component of **D** must also be continuous and since **D** = ε**E**, we must have

$$\varepsilon_1 E_{1x}+\varepsilon_1 E_{3x}=\varepsilon_2 E_{2x}$$

or
$$\varepsilon_1\left[E_{10}+E_{30}\right]\sin\theta_1=\varepsilon_2 E_{20}\sin\theta_2 \qquad (8.58)$$

Replacing for E_{20}, in equation 8.51 we have

$$\varepsilon_1\left[E_{10}+E_{30}\right]\sin\theta_1=\varepsilon_2\sin\theta_2\frac{\left[E_{10}-E_{30}\right]}{\cos\theta_2}\cos\theta_1$$

or
$$\left[\varepsilon_2\sin\theta_2\cos\theta_1+\varepsilon_1\sin\theta_1\cos\theta_2\right]E_{30}$$
$$=\left[\varepsilon_2\sin\theta_2\cos\theta_1-\varepsilon_1\sin\theta_1\cos\theta_2\right]E_{10}$$

Thus
$$r_{\|}=\frac{E_{30}}{E_{10}}=\frac{\varepsilon_2\sin\theta_2\cos\theta_1-\varepsilon_1\sin\theta_1\cos\theta_2}{\varepsilon_2\sin\theta_2\cos\theta_1+\varepsilon_1\sin\theta_1\cos\theta_2} \qquad (8.59)$$

where r_1 denotes the amplitude reflection coefficient, the subscript ‖ refers to the fact that we are referring to parallel polarization. By dividing the equation 8.57 by $\mathbf{E}_{10}$ and replacing the expression for $\frac{E_{30}}{E_{10}}$ from equation 8.59 we have the following

$$\left[1-\frac{\varepsilon_2\sin\theta_2\cos\theta_1-\varepsilon_1\sin\theta_1\cos\theta_2}{\varepsilon_2\sin\theta_2\cos\theta_1+\varepsilon_1\sin\theta_1\cos\theta_2}\right]\cos\theta_1=\frac{E_{20}}{E_{10}}\cos\theta_2$$

or

$$t_{\|} = \frac{E_{20}}{E_{10}} = \frac{2\varepsilon_1 \sin\theta_1 \cos\theta_1}{\varepsilon_2 \sin\theta_2 \cos\theta_1 + \varepsilon_1 \sin\theta_1 \cos\theta_2} \qquad (8.60)$$

where $t_{||}$ denotes the amplitude transmission co-efficient.

We will have to find the ratio of x- component, of the poynting vector associated with the reflected and transmitted wave in order to calculate the reflection co-effecient. The reason why we should take the ratio of the x-component can easily be understood by referring to Figure 8.9. If S_1 denotes the magnitude of the Poynting vector associated with incident wave then the energy incident on the area $d\mathbf{A}$ (on the surface $x = 0$) per unit time would be $S_{1x}\, dA = S_1\, dA \cos\theta$. Similarly, the energy transmitted through the area dA would be $S_{2x}\, dA = S_2 dA \cos\theta_2$ and the energy reflected from the adra dA would be $S_{3x}\, dA = S_3\, dA \cos\theta\, dA$. If reflection and transmission coeffecients are denoted by $R_{\|}$ and $T_{\|}$ then we get—

$$R_{\|} = \frac{S_{3x}}{S_{1x}} = \frac{S_3 \cos\theta_1}{S_1 \cos\theta_1} \qquad (8.61)$$

$$= \frac{\langle \mathbf{E}_3 \times \mathbf{H}_3 \rangle}{\langle \mathbf{E}_1 \times \mathbf{H}_1 \rangle} = \frac{\sqrt{\varepsilon_1 / \mu_1}\, |E_{30}|^2}{\sqrt{\varepsilon_1 / \mu_1}\, |E_{10}|^2}$$

$$= \left|\frac{E_{30}}{E_{10}}\right|^2$$

or

$$R_{\|} = \left[\frac{\varepsilon_2 \sin\theta_2 \cos\theta_1 - \varepsilon_1 \sin\theta_1 \cos\theta_2}{\varepsilon_2 \sin\theta_2 \cos\theta_1 + \varepsilon_1 \sin\theta_1 \cos\theta_2}\right]^2 \qquad (8.62)$$

and

$$T_{\|} = \frac{S_{2x}}{S_{1x}} = \frac{S_2 \cos\theta_2}{S_1 \cos\theta_1}$$

$$= \frac{\langle \mathbf{E}_2 \times \mathbf{H}_2 \rangle \cos\theta_2}{\langle \mathbf{E}_1 \times \mathbf{H}_1 \rangle \cos\theta_1} = \frac{\sqrt{\varepsilon_2 / \mu_2}\, |E_{20}|^2 \cos\theta_2}{\sqrt{\varepsilon_1 / \mu_1}\, |E_{10}|^2 \cos\theta_1}$$

$$= \sqrt{\varepsilon_2 / \varepsilon_1} \sqrt{\varepsilon_2 / \varepsilon_1} \frac{\sin\theta_2}{\sin\theta_1}$$

$$\left[\frac{2\varepsilon_1 \sin\theta_1 \cos\theta_1}{\varepsilon_2 \sin\theta_2 \cos\theta_1 + \varepsilon_1 \sin\theta_1 \cos\theta_2}\right]^2 \frac{\cos\theta_2}{\cos\theta_1}$$

where we have substituted for $\sqrt{\mu_2 / \mu_1}$ from Equation (8.54). Thus

$$T_{||} = \frac{4\varepsilon_1\varepsilon_2 \sin\theta_1 \sin\theta_2 \cos\theta_1 \cos\theta_2}{\left[\varepsilon_2 \sin\theta_2 \cos\theta_1 + \varepsilon_1 \sin\theta_1 \cos\theta_2\right]^2} \tag{8.63}$$

It can easily be seen that

$$R_{||} + T_{||} = 1 \tag{8.64}$$

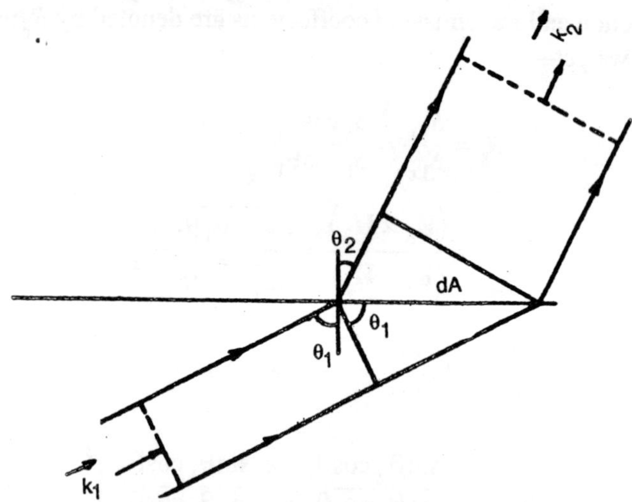

Figure 8.9 : *If the cross-sectional area of the incident beam is dA cos θ_1 then the cross-sectional area of the transmitted beam is dA cos θ_2.*

If the medium is not magnetic then $N_1 = N_2 \approx N_0 = 4\pi \times 10^{-7}$ *N/amp*^{2} and in that case the expression for the amptitude reflection co-effrcient will be

$$r_{||} = \frac{n_2^2 \sin\theta_2 \cos\theta_1 - n_1^2 \sin\theta_1 \cos\theta_2}{n_2^2 \sin\theta_2 \cos\theta_1 + n_1^2 \sin\theta_1 \cos\theta_2}$$

Since $$n_1 \sin\theta_1 = n_2 \sin\theta_2 \tag{8.65}$$

we get $$r_{||} = \frac{n_2 \cos\theta_1 - n_1 \cos\theta_2}{n_2 \cos\theta_1 + n_1 \cos\theta_2} \tag{8.66 (a)}$$

$$= \frac{\sin\theta_1 \cos\theta_1 - \sin\theta_2 \cos\theta_2}{\sin\theta_1 \cos\theta_1 + \sin\theta_2 \cos\theta_2} \tag{8.66(b)}$$

or

$$r_{||} = \frac{\sin 2\theta_1 - \sin 2\theta_2}{\sin 2\theta_1 + \sin 2\theta_2} = \frac{2\cos(\theta_1 + \theta_2)\sin(\theta_1 - \theta_2)}{2\sin(\theta_1 + \theta_2)\cos(\theta_1 - \theta_2)}$$

$$= \frac{\tan(\theta_1 - \theta_2)}{\tan(\theta_1 + \theta_2)} \tag{8.66 (c)}$$

If we start from equation (8.60) we could have the following

$$t_{||} = \frac{2\cos\theta_1 \sin\theta_2}{\sin(\theta_1 + \theta_2)\cos(\theta_1 - \theta_2)} \tag{8.67}$$

From Equations 8.66 and 8.67 we may deduce the following:

(a) *No reflection when $n_2 = n_1$.*

When $n_2 = n_1$, $\theta_2 = \theta_1$ and we get

$$r_{||} = 0 \text{ and } t_{||} = 1$$

Thus there is no reflection when the second medium has the same refractive index as the first medium (obviously!). So if a transparent solid is immersed in a liquid of same refractive index, we could not see the solid.

(b) *Polarization by reflection: Brewster's law*

Imagine that the angle of incidence in such a way that $\theta_1 + \theta_2 = \frac{\pi}{2}$, the angle of reflaction $r_{||} = 0$. It means there would be no reflected beams. Thus, if an unpolarized beam is incident at an angle such that $\theta_1 + \theta_2 = \frac{\pi}{2}$, then the parallel component of the

E-vector will not be reflected and the reflected light will be polarized with its **E**-vector perpendicular to the plane of incidence.

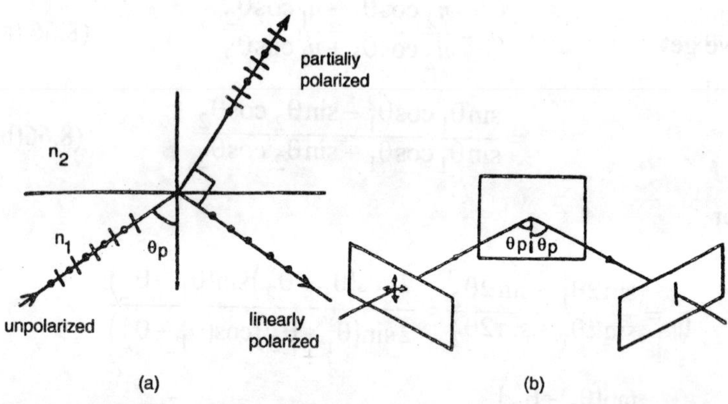

Figure 8.10: ***When an unpolarized beam of light is incident on a dielectric at the polarizing angle (i.e. the angle of incidence is equal to tan^{-1}(n_2/n_2)) then the reflected beam is plane-polarized with its E-vector perpendicular to the plane of incidence. The transmitted beam is partially polarized. The dashed line in (b) is the normal to the reflecting surface.***

This law is widely known as Brewster's law. The corresponding angle of incidence is known as the Brewster angle (or the polarizing angle) and is usually denoted by θ_p. Here, it is important to note that the angle of reflection is equal to $\frac{\pi}{2} - \theta_p$, in which case the Snell's law changes its form as

$$\frac{n_2}{n_1} = \frac{\sin\theta_1}{\sin\theta_2} = \frac{\sin\theta_p}{\sin\left(\frac{\pi}{2} - \theta_p\right)} = \tan\theta_p \tag{8.68}$$

or

$$\theta_p = \tan^{-1}\left(\frac{n_2}{n_1}\right) \tag{8.69}$$

If the angle of incidence in given by $\tan^{-1}\left(\frac{n_2}{n_1}\right)$ then the reflected beam is plane polarized. Further, the transmitted beam is partially polarized. It is easily seen that at the polarizing angle, the reflected ray is at right angles to the refracted ray.

(c) *Phase change on reflection and Stokes' relations*

But if the light is incident on *a* denser medium $\theta_2 < \theta_1$ and then $(\theta_1 + \theta_2) > \frac{\pi}{2}$. In that case $r_{||}$ is negative which indicates a phase change of *n*. However, no such phase change occurs when $\theta_1 < \theta_p$. Discussion on this point will be done in later part of the chapter

Storke's equation is satisfied by both complitude and transmission coefficient.

(d) *Reflection at grazing incidence*

In case of grazing incidence $\left(\theta_1 \approx \frac{\pi}{2}\right)$, we can rewrite equation 8.66 (b) as follow.

$$r_{||} = \frac{\frac{\sin\theta_1}{\sin\theta_2}\sin\alpha_1 - \sin\alpha_2}{\frac{\sin\theta_1}{\sin\theta_2}\sin\alpha_1 + \sin\alpha_2} = \frac{n\sin\alpha_1 - \sin\alpha_2}{n\sin\alpha_1 + \sin\alpha_2} \qquad (8.70)$$

where $n = \frac{n_2}{n_1}$, $\alpha_1 = \frac{\pi}{2} - \theta_1$ and $\alpha_2 = \frac{\pi}{2} - \theta_2$ and at grazing incide-nce both these angles will be small. Now

$$n = \frac{\sin\theta_1}{\sin\theta_2} = \frac{\cos\alpha_1}{\cos\alpha_2}$$

or

$$\sin\alpha_2 = \left[1 - \cos^2\alpha_2\right]^{1/2} = \left[1 - \frac{\cos^2\alpha_1}{n^2}\right]^{1/2}$$

Thus

$$r_{||} = \frac{n\sin\alpha_1 - \left[1 - \frac{\cos^2\alpha_1}{n^2}\right]^{1/2}}{n\sin\alpha_1 + \left[1 - \frac{\cos^2\alpha_1}{n^2}\right]^{1/2}} \approx \frac{n\alpha_1 - \left[1 - \frac{1}{n^2}\right]^{1/2}}{n\alpha_1 + \left[1 - \frac{1}{n^2}\right]^{1/2}} \qquad (8.71)$$

In order to retained terms proportional to α_1 but avoiding terms of higher order, we substituts sin α_1 by α_1 and cos α_1 by 1. Now we get

$$r_{||} \approx -\left[1-\frac{n\alpha_1}{\sqrt{(n^2-1)/n^2}}\right]\left[1+\frac{n\alpha_1}{\sqrt{(n^2-1)/n^2}}\right]^{-1}$$

$$\approx -\left[1-\frac{2n^2\alpha_1}{\sqrt{n^2-1}}\right] \to -1 \text{ as } \alpha_1 \to 0 \tag{8.72}$$

which shows that the reflection is complete at grazing incidence. The transmission coefficient tends to zero. When a glass plate is held horizontal to our eye level, the plate will act as a mirror as the angle of incidence become close to $\frac{\pi}{2}$.

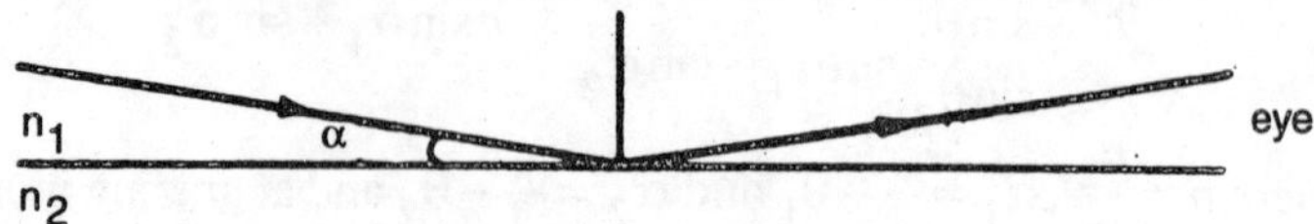

Figure 8.11: ***When light is incident at grazing angle (i.e., α ≈ 0) the reflection is almost complete.***

(e) *Total internal reflection*

If an electromagnetic wave is incident on a more rarer medium then $\theta_2 > \theta_1$ and the form of Snell's law changes its form as

$$\sin\theta_2 = \frac{n_1}{n_2}\sin\theta_1 = \sqrt{\varepsilon_1/\varepsilon_2}\,\sin\theta_1 \tag{8.73}$$

in the above case we have imagined the media to be non magnetic *i.e.*

$$n_1 = \sqrt{\varepsilon_1/\varepsilon_0} \text{ and } n_2 = \sqrt{\varepsilon_2/\varepsilon_0} \tag{8.74}$$

Clearly $\qquad \sin\theta_2 > 1$

where $\qquad \theta_1 > \theta_c$

$$\theta_c = \sin^{-1}(n_2 / n_1) = \sin^{-1}\sqrt{\varepsilon_2 / \varepsilon_1} \tag{8.75}$$

In the above equation angle θ_c is also known as the critical angle. Now, for non-magnetic media the amplitude reflection coefficient will be given by

$$r_{||} = \frac{\varepsilon_2 \cos\theta_1 \sin\theta_2 - \varepsilon_1 \sin\theta_1 \cos\theta_2}{\varepsilon_2 \cos\theta_1 \sin\theta_2 + \varepsilon_1 \sin\theta_1 \cos\theta_2}$$

$$= \frac{\cos\theta_1 - \sqrt{\varepsilon_1 / \varepsilon_2}\sqrt{1-\sin^2\theta_2}}{\cos\theta_1 + \sqrt{\varepsilon_1 / \varepsilon_2}\sqrt{1-\sin^2\theta_2}} \tag{8.76}$$

$$= \frac{\cos\theta_1 - (\varepsilon_1 / \varepsilon_2)\sqrt{\varepsilon_2 / \varepsilon_1 - \sin^2\theta_1}}{\cos\theta_1 + (\varepsilon_1 / \varepsilon_2)\sqrt{\varepsilon_2 / \varepsilon_1 - \sin^2\theta_1}}$$

$$= \frac{\cos\theta_1 - (\varepsilon_1 / \varepsilon_2)\sqrt{\sin^2\theta_c - \sin^2\theta_1}}{\cos\theta_1 + (\varepsilon_1 / \varepsilon_2)\sqrt{\sin^2\theta_c - \sin^2\theta_1}} \tag{8.77}$$

It shows that $\theta_1 > \theta_c$ means the quantity under square root becomes negative thus

$$\varepsilon_1 / \varepsilon_2\sqrt{\sin^2\theta_c - \sin^2\theta_1}\left(= (\varepsilon_1 / \varepsilon_2)\sqrt{(\varepsilon_1 / \varepsilon_2) - \sin^2\theta_1}\right) = i\gamma \tag{8.78}$$

where γ is a real number. Substituting this in Equation 8.77, we get

$$r_{||} = \frac{\cos\theta_1 - i\gamma}{\cos\theta_1 + i\gamma}, \tag{8.79}$$

and the reflection coefficient will be given by

$$R = |r_{||}|^2 = 1 \tag{8.80}$$

showing that the entire energy is reflected back into the first medium. This is a case showing total internal reflection. We may, however, note two points:

(i) since $r_{||}$ is a complex number, the phase change on reflection lies between 0 and π

(ii) The amplitude transmission coefficient is given by

$$t_{\|} = \frac{2\varepsilon_1 \sin\theta_1 \cos\theta_1}{\varepsilon_1 \sin\theta_1 \cos\theta_2 + \varepsilon_2 \cos\theta_1 \sin\theta_2}$$

which is not zero. It prooves that the field in the rarer medium is not zero

Thermal Raidation

When a body is heated it emits radiant energy, the quantity and quality is temperature dependent. Thus the rate at which an incandescent filament emits radiation increases rapidly with increase in temperature, and the emitted light becomes whiter. If this light is dispersed by a prism, a continuous spectrum is formed. These king of radiation is emitted by the objects which are thick enough to be opaque.

Body absorbed partially the radiation falling upon them. At all temperatures bodies are emitting and absorbing thermal radiation; if they are neither rising nor falling in temperature, this is because as much radiant energy is absorbed each second as is emitted. In order to deal with the simplest possible case of *thermal equilibrium,* consider a cavity whose walls and contents are all at a common temperature. The field of radiation in side an isothermal enclosure, possesses some remarkable simple properties.

Inside an isothermal enclosure, the stream of radiaton is same in all directins. It must be the same at every point inside the enclosure; and it must be the same in all enclosures at a given temperature, irrespective of the materials composing them. Furthermore, all these statements hold for each spectral component of the radiation taken separately.

The above statements can be prooved by showing that if any of the statement is wrong then it is possible to make a device which would not obey 2nd law of thermodynamics. For example, if the stream of radiation traveling west were greater than that traveling north, we could introduce two similar absorbers, one facing east and the other south. Then the observer which absorb stronger radiant stream will become hotter than other. We could, therefore, operate a Carnot engine, using the two absorbers as source and sink, and so could convert heat continuously into work without causing other changes in the system, in violation of the second law. Similarly, by using selective absorber, the radiation of any particular wavelength can be tested.

There is an important relation between the radiation field of an enclosure and the energy emitted by walls. An ideal black surface has the property that it absorbs completely all radiation falling upon it. Radiation leaving a black surface consists entirely of radiation emitted by it. Hence the *stream of radiation emitted by any black surface or body in any direction is the same as the stream of radiation that travels in one direction in an isothermal enclosure at the same temperature.* The amount of energy density produced by *a* black surface in front of itself by radiation of that surface is half as compared to the energy density produced by the same surface in an enclosure due to the fact that in open the radiantion is confined to a hemisphere but in enclosure it travells in all direction.

The phenomenon of blackbody radiation has properties, which have universal character being independent of the properties of any particular material substance. Several questions press at once for an answer. How does the energy density in blackbody radiation vary with the temperature? And what is the spectral distribution of the radiation? Another question arise in mind is thtat to understand the pattern of this partcular distribution by which it is brought into existence by the atomic processes.

During last century it was found possible to obtain further information from thermodynanics without making any assumption as to the atomic process. The method consisted in considering the effect of expanding or contracting an isothermal enclosure and taking. account of the work done on the walls by the radiation in consequence of *radiation pressure*.

Early Radiation Laws

Stephen discovered in 1879 that the power emitted by unit area by the black body can be empiracally related to the absolute temprature of the body which is as follow

$$R_B = \sigma T^4 \tag{8.81}$$

where R_B = blackbody radiant emittance = power radiated per unit area

σ = Stefan's constant = 0.56686×10^{-7} W/ m^2 – °K^4

T = absolute temperature

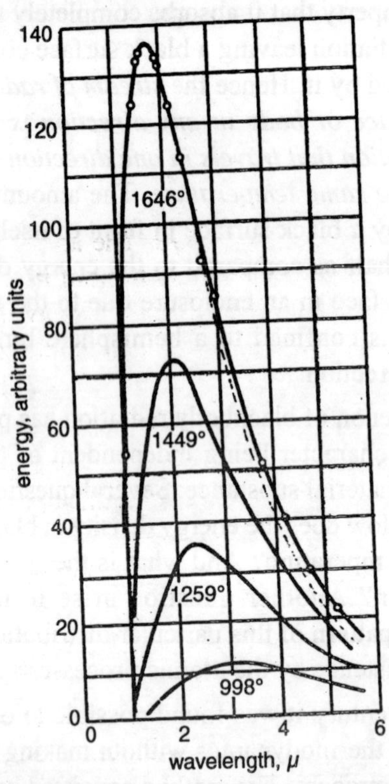

Figure 8.12 : ***Distribution of energy in constant wavelength interval $\Delta\lambda$ for the blackbody spectrum at four Kelvin temperatures.***

On the basis of thermodynamical considerations, Boltzman derived *a* Boltzman-Stephen equation or law. We define the *monochromatic emissive power* e_λ thus: the radiant power emitted per unit area in the spectral range λ to $\lambda + d\lambda$ is $e_\lambda\, d\lambda$ is clearly, the radiant emittance $R = \int_0^\infty e_\lambda\, d\lambda$. e_λ is a function of bath wavelength and temprature. In 1893 Wien showed that for any blackbody

$$e_\lambda = T^5 f(\lambda T) = \lambda^{-5} F(\lambda T) \tag{8.82}$$

where $F(\lambda T) = (\lambda T)^5 f(\lambda T)$. Thus e_λ / T^5 is the same for all blackbodies as is shown in, where the data cf Figure 8.12, obtained in 1899 by Lummer and Pringsheim, are presented in a different form. A single curve serves to represent blackbody radiation at all emperatures. Thus, if e_λ is maximum at the wavelength λ_m for all temprature then the equation is as follows—This is an unique case of wien's displacement law.

$$\lambda_m T = \text{const} = 2.898 \times 10^{-3} \text{ m-°K} \tag{8.83}$$

Now, the problem of blackbody radiation is depends on the determination of *a* single unknown function : $f(\lambda T)$ or $F(\lambda T)$ according to the reasoning based on thermodynamics. *All attempts to obtain the correct form for this function from classical theory failed.*

However, both the formulas based on *a* classical basis nevertheless deserve mention.

By assuming special conditions regarding emission and absorption of radiation, wein derived a formula and he found that the energy density $U \lambda \, d\lambda$ for wavelengths between λ and $d\lambda$ is given by

$$U_\lambda \, d\lambda = \frac{1}{4} c e_\lambda \, d\lambda = C_1 \lambda^{-5} e^{-C_2/\lambda T} \, d\lambda \tag{8.84}$$

where c is the speed of light in free space. By proper choice of the undetermined constants $C_1$and C_2 Wien's radiation law can be made to, fit the data of Figure 8.13 except at higher values of λT, where it predicts too low a value. In 1900 Rayleigh applied the principle of equipartition of energy to blackbody radiation. This led, with a contribution from Jeans, to a formula for e_λ which fitted the high-λT part of the e_λ curve but rose to infinity with decreasing λT. We will now discuss the reasoning used by Rayleigh and Jeans which has applications in modern physics.

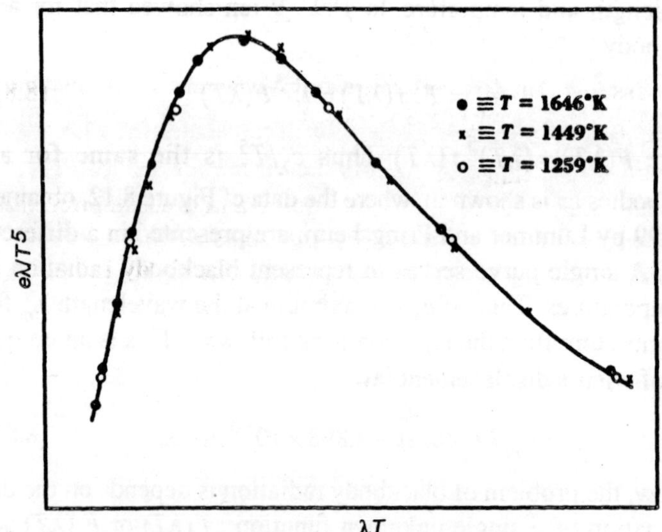

Figure 8.13 : ***Experimental verification of the Wien displacement law for blackbody radiation.***

Degrees of Freedom in an Enclosure

Many modes of vibration can be produced by *a* violin string or organ pipe. We are interested in the number of modes within a given frequency range for a cavity, but it is instructive to apply the method first. to a one-dimensional case. Let a string of great length *L* be stretched between two fixed points *A* and *B*. Standing waves can be set up only for those frequencies which give a whole number of loops between *A* and *B*. As each one loop represent one half of a wavelength, so standing waves occurs only for those wave lengths λ for which

$$\frac{L}{\lambda/2} = n_x = \frac{2L}{\lambda} \tag{8.85}$$

where n_x is an integer.

Now let us take look that how many of these modes have wavelength falling in the range between λ and $\lambda + \Delta\lambda$. If we decrease n_z by some integer Δn_z, the value of λ given by Equation (8.85) increases to $\lambda + \Delta\lambda$, where

$$\frac{2L}{(\lambda + \Delta\lambda)} = n_x - \Delta n_z$$

If we assume that L is very large and $\Delta\lambda$ is very smal, then we can write $\Delta n_z = -2L\Delta\lambda / \lambda^2$, where Δn_z represent the modes with wavelength in the range of $\Delta\lambda$. As the vibrations are transverse and any point on the string is free to move in a plane right angle to the string, so there are two degrees of freedom are associated with each mode. The total number of degrees of freedom per unit length of string in the wavelength range between λ and $\lambda + \Delta\lambda$ is therefore

$$\Delta n_l = \frac{4\Delta\lambda}{\lambda^2} \tag{8.86}$$

The situations, which is two dimensional, are more complex but the principals remain identical. Let us first discuss the system of waves in a square of side L. A set of waves moving initially in direction OM_1 will after reflection move in the direction M_1M_2. After reflection at M_2 the direction of propagation is M_2M_3, which is parallel but opposite to $OM_{1,}$ etc. Four directions of motions are possible for this group of waves, these directions are $\pm\ OM_1$ and $\pm\ M_1M_2$.

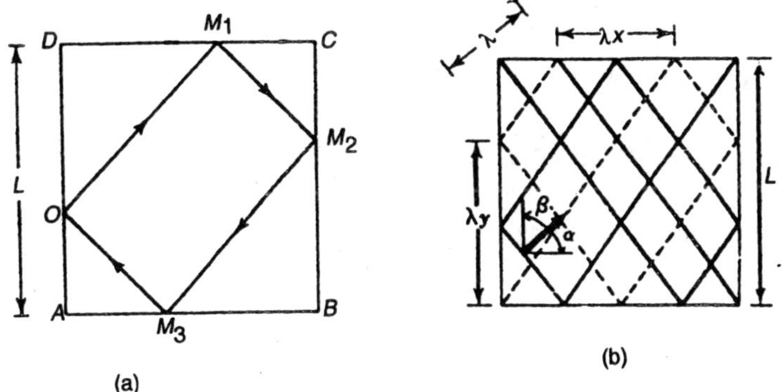

Figure 8.14 : (a) Reelection of waves inside a square.
(b) Standing waves in the xy plane of a cube.

All these four waves combine to produce a group of standing waves but this is possible only when in both n and y directions there is a whole number of half waves. Consider a cube of side L with perfectly reflecting walls in which there are electromagnetic waves polarized with the electric intensity in the z direction. Since the z component E_z of the electric intensity must be zero at the bounding planes when $x = 0$ or L and $y = 0$ or L, the allowed solutions of the wave equation are of the form

$$E_z = f(z, L)\sin\frac{n_x\pi}{L}x\sin\frac{n_y\pi}{L}y$$

where n_x and n_y are integers; this is the equation for a standing wave. The condition that there be nodes at the walls of puts a limitation on the allowed wavelengths; specifically, the projection of any side along the propagation direction must be an integral number of half wavelengths. In the above figure *a* thick arrow is representing one of the propagation directions this arrow makes angle $\alpha = 37°$ with the axis. Four half wavelengths associated with the z direction and three with y direction are present in this case.

From the figure it is clear that $\lambda_z = \lambda / \cos\alpha$ and

$$\lambda_y = \frac{\lambda}{\cos\beta} = \frac{\lambda}{\sin\alpha}$$

Since $\sin(n_x\pi x/L) = \sin(2\pi x/\lambda_x)$, we have $n_x\pi x/L = (2\pi\cos\alpha)/\lambda$; similarly $n_x\pi/L = (2\pi\cos\beta)/\lambda$. If we solve for n_x and n_y, square each and add, we have

$$n_x^2 + n_y^2 = \frac{4L^2}{\lambda^2}\left(\cos^2\alpha + \cos^2\beta\right) = \frac{4L^2}{\lambda^2}\left(\cos^2\alpha + \sin^2\alpha\right) = \frac{4L^2}{\lambda^2}$$

A three dimensional cube can be used for constructing these standing waves. The condition $n_z\pi/L = (2\pi\cos\gamma)/\lambda$ must be applied, and since

$$\cos^2\alpha + \cos^2\beta + \cos^2\lambda = 1, \text{ we find}$$

$$n_x^2 + n_y^2 + n_x^2 = \frac{4L^2}{\lambda^2} \qquad (8.87)$$

With the help of above equation, the number of wavelength which exceed a given value λ_m can be found. This number is equal to the number of possible combinations of the positive integers $n_{x,}$ n_y, n_z which make the left-hand member of less than $4L^2/\lambda_m^2$. To find this number, imagine each set $n_{x,}$ $n_{y,}$ n_z represented by a point on the threedimensional plot of these points lie at the corners of cubic cells of unit length. So the number of values for whom $\lambda > \lambda_m$ is therefore just the number of points inside sphere of which has a radius as $\frac{3L/\lambda_m}{8}$.

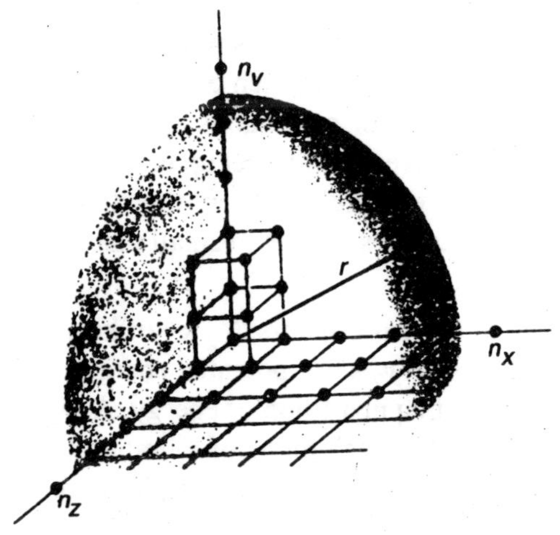

Figure 8.15: ***The Rayleigh scheme for counting the number of allowed values of n_z, n_y, n_z for which $n_x^2 + n_y^2 + n_z^2$ is less than r^2.***

$$n = \left(\frac{4\pi}{3}\frac{8L^3}{\lambda_m^3}\right)\frac{1}{8} = \frac{4\pi}{3\lambda^3}L^3$$

By differentiating the above equation we will find the number of wavelength in a range from λ to $\lambda/d\lambda$: so

$$dn = \frac{4\pi\, d\lambda}{\lambda^4}(\text{vol of cube}) \tag{8.88}$$

where both dn and $d\lambda$ are taken as positive for. convenience. There are two polarization per allowed wavelength for electromagnetic waves. And so we finally obtain for the number of degrees of freedom per unit volume for standing waves in the enclosure with wavelength from λ to $\lambda + d\lambda$ or frequency from v to $v + dv$

$$g(\lambda)\,d\lambda = \frac{8\pi\, d\lambda}{\lambda^4} \quad g(v)\,dv = \frac{8\pi v^2 dv}{c^2} \tag{8.89}$$

The Rayleigh-Jeans Radiation Law

Inside an isothermal enclosure, the quantity and spectral characteristics of a radiation remain constant because the any amound of energy absorbed is again radiated back. The result is the same as if all the energy were reflected at the walls. Rayleigh and Jeans assumed that oscillators in the wall absorbed and emitted radiation constantly, with each oscillator having its own characteristic frequency. For continuous operation of any given oscillator, standing waves must be set up in the enclosure. If the size of enclosure is reasonable, then the differences between neighboring frequencies appear so small that the radiation seems to be continuous.

Assigning of these oscillators, Y_2 k+ of kinetic energy per degree of freedom Mus y_2 k+ of potential energy, is the requirement by the principal of equipartition of energy. Assignment of an average energy of kT to each mode of vibration leads to an energy density $U_\lambda\, d_\lambda$ for waves with wavelength between λ and $\lambda + d\lambda$ given by

$$U_\lambda\, d\lambda = kTg(\lambda)\, d\lambda = 8\pi kT\lambda^{-4} d\lambda \tag{8.90}$$

The above equation represent the formula given by Rayleigh for the black body radiation. It will be noted that it contains *no new constants*. At *long* wave*lengths this formula agrees with observation,* but near the maximum in the spectrum and at short wavelengths it gives much too large values. Furthermore, it assigns no maximum at all to U_λ. Thus, U *i.e.* the energy density inside the enclosure remain unlimited *i.e.* infinite as there is no lower limit for wavelength and $U = \int_0^\infty U_\lambda d_\lambda \to \infty$. In reality, the radiation density in an isothermal enclosure is ordinarily much less than the energy density due to thermal agitation of the molecules of a solid body; e.g., at 1000°K U is about 7.5×10^{-4} J/m^3, as compared with a total density on the order of 10^9 J/m^3 in iron. At temperatures above 10^{60} K the comparison would be reversed. Due to presence of infinite energy inside the electromagnetic field, specific heat of all material bodies would also become infinite.

Planck's Investigation of Blackbody Radiation

Max planck was the scientist who first of all gave a correct blackbody formula in 1900. He found a formula in complete agreement with experiment by introducing a radical innovation quite at variance with previous concepts. This was the first step in the history of quantum theory.

The electromagnetic theory was confirmed by experiment per formed by Hertz, after which Planck was convinced that the key point to the black body radiation would be found in the laws governing absorption and emission of radiation by electric oscillators. We may imagine that the walls of an isothermal enclosure contain oscillators of all frequencies, essentially similar to the hertzian oscillator, and that the emission and absorption of radiation by the walls are caused by these oscillators.

When Planck investigated with the help of the electromagnetic theory, he concluded that an oscillator would affect, in due course of time, only radiation of the same frequency. In the state of equilibrium, there would be a definite ratio between the density of radiation of any frequency ν and the average energy of the oscillators of that frequency. The problem of the blackbody spectrum was thus reduced to the problem of the average energy of an oscillator at a given temperature. Then, if Planck considered the value of the average energy to be kt, as suggested by the equipartition energy formula, then he would return back to the formula of Rayleigh and Jeans.

But he dismissed the principle of equipartition of energy. On the basis of a different assumption he was led at first to Wien's formula for the radiation density, Equation (8.4). In 1900 new measurements of the blackbody spectrum by Lummer and Pringsheim and by Rubens and Kurlbaum showed definitely that

Wien's formula is not correct. Then Planck modified the weien's formula empirically and the modified form suited the observations. Then he sought to modify the statistical theory of the distribution of energy among a set of oscillators so that the theory would lead to his new formula. He succeeded in doing so only after making a new assumption that broke drastically with classical principles. A brief description of the statistical theory is needed in order to understand his assumption.

Distribution of Energy among Oscillators in Thermal Equilibrium

Let as assume a linear oscillator which perform simple harmonic motion (SHM) and the mass of the oscillator be *m*, its displacement be *x* and its momentum be $p = m\,dx/dt$. The potential energy of the oscillator can be written in the form $\frac{1}{2}bx^2$, *b* denoting the force constant; its kinetic energy is $\frac{1}{2}m(dx/dt)^2 = p^2/2m$.

So, the total energy of the oscillator is given by

$$\varepsilon = \frac{p^2}{2m} + \frac{1}{2}bx^2 \qquad (8.91)$$

and by elementary theory its frequency of vibration ν has the value

$$\nu = \frac{1}{2\pi}\sqrt{\frac{b}{m}} \qquad (8.92)$$

If a number of oscillators are in thermal equilibrium then, the energy of *a* single oscillator differs widely but the energies of the whole group are distributed in energy according the Maxwell-Boltzman distribution. By a contraction of Equation to two phase-space dimensions we find the number of oscillators *dN* with co-ordinates between *x* and *x* + *dx* and momentum between *p* and *p* + *dp* to be

$$dN = N A e^{-\varepsilon/kT}\, dx\, dp \qquad (8.93)$$

where *A* is a proportionality constant such that $\int dN = N$.

As we want to know only the energies related to the oscillators, So we can change the form of the formula. Let us construct a plot on which x and p are taken as cartesian coordinates (Figure 8.17),. Each oscillator is represented by a point on this plot which moves about as the oscillator vibrates. So long as the oscillator is free: from disturbing forces, x and p change in such a way that the energy remains constant: hence the representative point moves along an ellipse given by Equation (8.91; with a fixed value of ε. Let two such ellipses be drawn for slightly different energies ε and ε + Δ*e*. If Δε is very small, then through out the elliptical ring between these ellipses the amount $e^{-e/kt}$ would remain constant. So, if the state of thermal equilibrium is maintained, then the number of oscillators represented by points in this elliptical ring is given as above equation

$$\Delta N = N\,Ae^{-\varepsilon/kT}\iint dx \quad dp \tag{8.94}$$

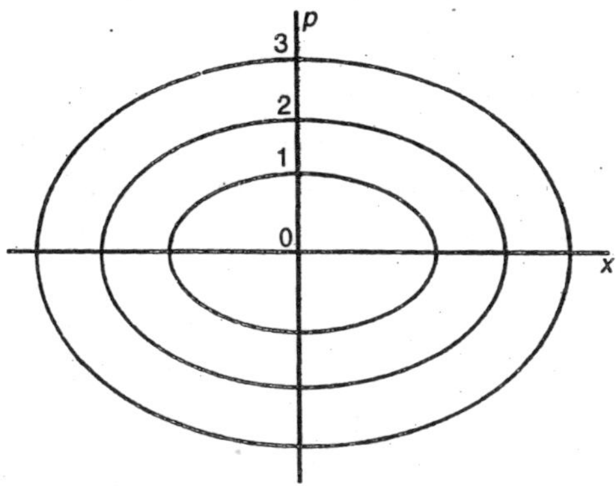

Figure 8.16 : ***The momentum-coordinate plane for a harmonic oscillator.***

The area of the elliptical ring is $\iint dx \quad dp$ and its value can be easily determined in terms of Δ*e*. We first note that the area inside an ellipse whose semiaxes are x_m and p_m is

$$S = \pi p_m x_m \tag{8.95}$$

In the above equation x_m represents maximum value for x and p_m represents maximum value for p during a vibration of an oscillator with energy e, putting the maximum value of x and $p = 0$ at first and maximum value of p and $x = 0$ at 2nd chance we get

$$x_m = \sqrt{\frac{2\varepsilon}{b}} \quad p_m = \sqrt{2m\varepsilon} \quad S = 2\pi\varepsilon\sqrt{\frac{m}{b}} = \frac{\varepsilon}{\nu} \tag{8.96}$$

by (8.92). The area of a ring corresponding to an increment $\Delta\varepsilon$ is thus $\iint dx\ dp = \Delta S = \Delta\varepsilon / \nu$. Hence (8.94) can be written in the form

$$\Delta N = N A_1 e^{-\varepsilon/kT} \Delta\varepsilon \tag{8.97}$$

where $A_1 = A/\nu$.

Adding up the individual energy of each oscillator and dividing the sum with N, we get the average energy of oscillators. Ordinarily this would be done by means of integrals, but for our present purpose we employ discrete sums. Let ellipses centered at the origin be drawn so as to divide the xp plane into rings of equal area h and let these rings, of which the innermost is actually an elliptical area, be numbered 0, 1, 2,... from the center outward. So inside ring number r the total area is $S = rh$; and the energy of an oscillator represented by *a* point on the inner boundary of ring number r is given by

$$\varepsilon = S\nu = \tau h\nu$$

The possible number of oscillators *i.e.* N_τ represented by points on ring τ, is written as

$$N_\tau = N_0 e^{-\tau h\nu/kT}$$

where N_o replaces $NA_1\Delta\varepsilon$. The total energy E of all oscillators is then, approximately,

$$E = \sum_0^\infty \tau h\nu N_\tau = N_0 h\nu e^{-h\nu/kT}\left(1 + 2e^{-h\nu/kT} + 3e^{-2h\nu/kT} + \cdots\right)$$

$$= N_0 h\nu e^{-h\nu/kT}\left(1 - e^{-h\nu/kT}\right)^{-2}$$

since by the binomial theorem the last series is of the form

$$+ 2x + 3x^2 + \cdots = (1 - x)^{-2}$$

Similarly, it can be written

$$N = \sum N_\tau = N_0\left(1 + e^{-h\nu/kT} + \cdots\right) = N_0\left(1 - e^{-h\nu/kT}\right)^{-1}$$

For the *average energy* per oscillator we have then, finally.

$$\bar{\varepsilon} = \frac{E}{N} = \frac{h\nu}{e^{h\nu/kT} - 1} \tag{8.98}$$

This expression represents also the energy of a particular oscillator averaged over any length of time that is not too short. Due to the fact that the oscillators are similar, they have same average energy which must be *E*/*N*.

If becomes essential to let $h \to 0$ in classical theory. The approximations made then disappear. Using the series $e^z = I + x + x^2/2 + \ldots$, we find that, to the first order in *h*,

$$e^{h\nu/kT} - 1 = \frac{h\nu}{kT}$$

Hence, in the limit as $h \to 0, \varepsilon = kT$. So result of average energy in this equation is equal to the equipartition of energy and this result leads us back to Rayleigh-Jeans equation.

Planck's Quantum Hypothesis

Planck introduced a new assumption which is equivalent to keeping *h* finite in following formulas. In the first formulation of the new theory, Planck assumed that the oscillators associated with a given ring all have the energy proper to the inner boundary of that ring. Then Equation (8.99) for ε holds exactly. Depending upon this theory it can be said that energy of an oscillator cannot differ continuously, rather it must take on one of the discrete set of values 0, *hν*, 2*hν*. . ., *rhν*. . .. The actual original form of Planck's assumption was that the energy of the oscillator must always be an integral multiple of a certain quantity ε_0 but he then showed that for oscillators of different frequencies, ε_0 must be proportional to *ν* if the radiation law is to harmonize with the Wien displacement law. Thus he assumed that $\varepsilon_0 = h\nu$, where *h* is a constant of proportionality. Later, planck discovered the relation between h and area on the xp plane on the oscillator.

If is important to note that the assumption of *a* discrete set of posible energy values is not in agreement with classical ideas. According to this assumption, if the energies of a large number of oscillators were measured, some might be found to have zero energy, some $h\nu$ each, others $2h\nu$, and so on. But not a single oscillator would be found which had energy, say, $1.73h\nu$. Therefore, it is said that if the energy of a oscillator changes then it must change suddenly and discontinuously. According to the older conceptions, the interchange of energy between two systems, e.g., between one gas molecule and another or between radiation and oscillators, is a continuous process, and the energy of an oscillator would likewise vary continuously. Such continuity of energy values is demanded by classical physics. For example, the electric and magnetic vectors in a light wave may have any values whatsoever, *from zero up;* and, accordingly, the wave may have any intensity, from zero up. So, the process of emission and absorption of energy by the walls of enclosure should be perfect and continuous.

Some serious challenges were offered, by the absorption and emission of radiation, to the new theory. If the energy of an oscillator can vary only discontinuously, the *absorption and emission of radiation* must be *discontinuous processes,* As long as the oscillator remains in one of its *quantum states,* as we now call them, with its energy equal to one of the allowed discrete values, it cannot be emitting or absorbing radiation according to the laws of classical physics, for then the conservation of energy would be violated. However, it is contrast to the classical theory, which needs an isolated and accelerated electric charge to radiate energy.

Planck suggested a new theory, according to which the emission of energy is possible only when oscillators jump from one energy level to another; if the oscillator travells from a energy level towards a low nergy level then the energy $h\nu$, lost by the oscillator will be emitted in form of a short pulse of radiation. *Absorption* was also assumed at first to be discontinuous. An oscillator, Planck assumed, can absorb a quantum $h\nu$ of radiant energy and jump up to its next higher energy level. However, there were also some difficulties faced by this assumption. For the quantum of radiant energy emitted by an oscillator, according to the classical wave theory, would spread out over an ever expanding wavefront, and it is hard to see how another oscillator could ever, gather this energy together again so as to absorb it all and thereby acquire the energy for an upward quantum jump. So, according to Planck's theory, the process of absorption seem impossible.

Due to such difficulties, Planck modified this theory. In his new theory absorption in a continuous process become possible, but the process of emission remain discontinuous. The energy of an oscillator could then take on all values, as in classical theory; but every time the energy passed one of the critical values $\tau h\nu$, there was assumed to exist a certain chance that the oscillator would jump down to a lower energy level, emitting its excess energy as a quantum of radiation. This came to be known as the second form of Planck's quantum theory. According to this theory, the oscillators remain distributed evenly on each ring on the p^n plane, instead of being all on the inner boundary the average energy of all oscillators given by

$$\bar{\varepsilon} = \frac{h\nu}{e^{h\nu/kT} - 1} + \frac{1}{2}h\nu \qquad (8.99)$$

At this point, it seems quite confusing to determine the essential assumption of Planck's theory. This confusion can be no worse than that which existed in the minds of most physicists in the year, say, 1911. The situation was made still more puzzling. By the success of Einstein's theory of the photoelectric effect the situation was made still more puzzling; for Einstein assumed not only that radiation came in quantized spurts but that each spurt was closely concentrated in space, contrary to the wave theory. While moving ahead, the confusion become more complex but later on it become clear cut to laid a path leading straight to the goal.

One of our main aim is to know exactly how the theory become clear of all confusions. It will he found that the following two new ideas introduced by Planck have been retained permanently and form a part of modern wave mechanics.

1. *An oscillator or any similar physical system has a discrete set of possible energy; values or levels; energies intermediate between these allowed values never occur.*

2. *The emission and absorption of radiation are associated with transitions, or jumps, between two of these levels, the energy thereby lost or gained by the oscillator being emitted or absorbed, respectively, as a quantum of radiant energy,* of *magnitude $h\nu$, ν being the frequency of the radiation.*

When the theory of wave mechanics was used in the electromagnetic field itself, the difficulties will be overcome.

If should be noted that Planck's assumption were not made on the extension of classical physics. Quite the contrary; they represented an *empirical modification* of classical ideas made in order to bring the theoretical deductions into harmony with experiment. If the magnitude of the quantum energy would not be $h\nu$ then the new theory will be in the form of a simple atomicity of energy. Such is not the case, however. Rather, it is the new universal constant h that represents the essentially new element introduced into physics by the quantum theory. As 'h' is playing a vital role in the wide variety of atomic phenomenon, so we should find out the value of 'h'.

Planck's Radiation Law

The new theory, given by Planck, was based upon the interaction between the radiation inside an isothermal enclosure and electric oscillators, which he thought to be present in the wall of enclosure. A more direct and equally satisfactory procedure is to combine Planck's new expression for the mean energy of an oscillator with the analysis of the electromagnetic field by the method of Rayleigh and Jeans, in which the various modes of oscillation of the field inside an enclosure are treated as if they were oscillators. We found that there would be $8\pi \, d\lambda / \lambda^4$ such modes of oscillation or degrees of freedom per unit volume in the wavelength range λ to $\lambda + d\lambda$. Multiplying this number by $\bar{e}$ we will get the following equation.

$$U_\lambda d\lambda = 8\pi \frac{d\lambda}{\lambda^4} \frac{h\nu}{e^{h\nu/kT} - 1}$$

Let us substitute here $\nu = c/\lambda$, c, being the speed of light in vacuum. Thus we obtain, as Planck's new radiation law,

$$U_\lambda = \frac{8\pi ch}{\lambda^5} \frac{1}{e^{ch/\lambda kT} - 1} \tag{8.100}$$

In fact, instead of $\bar{e}$, we should have used its value as the value agrees with the wave mechanics. The effect of this change would be to add in U_λ a term independent of temperature. But this term would be without physical effect as only changes in U are perceptible.

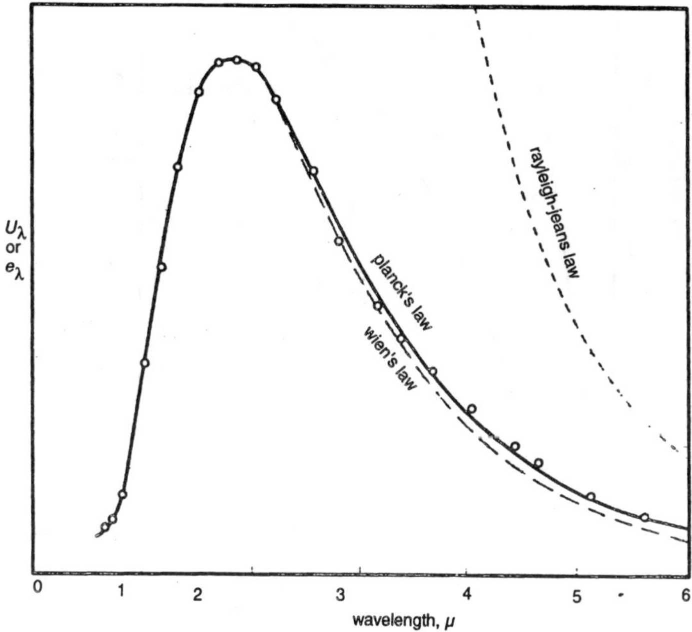

Figure 8.17: ***Comparison of the three radiation laws with experiment at 1600°K.***

When planck's formula applied to *a* spectrum it leads to Wien's formula at one end and Rayleigh - Jeans formula at the other end.

In Figure 8.17 is shown a comparison between the several spectral-energy distribution formulas and the experimental data. The circles show observations by Coblentz on the energy distribution in the spectrum of a blackbody at 1600°K. The reason that the curve for Planck's formula drops below the Rayleigh-Jeans formula is the failure of the classical principle of the equipartition of energy. The oscillation of the electromagnetic field, are of high frequency modes and their average energy should be *kt*, according to classical theory. These oscillations remain in their least possible quantum states and hence contribute little to the density of radiant energy.

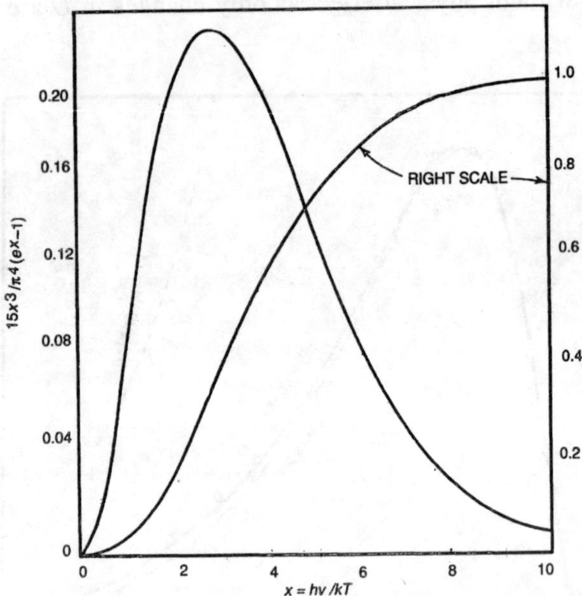

Figure 8.18: ***Spectral distribution*** **Uv** ***of blackbody radiation normalized to make U unity*** **(left scale)** ***and fraction of energy radiated with x (= hv/kT) less than value given on abscissa*** **(right scale).**

We see that the form of the Planck's formula is in agreement with the requirement of the thermodynamics principles can be written as

$$U_\lambda = T^5 f(\lambda T) \quad f(\lambda T) = \frac{8\pi ch}{(\lambda T)^5} \frac{1}{e^{ch/k(\lambda T)} - 1} \quad (8.101)$$

The formula given by Boltzman and Wien displacement law, follow as mathematical consequences. By comparing the observed value of U_λ with the formula, the value of new constant '*h*' can be determined. In his original paper (1901) Planck obtained in this way the value $h = 6.55 \times 10^{-34}$. But the thermal data can be improved in terms of accuracy by using better methods.

$$h = 6.6256 \times 10^{-34} \ \text{J-s}$$

Preferably planck's laws is written in terms of frequency rather than λ. In this case the energy density U_v in the range between v and $v + dv$ is

$$U_v \ dv = \frac{8\pi h v^3 dv}{c^3\left(e^{hv/kT} - 1\right)} \tag{8.102}$$

The total energy density U is

$$U = \int_0^\infty U_\lambda \ d\lambda = \frac{8\pi k^4 T^4}{c^3 h^3} \int_0^\infty \frac{x^3 \ dx}{e^x - 1} = \frac{8\pi^5}{15} \frac{k^4 T^4}{c^3 h^3} \tag{8.103}$$

where $x = ch/\lambda T$ and the integral has the value $\pi^4/15$ Figure shows the normal spectral distribution for a blackbody along with *a* curve showing the fraction of the energy radiated at *n* which is less than the allowed value.

As is shown in previous section, the power radiated per unit area, total, or in a given λ or v interval, is just $c/4$ times the appropriate energy density. So, the power radiated per unit area is $R_a = e\ U/4$.

Classical Radiation Theory

Pressure and Energy Flux Due to Isotropic Radiation

Let us imagine that in a vaccum, the radiation falls normally on the surface of a body. Then, if w is the mean energy density in the oncoming waves, they carry also w/c units of momentum per unit volume. Thus the waves bring up to each unit area of the surface, along with cw J of energy, w units of momentum per second, the momentum as a vector being directed normally toward the surface. If the surface absorbed the waves then it recieves momentum and will experience a pressure equal to w.

Imagine, that the waves are not falling normally rather they are making an angle θ. Then the energy that crosses a unit area drawn perpendicular to the rays (PQ in Figure 8.19) is received by a larger area of magnitude 1/cos 0 on the surface PR. Furthermore, the component of the momentum normal to the surface is less than in the case of normal incidence in proportion to $\cos\theta$. Thus the momentum in the direction of the normal that is delivered to unit area of the surface per second is decreased. by the obliqueness of incidence in the ratio $\cos^2\theta$ on the resulting pressure, if the radiation is entirely absorbed, is

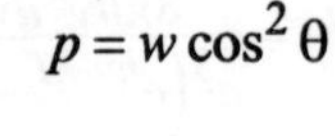

$$p = w\cos^2\theta$$

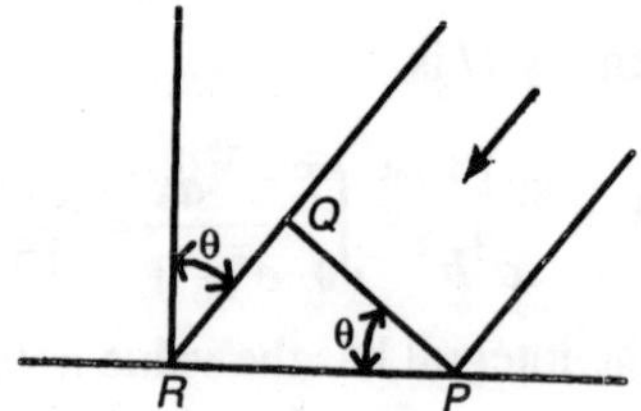

Figure 8.19

The above equation will be used for the pressure caused by emission of *a* beam at an angle θ or for the additional pressure caused by occurence of a reflected beam. If an incident beam is specularly reflected from a surface at the angle of incidence θ, the total pressure on the surface is $2w\cos^2\theta$.

Imagine, now, that like in an isothermal enclosure radiation is streaming towards a surface and also away from it with equal intensities in all directions. Such a distribution of radiation is equivalent to a large number of beams of plane waves, all of equal intensity, with their directions of propagation distributed equally in direction. Let there be N beams in all, and let the energy density due to any one of them be w. So, the pressure p and the total energy density U just in front of the surface, are given as follows

$$U = Nw \tag{8.104)(a}$$

$$p = \sum w \cos^2 \theta = w \sum \cos^2 \theta \qquad (8.104)(b)$$

In order to get the latter sum in above equation imagine lines are drawn out ward from *a* point O on the surface to represent the various directions of the beams wheather moving towards the surface or away from it and then about O as *a* centre draw *a* hemispherical surface of unit radius with its base on the surface. From the hemisphere cut out a ring-shaped element of area QS by means of two cones of semiangle θ and $\theta + d\theta$, drawn from O as apex and with the normal OP as axis. The edge of this element *is* a circle of perimeter $2\pi \sin \theta$, and its width is $d\theta$; hence its area is $2\pi \sin \theta \, d\theta$, whereas the area of the whole hemisphere is 2π. If drawn through O, then the lines of approach of the N beam of radiations will cut the hemisphere equally in points distributed all over the surface. Hence, if we let dN denote the number of these lines that pass through the ring-shaped element, dN will be to N in the ratio of the area of the ring to the area of the hemisphere, so that

$$\frac{dN}{N} = \frac{2\pi \sin\theta \, d\theta}{2\pi} = \sin\theta \, d\theta$$

For all the dN beams, the value of $\cos^2\theta$ remain same. hence their contribution to $\Sigma \cos^2 \theta$ is $\cos2 \theta \, dN$ or, from the last equation, $N \cos^2 \theta \sin \theta \, d\theta$. Thus

$$\sum \cos^2 \theta = \int \cos^2 \theta \, dN = N \int_0^{\pi/2} \cos^2\theta \sin\theta \, d\theta = \tfrac{1}{3} N \quad (8.105)$$

(The limit is $\pi/2$ because directions all around the normal OP are included in the ring.) For the pressure we thus obtain, from Equation (8.104), $p = \frac{1}{3} wN$, or

$$p = \tfrac{1}{3} U \qquad (8.106)$$

It gives that the pressure on the wall of an isothermal enclosure is equal to the one third of the radiant energy density at any point inside the enclosure.

To calculate the total energy brought up to the surface will also be useful. Since half of the N waves are moving toward the surface and each wave delivers energy $cw \cos\theta$, the total energy brought up to unit area per second is

$$\frac{1}{2}N\int_0^{\pi/2} cw\cos\theta\sin\theta\, d\theta = \frac{1}{4}cwN = \frac{1}{4}cU$$

by (8.104*a*). An equal amount of energy is carried away from the surface in an isothermal enclosure. Similarly, it can also be prooved that in an isothermal enclosure with an energy density u, energy $\frac{1}{4}Cu$ per second passes in each direction across unit area of any imaginary surface drawn inside the enclosure.

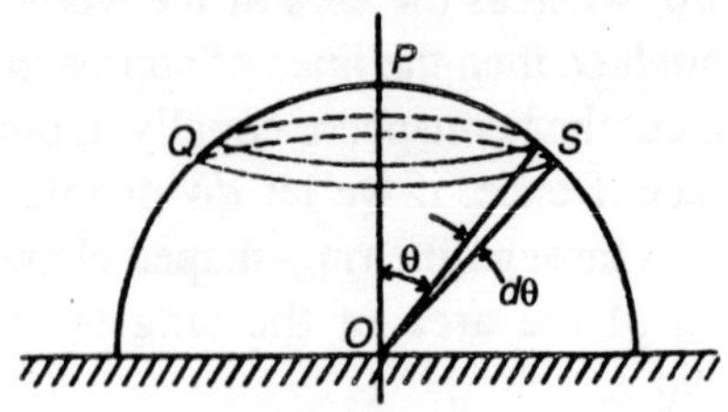

Figure 8.20

Again, the energy emitted per second by unit area of any black surface *i.e.* its radiant emittance R, is related to the energy density U of an isothermal enclosure at the same temprature is given by

$$R = \frac{1}{4}cU \tag{8.107}$$

The same relation holds for each wavelength separately. Imagine that the energy emitted from the blackbody with *a* range of wavelength $d\lambda$ is represented by $e\lambda\, d\lambda$ and the density of radiant energy in the enclosure within the same range of wavelength is denoted by $U_\lambda\, d_\lambda$. thus, we have

$$e_\lambda = \frac{1}{4}cU_\lambda \tag{8.108}$$

The Stefan-Boltzmann Law

Boltzman derived a theoretical law for the variation of the total intensity of black body radiation with temprature. For this purpose, he applied the laws of the Carnot cycle to an engine in which the radiation played the part of the working substance.

He suggested that an ideal carnot engine has an evacuated cylinder with wall impervious to heat, a piston which is also impervious to heat and moving without friction, and a base which permit the transfer of heat. Let the walls, piston, and base be perfectly reflecting except for a small opening O in the base, which can be covered at will by a perfectly reflecting cover. Let this cylinder be placed with the opening O uncovered and opposite an opening in an evacuated isothermal enclosure B_1, which is maintained at temperature T_1 (Figure 8.21). As *a* result the radiation entering through O from B_1 will fill the cylinder until there is the same density U of radiation in the cylinder as there is in B_1 at which time radiation will be passing at the same rate from O to B_1 as from B_1 to O.

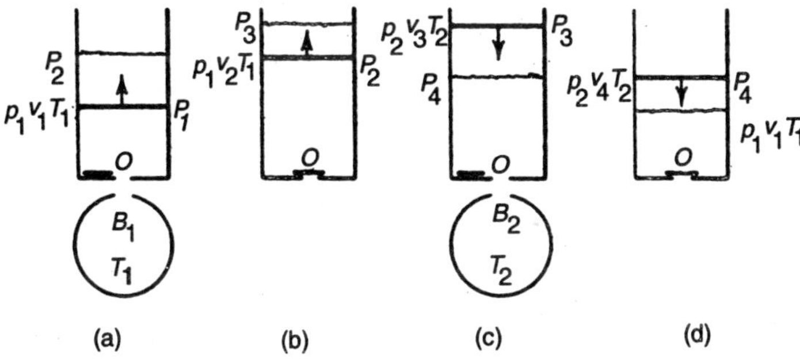

Figure 8.21: ***Boltzmann's radiation engine.***

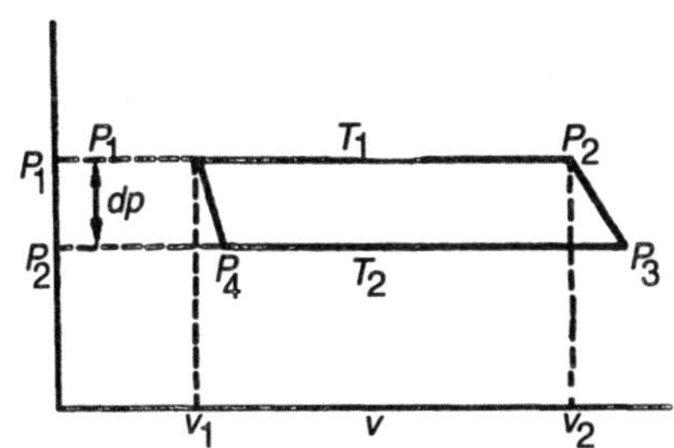

Figure 8.22: ***The p-v diagram for the Carnot cycle of the radiation engine.***

We may now consider the following cycle of events:

Initially the piston is at position P_1 and the initial volume of the cylinder being v_1 and the initial pressure due to the radiation P_1 = 1/3 U_1, we cause the piston to move upward slowly, until position P_2 is reached, the volume increasing to v_2. During this process the radiation density within the cylinder remains constant at U_1. To keep it constant, additional radiation must enter the opening *0* from the enclosure B_1, for two reasons:

(*a*) Work W_e. is done by the radiation on the piston. If T_1 remains constant, so do U_1 and p_1 and

$$W_e = p_1(v_2 - v_1) = \frac{1}{3}U_1(v_2 - v_1)$$

(*b*) The increase of the volume of cylinder from v_2 to v_1 will require an additional influx of energy equal to U_1 (v_2-v_1). Thus the total influx H_1 of radiation from B_1 must be

$$H_1 = \frac{4}{3}U_1(v_2 - v_1) \tag{8.109}$$

In figure 8.22, the horizontal line P_1 P_2 represents the isothermal process on the *p-v* diagram. The energy H_1 is equivalent to heat supplied to the space within the cylinder, just as in an ordinary Carnot cycle the first isothermal expansion is accompanied by an absorption of heat. In order to keep the temprature of B_1 constant, an amount of heat equal to H_1 must also be supplied from an external source to B_1.

As the perfectly reflecting cover is placed over the opening *O* at the time when piston reached P_2 complete thermal isolation of the interior of the cylinder will be achieved and *a* further expansion of position P_2 is made. External work is done, as before, on the piston, the energy required for this external work being supplied by the radiation. *The energy density of the radiation within the cylinder must decrease* from U_1 to some smaller value U_2 party because of this work and partly because of the increase in volume. The pressure, likewise, has decreased. This is obviously an adiabatic process. The line P_2 P_3 of fig. 8.22 is representing it.

At a certain new temprature T_2, the new energy density U_2 is now wqual to the energy density in an enclosure. If the expansion during this second process was very small, we may represent the change in temperature $T_1 - T_2$ by dT and the corresponding change in energy density $U_1 - U_2$ by dU. Since $p = \frac{1}{3}U$, we have then

$$dp = \frac{1}{3}dU \tag{8.110}$$

In above equation *dp* represents the chage in pressure.

At this time the engine is placed opposite to an isothermal enclosure B_2 at temprature T_2, the slide is removed from the opening *O* and by applying force the piston is moved from P_3 to P_4. On account of this compression, there is a tendency for the density of radiation within the cylinder to rise and for radiation to pass through *O* into B_2. The compression is supposed to take place so slowly, however, that the radiation density remains constant at a value only infinitesitimally in excess of U_2. In this isothermal process, radiant energy in amount H_2 leaves the engine.

After the piston is reached at P_4, the opening *O* is closed and the radiation is then compressed adiabatically until the piston comes back to its initial position P_1.

The area $P_1\ P_2\ P_3\ P_4$ of figure 8.21 (c) represents the net externat work done during this cycle. If we assume the change of pressure to have been very small, this area equals $(v_2 — v_1)\ dp$. Calling the net. external work dW, we have, therefore,

$$dW = (v_2 - v_1)dp = \frac{1}{3}(v_2 - v_1)\,dU$$

Thus according to usual rule of carnot cycle.

$$\frac{dW}{H_1} = \frac{T_1 - T_2}{T_1} = \frac{dT}{T_1}$$

and, using the value found for dW and also Equation (8.109), we have

$$\frac{dU}{U_1} = \frac{4dT}{T_1}$$

Thus, dropping the subscript,

$$\frac{dU}{U} = 4\frac{dT}{T}$$

Integrated, this equation gives log $U = 4$ log T + const or

$$U = aT^4 \tag{8.111}$$

where a is a constant, not yet known. From Equation (8.107) we have then also, for the emissive power or radiant emittance

$$R = \sigma T^4 \tag{8.112)(a}$$

$$\sigma = \frac{1}{4}ca \tag{8.112)(b}$$

The above equation represents the Stefan-Boltzman law. According to this law both the energy density of the radiation within an isothermal enclosure and the total emissive power of *a* black body are proportianal to the fourth power of the absolute temprature T.

Reflection from a Moving Mirror

Till now, we have not given our attention to spectral distribution of radiation. The question presents itself, however, whether the same law can be applied also to the separate wavelengths. To know the answer we must know that what happened to spectral distribution of a beam of radiation when it is reflected from a mirror in motion.

First of all, now let us consider the effect of such motion upon *a* monochromatic beam. For this purpose we employ Huygens' principle. In Figure 8.23(a) *MM* represents a mirror moving with a component of velocity V perpendicular to its plane. *AB* and *DE* represent parts of two incident waves which are one wavelength, or a distance λ, apart and are falling on the mirror at an angle of incidence θ; *CB* and *FE* are parts of the same waves which have been reflected and are now leaving the mirror at an angle of reflection θ' a distance λ' apart. From the Figure, it becomes clear that $\lambda = BE \sin\theta$, $\lambda' = BE \sin\theta'$, thus

$$\frac{\lambda'}{\lambda} = \frac{\sin\theta'}{\sin\theta} \tag{8.113}$$

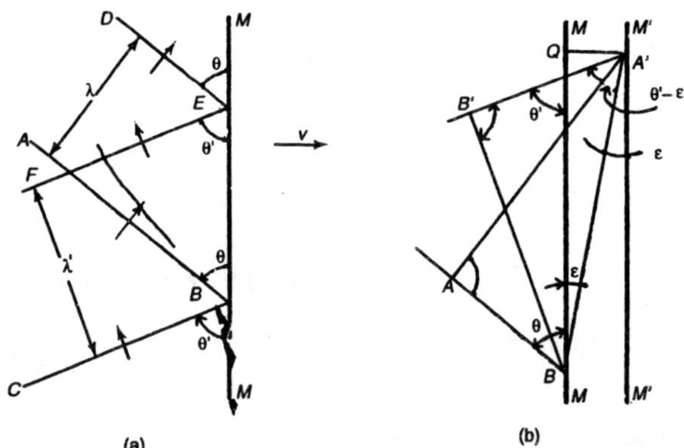

Figure 8.23: ***Reflection from a moving mirror.***

We can get a second relation between λ' and θ' by assuming two successive position of the same wave. In Figure 8.23(b) the part *AB* of a wave is just beginning to fall on the mirror, whose instantaneous position is *MM*. The same portion of the wave at a later instant, after it has been reflected at the angle θ is shown by *A'B'*, the point *A* being now in contact at *A'* with the mirror, which is in the new position *M'M'*. While *A* traveled along the ray *AA'*, *B* traversed *BB'*; hence *AA'* = *BB'*. The angles *BAA'* and *BB'A'*, being angles between ray and wave, are right angles.By similar triangle theorem, it follws that the angle.

$$B'A'B = ABA'$$

or

$$\theta' - \varepsilon = \theta + \varepsilon \tag{8.114}$$

where ε is the angle *QBA'*. Furthermore, while *A* went from *A to A'* at the speed *c* of light, the mirror moved from *MM* to *M'M'* at the speed *u*.So, if *A'Q is* a perpendicular from *A'* to *MM*, then.

$$\frac{V}{c} = \frac{A'Q}{AA'} = \frac{BA'\sin\varepsilon}{BA'\sin(\theta+\varepsilon)}\ \frac{\sin\varepsilon}{\sin(\theta+\varepsilon)} \tag{8.115}$$

By removing ε between the last two equations, the equation

$$\tan\frac{1}{2}\theta' = \frac{c+V}{c-V}\tan\frac{1}{2}\theta \tag{8.116}$$

The above equation is the law of reflection from a moving mirror.

For the change of wavelength Δλ, caused due to reflection by equation 8.113. we have

$$\frac{\Delta\lambda}{\lambda} = \frac{\lambda' - \lambda}{\lambda} = \frac{\sin\theta' - \sin\theta}{\sin\theta}$$

We need the value of $\Delta\lambda$, however, only for an infinitesimal value of V and hence of $\theta' - \theta$. For such a value

$$\sin\theta' - \sin\theta = (\theta' - \theta)\frac{d}{d\theta}\sin\theta = (\theta' - \theta)\cos\theta$$

Also, from Equations (8.113) and (8.114), in which a is an infinitesimal,

$$\theta' - \theta = 2\varepsilon = 2\sin\varepsilon = 2\frac{V}{c}\sin(\theta + \varepsilon)$$

to the first order; and here sin $(\theta + \varepsilon)$ may be replaced by sin θ.Thus, we find the first order in v.

$$\Delta\lambda = 2\frac{V}{c}\lambda\cos\theta \qquad (8.117)$$

Effect of an Adiabatic Expansion upon Blackbody Radiation

Now, lets get back to the sequence of operations we discussed earlier, let us assume the effect of the adiabatic process upon the spectral distribution of the radiation. In this process, blackbody radiation initially at temperature T_1 imprisoned in a cylinder with perfectly reflecting walls is slowly expanded from an initial energy density U_1 to a new energy density U_2. Let the former restriction to a small expansion be dropped. The change in direction of the rays that is produced by the moving piston, according to Equation (8.112), will tend to make the radiation no longer isotropic. We can obviate this inconvenient effect, however, by letting part of the walls of the cylinder reflect perfectly but diffusely. A surface of magnesium oxide does this very well. As all rays strike the diffusing surface repeatedly, then if the expansion is made very slowly, the radiation will be kept effectively isotropic and the pressure on the piston, will be at all times equal to 1/3 U.

Imagine that the length of the cylinder is l and cross section A.

Then, if U is the energy density at any moment, when the piston moves outward a distance dl, work $p\,dv = \frac{1}{3} UA\,dl$ is done on it by the force due to radiation pressure. The enclosed energy is used during this work done, the total amount of the energy is lAU thus

$$\frac{1}{3} UA\,dl = -d(lAU) = -AU\,dl - Al\,dU$$

$$\frac{dU}{U} = -\frac{4}{3}\frac{dl}{l}$$

and, after integration,

$$\log U = -\log l^{\frac{1}{3}} + \text{const} \qquad 8.118(a)$$

$$U = Cl^{-\frac{4}{3}} \qquad 8.118(b)$$

C denoting a constant.

By using thermodynamic reasoning, it can be shown that an expansion of the type considered here cannot destroy the blackbody property of the radiation. For, at a certain instant, suppose that the expansion has reduced the total energy density to U_2, and let T_2 be the temperature of an enclosure in which the density has this same value. Suppose that in the cylinder there were more radiation per unit volume of wavelengths near some value λ' than at the same wavelengths in the enclosure and less radiation near some other wavelength λ''. By convering the opening in the base of the cylinder, it is possible to cause a little radiation to pass from the cylinder into second enclosure at *a* temprature T'_2 slightly above T_2. The opening of the cylinder must be closed with a plate which tensmit wavelength near λ' but reflecting all others and putting the cylinder into communication with the second enclosure. In a similar way, enough radiation near λ'' could be passed into the cylinder from an enclosure at a slightly *lower* temperature T''_2 to restore the total energy to U_2. Then the radiation could he compressed back to U_1, the changes in U and l and the amount of the work done being just the reverse of these quantities during the expansion. Finally, putting the cylinder again into communication with the enclosure at T_1 we could allow the spectral distribution to be restored to that proper to a blackbody at T_1 but without any net transfer of energy between' cylinder and enclosure, since the total energy density has already been restored to that corresponding to T_1 Thus we should have performed a cyclic operation, the only effect of which is to transfer heat energy from an enclosure at T_2' to one at a higher temperature T_1. However, this does not remain constant with second law of thermodynamics.

During any slow adiabatic expansion or compressin the black body radiation must remain black. Its density and temperature, however, decrease. If we commine the last (equation 8.111) with equation we will get

$$T \propto \frac{1}{l^{\frac{1}{3}}} \tag{8.119}$$

It is necessary to find the average rate at which wavelength is increased in order to determine the effect on spectral destribution. Suppose, first, that the walls of the cylinder and the piston reflect specularly. Then any ray preserves its angle of inclination to the axis of the cylinder θ in spite of repeated reflections and has, therefore, a constant component of velocity c cos θ perpendicular to the piston. *v* being the speed of the piston, the ray strikes the piston $(c \cos \theta)/2l$ times a second and its wavelength increase each time by $(2 v \lambda \cos \theta)/c$. Thus its wavelength increases at a rate

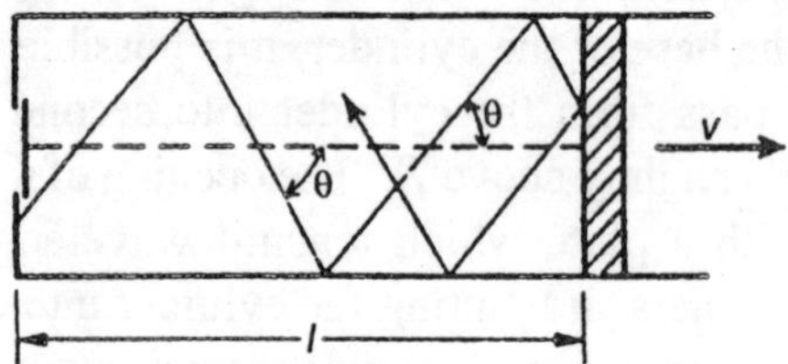

Figure : 8.24

$$\frac{d\lambda}{dl} = \frac{c \cos \theta}{2l} \frac{2V\lambda \cos \theta}{c} = \frac{V\lambda}{l} \cos^2 \theta$$

According to equation 8.105, if the rays of wavelength λ are equally distributed in directions, the average value of $\cos^2\theta$ for all of them is 1/3.

$$\frac{d\lambda}{dt} = \frac{V_\lambda}{3l} = \frac{\lambda dl}{3ldt}$$

since $V = dl/dl$.

We can assume, that there is a diffusely reflecting spot on the walls in order to simplily the caloulations. Then all the rays will take turns moving in the various directions, and all rays of wavelength λ will undergo the average change of wavelength just calculated. By integrating, thus, the last equation we have

$$\frac{d\lambda}{\lambda} = \frac{1}{3}\frac{dl}{l} \tag{8.120)(a}$$

$$\log\lambda = \log l^{\frac{1}{3}} + \text{const} \tag{8.120)(b}$$

$$\lambda \propto l^{\frac{1}{3}} \propto \frac{1}{T} \tag{8.120)(b}$$

by (8.118). It is concluded that each spectral component of black body radiation must change in wave length in such a way that $\lambda \propto \frac{1}{T}$.

The Wien Displacement Law

Lit us consider *a* particular spectral range from λ_1 to $\lambda_1 + d\lambda_1$ containing energy $U\lambda_1\ d\lambda_1$ per unit volume in an enclosure at temperature T_1. An adiabatic expansion which lowers the temperature to T_2 changes the limits of this range, according to the result just reached, to λ_2 and $\lambda_2 + d\lambda_2$, where

$$\frac{\lambda_2}{\lambda_1} = \frac{\lambda_2 + d\lambda_2}{\lambda_1 + d\lambda_1} = \frac{T_1}{T_2} \tag{8.121)(a}$$

Therefore

$$\frac{d\lambda_2}{d\lambda_1} = \frac{T_1}{T_2} \tag{8.121)(b}$$

At the same time the energy in $d\lambda_1$ decreased and in the same ratio as the total energy is; for in the argument that led upto equation 8.118, we might have started with only the radiation in $d\lambda_1$

present in the eylinder. Hence, $U_{\lambda 2}$ being the new value of U_λ

$$\frac{U_{\lambda 2} d\lambda_2}{U_{\lambda 1} d\lambda_1} = \frac{U_2}{U_1} = \frac{T_2^4}{T_1^4}$$

and using (8.121(b)),

$$\frac{U_{\lambda 2}}{U_{\lambda 1}} = \frac{e_{\lambda 2}}{e_{\lambda 1}} = \frac{T_2^5}{T_1^5} \qquad (8.122)$$

In the above equation e_λ is the emissive power of blackbody given in terms of U_λ in equation 8.108.

So, when the values of either U_λ or e_λ at two different tempratures are compared not at the same wavelength but at wavelength inversly proportional to T then their values will be proportional to the fifth power of absolute temprature. This conclusion is known as the *Wien displacement law.* It can also he expressed by saying that at a wavelength varied so that the product λT is held constant, U_λ/T^5 or e_λ/T^5 has the same value at all temperatures. We will obtain a single curve if we plot either of these ratio against λT as abscissa, and this curve will valid at all temprature.

9

Photons

A New Direction

The Einstein's theory of relativity deals with the objects moving at a speed close to speed of light. Among other surprises, Einstein's theory predicts that the rate at which a clock runs depends on how fast the clock is moving relative to the observer: the faster the motion, the slower the clock rate. Theory of relativity has passed every experimental test and with the help of this theory we have explored more satisfying view regarding space and time. Now we will move ahead to explore another extraordinary world *i-e* sub atomic world. You will encounter a new set of surprises that, though they may sometimes seem bizarre, have led physicists step by step to a deeper view the nature of reality.

The theory dealing with subatomic world is called Quantum mechanics which will answer some questions as why do the star shine. Why do the elements exhibit the order that is so apparent in the periodic table? How do transistors and other microelectronic devices work? Why does copper conduct electricity but glass does not? Quantum mechanics includes study of chemistry, including biochemistry, So we have to understand deeply.

A few predictions of quantum theory feel strange even to physicists. Nevertheless, this theory too. has passed every experimental lest-and there have been many—with flying colors. But its prediction stands always right.

Light Waves and Photons

Light is always described as a wave with a wavelenght λ speed

c and frequency f_0.

$$c = \lambda f \tag{9.1}$$

In previous chapter used Maxwell's equations to show that a light wave is an interdependent combination of electric and magnetic fields, each alternating at frequency f. It is well known that visible light is part of electro magnetic spectrum which extends along *a* wide range of wavelengths.

A characteristic property of visible light is proposed by Einstein in 1905. According to this theory when an atom absorbs or emits light energy transfered or travel in *a* small boxes, which are known as photons.

Eeinstein also proposed the energy E, which transferred by *a* single photon, associated with *a* light wave of frequency f as

$$E = hf \text{ (photon energy).} \tag{9.2}$$

Here h is the **Planck constant,** which has the value

$$h = 6.63 \times 10^{-34} \text{ J.s} = 4.14 \times 10^{-15} \text{ eV.s.} \tag{9.3}$$

The Planck constant is the basic constant of quantum mechanics, much as the speed of light c is the basic constant of relativity. If we assume that 'c' is infinite and 'h' is zero then quantum mechanics would not exists.

The Photoelectric Effect

If a beam of light of short wavelength is reflected on a metal surface, then electrons will be released from that metal piece. This photoelectric effect is used in many devices, including TV cameras, camcorders, and night vision viewers. Eeinstein used his theory to explain this photoelectric effect in order to support his theory of photon.

Consider the case shown in fig 9.1 where light of frequency f falls on the metal piece 'T' and electrons are released from the metal by light. A potential difference V is maintained between target T and collector cup C to sweep up these electrons, said to be photoelectrons. A photoelectric current i is produced meter A due to the potential difference.

First Photoelectric Experiment

By adjusting the potential difference V with the help of sliding contact, we will make C negative with respect to T. This potential difference acts to slow down the ejected electrons. We then vary V until it reaches a certain value, called the **stopping potential** V_{stop}, at which the reading of meter A has just dropped to zero. When $V = V_{stop}$, the most energetic ejected electrons are turned back just before reaching the collector. The kinetic energy of the most energetic electron is given by

$$K_{max} = eV_{stop}, \tag{9.4}$$

where e is the elementary charge.

The equation indicates that for light of a given frequency, K_{max} is independent of the intensity of light. Whether the source is dazzling bright or so feeble that you can scarcely detect it (or has some intermediate brightness), the maximum kinetic energy of the ejected electrons always has the same value.

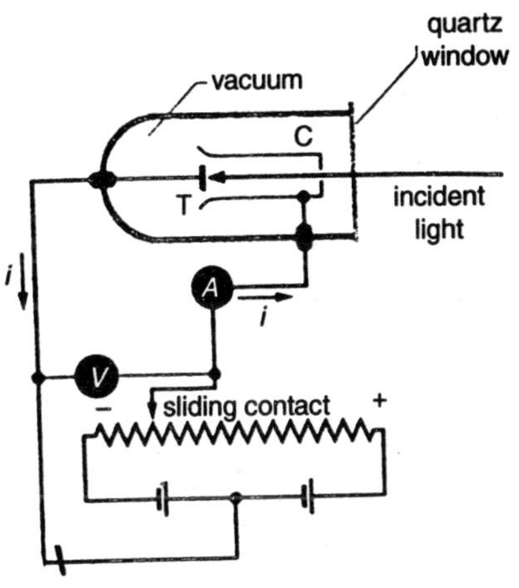

Figure 9.1: ***An apparatus used to study the photoelectric effect. The incident light falls on target T, ejecting electrons, which are collected by collector cup C. The electrons move in the circuit in a direction opposite the conventional current arrows. The batteries and the variable resistor are used to produce and adjust the electric potential difference between T and C.***

Again, this seem quite strange for classical physics. If we view the incident light as a classical electromagnetic wave, we have in mind the image of an electron in the target oscillating back and forth under the influence of the alternating electric field of the incident light wave. Under certain conditions, the oscillating electron will pick up enough energy to break through the surface of the target. If increase the intensity of the incident light beam, we increase the amplitude of the alternating electric field, and it seems reasonable to suppose that this stronger alternating field will give a more energetic "kick" to the ejected electron. *That is not what happens*. If an intence light ray and a table light ray both of same frequency, used one by one then both will produce same K_{max} to knock out electron.

If we think in terms of photons natural result follows. Now the maximum energy that an electron of target T in Figure 9.1 can pick up from the incident light is only that of a single photon. Increasing the light intensity increases the number of photons at the target surface, but the *energy per photon,* given by Equation 9.2, remains unchanged. Thus, K_{max} for an electron remain same.

Second Photoelectric Experiment

Now, we will plot a graph between f *vs* Y_{stop} after measuring V_{stop} for different frequencies f. Photo electric effect will not occur if frequency is less that of cut off frequency of this is also independent of itensity of light.

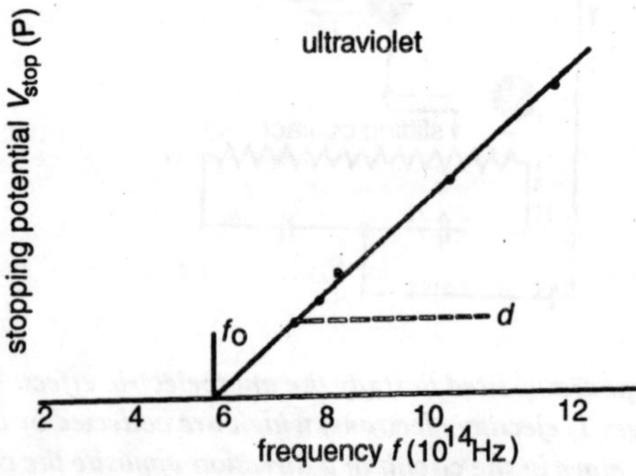

Figure 9.2: ***The stopping potential*** V_{stop} ***as a function of the frequency* f *of the incident light for a sodium target T in the apparatus of Figure 9.1.***

Thin seem like a puzzle for classical physics. If you view light as an electromagnetic wave, you must expect that no matter how low the frequency, electrons can always be ejected if you supply them with enough energy—that be if you use a bright enough light source. *That is not what happens*. The photoelectric effect will not occur if the frequency is below a certain frequency.

As, the energy travels through photons, so cut off frequency is a point which we should expect. The electrons within the target are held there by electric forces. (If they weren't, they would drip out of the target under the influence of gravity!) To just escape from the target. an electron must pick up a certain minimum energy Φ where Φ is a property of the target material called it is work function. If the energy hf transferred to an electron by a photon exceeds the work function of the material. that is, if $hf > \Phi$, the electron can escape through the target surface. The electron will not escape if the energy transfered less that work function.

The Photoelectric Equation

To get the photoelectric equation, Einstein added the results of the two experiment discussed above.

$$hf = K_{max} + \Phi \text{ (photoelectric equation)}, \quad (9.5)$$

which is a statement of the conservation of energy for a single interaction between a photon of frequency f and an electron in a target made of a material with work function Φ. If the electron is to escape from the target, it must pick up energy at least equal to Φ. Any remaining energy $(hf - \Phi$ that the electron acquires from the interaction appears as kinetic energy of the electron. If the electron escape the surface without loosing its kinetic energy then the electron has maximum kinetic energy in it.

By sustituting K_{max} by V_{stop} and a little arrangment in equation 9.5, we get

$$V_{stop} = \left(\frac{h}{e}\right) f - \frac{\Phi}{e}. \quad (9.6)$$

The ratios h/e and Φ/e are constants, so we expect a plot of the measured stopping potential V_{stop} against the frequency f to be a straight line, as in Figure 9.2. Further, the slope of that straight line should be h/e. As a check. we measure ab and bc in Figure 9.2 and write

$$\frac{h}{e} = \frac{ab}{bc} = \frac{2.35\text{ V} - 0.72\text{ V}}{\left(11.2 \times 10^{14} - 7.2 \times 10^{14}\right)\text{ Hz}}$$

$$= 4.1 \times 10^{-15}$$

Multiplying this result by the elementary charge *e*, we find

$$h = \left(4.1 \times 10^{-15}\text{V} \cdot \text{s}\right)\left(1.6 \times 10^{-19}\text{C}\right)$$

$$= 6.6 \times 10^{-34}\text{ J} \cdot \text{s}.$$

This measurment of plank constant is also proved by many other methods.

Photons Have Momentum

Einstein included linear moment in his photon concept by stating that linear momentum also transfered in photons when light interacts with metal.

When the photon is associated with a wave of frequency *f*, then the magnitude of momentum (P) is given by

$$p = \frac{hf}{c} = \frac{h}{\lambda}\text{(photon momentum)} \qquad (9.7)$$

Equation 9.2 ($E = hf = hc/\lambda$) and equation 9.7 tell us, for example, that photons associated with a beam of x rays ($\lambda \approx 50$ pm) have much greater values of both energy and momentum than do photons associated with a beam of visible light ($\lambda \approx 500$ nm $= 5 \times 10^5$ pm.)

At Washington university Arthur Compton performed an experiment in 1923, support the view that both light and momentum transferred via photons. He arranged for a beam of x rays of wavelength λ to fall on a target made of carbon, as shown in Figure 9.3.The wavelength and intensities of the x ray scattered by the target were also meaused by compton.

The experiment and result of compton is shown is figure 9.4 Although there is only a single wavelength ($\lambda = 71.1$ pm) in the incident x-ray beam, we see that the scattered x-rays contain a range of wavelengths with two prominent intensity peaks. One peak is centered about the incident wavelength λ, the other about a wavelength λ′ that is larger than λ by an amount Δλ, which is called the **Compton shift**. The magnitude of Compton shifts depend on the angle at which the scattered x-ray detected.

The experiment shown in figure 9.4 creats another complicated problem for classical physics. If you think of the incident x-ray beam as an electromagnetic wave, you must imagine an electron in the carton target oscillating back and forth under the influence of the alternating electric field of the incident way. The electron will oscillate at the frequency of the alternating electric field and-like a tiny radio transmitting antenna it will radiate *at this same frequency*. The frequency and wavelength of scattered x-ray should be same with the incident light but it never remain the same.

Let us see, first conceptually and then quantitatively, how this quantum mechanical picture leads to an understanding of Compton's results.

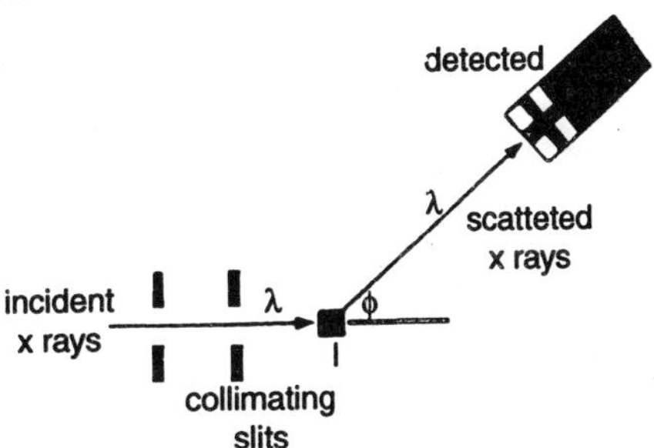

Figure 9.3: ***Compton's apparatus. A beam of rays length λ = 71.1 pm falls on a carbon target T. The x rays entered from the target are observed at various angles direction of the incident beans. The detector measures are the intensity of the scattered x rays and their wavelength.***

Compton Scattering

By classical theory of scattering, electric intensity of the incident waves drives electron in a simple harmonic motion whereas the accelerated electron radiate at the same frequency. Such scattering at the frequency of the incident radiation is observed over the entire electromagnetic spectrum. However, it is weak at high frequencies. It had been proved by Gray and others that scattered

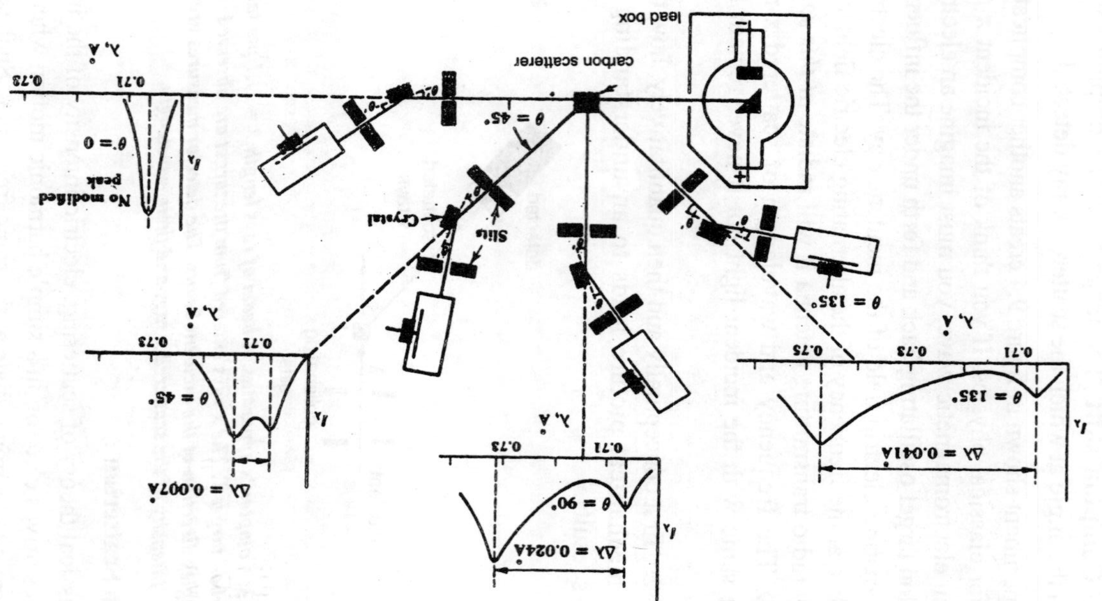

Figure 9.4: ***The scattering of Mo kα x-rays (l = 0.707 Å) at an angle q gives rise to two peaks in the scattered radiation. one at the incident wavelength and the second at a wavelength greater by $\Delta\lambda$ = 0.024 (1 – cos θ) Å.***

x-ray absorbed faster than incident beam. From this we conclude that scattered photons are of lower energy. Compton published result of this experiment in 1923. He meausred the x-ray frequency scattered by carbon atoms upon which monochromatic x-rays wer incident. He found that the scattered beam contained *two frequencies,* one the same as that of the incident beam and the second somewhat lower. On the basis of the quantum theory of radiation, Compton derived a relation which quantitatively predicted scattering at the observed lower frequency. One of the most unique evidence of the particle properties of electromagnetic radius is represented by this Compton effect. This effect supported Quantum mechanics very strongly, and play major role in convincing physicist in favour of Quantum mechanics.

To scatter, Mo $K\alpha$ radiation, Compton used a Graphite block and to measure the wavelength of the scattered photon he used *a* Bragg spectrometer. At each finite scattering angle, he observed two peaks in the scattered beam, one at the incident wavelength 0.707 Å and the second at a wavelength longer by an amount $\Delta\lambda$ dependent on the scattering angle θ, according to the relation $\Delta\lambda = 0.024(1 - \cos\theta)$ Å. The increase in the wavelength $\Delta\lambda$ is independent of the scattering material.

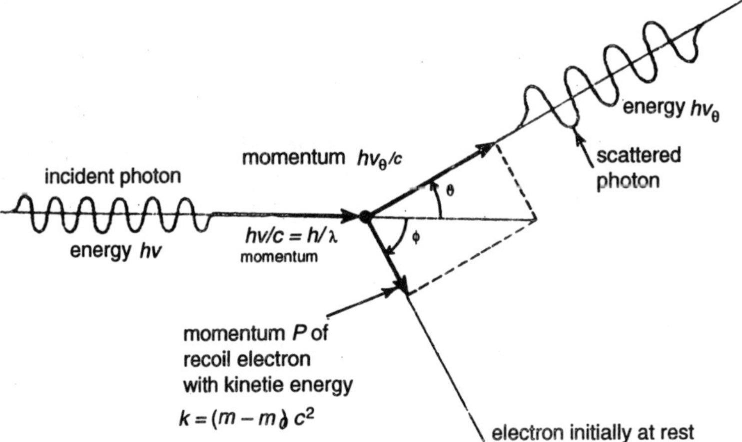

Figure 9.5 : ***Elastic collision of a photon with an electron initially at rest.***

Compton used the quantum picture of radian energy only to explain the occurence of the shifted component. He assumed that the scattering process could be treated as an elastic collision between a photon and an electron, governed by the two laws of mechanics, the conservation of energy and the conservation of momentum. Let an incident photon of energy $h\nu$ collide with an electron initially at rest. The photon is scattered through an angle θ, while the electron recoils in a direction φ (Figure 9.5) The kinetic energy K given to the electron is $(m - m_0)c^2$ According to conservation of energy law, if the frequency of the scattered photon is ν_0, then the sum of the kinetic energy of the electron and the energy of the scattored photon should be equal to the energy of the incident photon.

$$h\nu = h\nu_\theta + \left(m - m_0\right)c^2 \tag{9.8}$$

The amount of momentum carried by *a* photon is equal to its energy content $h\nu$ devided by c, the speed of light. Since momentum is a vector quantity and is conserved, the x and y components (Figure9.5) must obey the equations

$$\frac{h\nu}{c} = \frac{h\nu_\theta}{c}\cos\theta + p\cos\varphi \qquad 9.9(a)$$

$$0 = \frac{h\nu_\theta}{c}\sin\theta + p\sin\varphi \qquad 9.9(b)$$

where p is the momentum of the recoil electron. (The plus sign appears with sin φ because it is assumed all angles are positive when measured counterclockwise from the x axis; thus in Figure 9.5 φ is a negative angle.). By using wavelength of two photons as $\nu/c = 1/\lambda$ and replacing magnitude of p as $\sqrt{1-(\nu/c)^2}$, we get

$$\frac{h}{\lambda} - \frac{h\cos\theta}{\lambda_\theta} = \frac{m_o \nu\cos\varphi}{\sqrt{1-(\nu/c)^2}} \quad \frac{h}{\lambda_\theta}\sin\theta = -\frac{m_0 \nu\sin\varphi}{\sqrt{1-\nu^2/c^2}}$$

Adding the squares of these two equation will eleminates φ

$$\frac{h^2}{\lambda^2} + \frac{h^2}{\lambda_0^2} - \frac{2h^2}{\lambda\lambda_\theta}\cos\theta = \frac{m_o^2\nu^2}{1-(\nu/c)^2} = \frac{m_o^2c^2}{1-(\nu/c)^2} - m_o^2c^2 \tag{9.10}$$

Likewise, by deviding equation 9.8 by *c*, we get

$$\frac{h}{\lambda} - \frac{h}{\lambda_\theta} + m_0 c = \frac{m_0 c}{\sqrt{1-(v/c)^2}} \tag{9.11}$$

and squared to give

$$\frac{h^2}{\lambda^2} + \frac{h^2}{\lambda_\theta^2} - \frac{2h^2}{\lambda\lambda_\theta} + 2m_0 ch\left(\frac{1}{\lambda} - \frac{1}{\lambda_\theta}\right) + m_0^2 c^2 = \frac{m_0^2 c^2}{1 - v^2/c^2} \tag{9.12}$$

Subtracting Equation (9.10) from (9.12) gives

$$\frac{2h^2}{\lambda\lambda_\theta}(1-\cos\theta) - 2m_0 ch\left(\frac{1}{\lambda} - \frac{1}{\lambda_\theta}\right) = 0$$

or

$$\Delta\lambda = \lambda_\theta - \lambda = \frac{h}{m_0 c}(1-\cos\theta)$$

$$= 0.02426(1-\cos\theta)\ \text{Å} \tag{9.13}$$

Thus, the wavelength shift predicted by Compton's theory is in excelient agreement with the experimental results. Partick nature of photon is revealed by the compton scattering.

The Compton wavelength of the electron is represented by the quantity h/m_0c which is equivalent to wavelength of a photon which has energy equal to m_0c^2. this idea can alos applied to other particle. *e.g.* the Compton wavelength of proton is $\frac{\lambda}{M_p c}$.

The change in wavelength *i.e* Compton shift, is dependent on the scattering angle but it does not depend on the energy of photon. On the other hand, the energy difference between incident and scattered photon increases rapidly as the energy of the incident photon is raised, as can be readily found by rewriting Equation 9.13 in the form

$$\frac{1}{h\nu_\theta} - \frac{1}{h\nu} = \frac{1}{m_0 c^2}(1-\cos\theta) \tag{9.14}$$

A modified line same with the wavelength of primary line appears in the Compton experiment. This unmodified line can be explained in terms of the photon concept as the result of a collision of a photon with a bound electron, held sufficiently tightly for an entire atom (or even an entire crystal) to recoil. The shift in wavelength, in these collisions can be obtained by replacing m_o by the mass of recoiling atom in equation 9.13 and this is very small, so impossible to be observed.

The cause, due to which the modified line is more broader than the unmodified line, can be explained as the inifial scattered electrons are not in rest rather they have energies & momentum.

Compton Recoil Electrons

C. T. R. Wilson & Bothe, later, found the predicted recoiling electrons of Compton. The kinetic energy $K = (m - m_0)c^2$ of the recoil electrons is given by Equation 9.8

$$K = \left(m - m_0\right)c^2 = hv - hv_\theta \tag{9.15}$$

If ε represents hv/m_0c^2, it can be shown that

$$K = hv\frac{2\varepsilon\cos^2\varphi}{(1+\varepsilon)^2 - \varepsilon^2\cos^2\varphi} \tag{9.16}$$

$$K = hv\frac{\varepsilon(1-\cos\theta)}{1+\varepsilon(1-\cos\theta)} \tag{9.17}$$

By 1927 Bless, using the magnetic spectrograph, showed that the observed values of K for the recoil electrons are in agreement with the theory. The scattering angle φ for the electron and scattering angle θ for the photon share a relation

$$\cot\varphi = (1+\varphi)\tan\frac{\theta}{2} \tag{9.18}$$

The base for the Compton theory is the assumption that recoiling of electrons takes place simultaneously with the scattering of photon and also that during the process of collision both energy and momentum are conserved. The simultaneity and *conservation* requirements have been tested experimentally with great care by several observers by use of scintillation detectors and coincidence techniques. One detector fixes θ and measures $h_{v\theta}$ for the photon, while a second detector gives φ and K for the recoil electron.

Within experimental error, for every Compton photon scattered an angle θ there is a recoil electron at angle φ given by Equation 9.18 with kinetic energy given by Equation 9.17. The time limit for the simultaneous reaction *i.e.* recoil of electrons and scatering of photon, is less than 2×10^{-8}s in many experiments and in few cases less than 10^{-11}s.

We have made an assumption that the recoil electrons are initialy at rest and unbound and it is possible only when the kinetic energy acquired by recoiling electrons must be greater than the energy with which the electron are binding to its atom. The effect of the binding is to reduce $\Delta\lambda$ slightly. In addition to the Compton process, in which the electron is freed from the atom, there are also *Smekal-Raman* processes, in which the frequency of the scattered radiation is changed by an amount which corresponds.to transition between two bound states of the scattering atom.

At low frequencies, the Compton's modified line is more broader, This breadth can be thought of as caused by the motion of the electrons in the atom. In our simple deduction of the Compton effect, the electron was assumed to he initially at rest: if it assumed to have a component of velocity, positive or negative in the direction of the incident radiation, the wavelength shift is different. A probability distribution for the velocities of the atomic electrons is given by wave mechanics. The broadening can also be calculated from this distribution. As v is then increased further, the Compton line becomes narrower. Eventually, in any direction of scattering other than that of the incident beam, the unmodified line becomes weaker than the Compton line, sooner at large angles of scattering than at small angles, and sooner for heavy atoms than for light ones. At the end, only the Compton's line is in the appreciable intensity.

Electrons and Matter Waves

By assuming symmetry of nature, Pysicist have done quite appreciable work. Thus, when you learn that a changing magnetic field produces an electric field, you might guess as both *Faraday and Maxwell* did-that a changing electric field produces a magnetic field. And it is fact.

Louis de Broglie, a french physicist, made some appeal to symmetry. A beam of light is a wave, but it transfers energy and momentum to matter in photonsized lumps. Why can't a beam of particles have the same properties? So we can give a chance to the thought that why a moving electron or any other particle cannot produce a matter wave.

De brglies give a suggestion that equation $p=\frac{h}{\lambda}$ can also applied to other particles of matter *e.g.* electron. We used that equation in previous section to assingn a momentum p to a photon, knowing the wavelength λ of its associated wave. we now use it, in the form

$$\lambda=\frac{h}{p} \text{ (de broglie wavelength)} \qquad (9.19)$$

to assign a wavelength λ to a particle whode momentum is p. Above equation represents the equation for *De broglie* wavelength of matter.

C.J. Davisson and L.H. Germer of Bell telephone Laboratories and *George P thomson* of the university of A berdeen Scotland verified the theory of *De Broglie* wavelength of moving particle. More recently, the wave nature of a beam of electrons was demonstrated in a 1989 double-slit experiment like that used to demonstrate the wave nature of light. Figure 9.6 suggests how the fringe pattern builds up with time in this experiment as individula electrons strike the detecting screen. Later in 1994, Beams of eodine were used to produce interference fringes. These molecules were 500000 times heavier than electron.

According to the figure 9.7(a) beam of either x ray or electron allowed to fall on the tzarget having alumunium crystals. Scattered by the crystals, the beam emerges from the target with circular symmetry about its initial direction and because of Bragg reflections on the atomic planes of the aluminum crystals, forms concentric rings on a sheet of photographic film placed as indicated. The geometries of the rings are identical, showin that both x rays and electrons behave like waves in this experiment. In the above experiment the reason for choosing the energy of x-ray photons and momentum of electron, is that both have same wavelength.

Let us consider the wave natue of matter for granted. Diffraction studies involving beams of electrons or electrons are used routinely to study the atomic structures of solids and liquids. Matter waves are a valuable supplement to x-ray in such studies. Electrons, for example are penetrating than x rays and so are particularly useful in studying surface features. Also, x-ray interact largely with electrons in a target and for that reason are not effective in locating low-mass atoms—particularly hydrogen—that have few electrons. Neutrons are useful in compare to x-ray because neutrons interact with the of the target atom but x-ray are not.

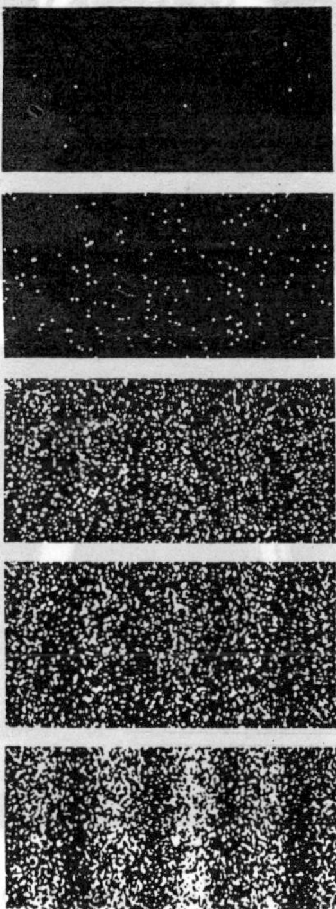

Figure 9.6: *The buildup of an interference pattern by a beam of electrons in a two slit interference experiment like that of figure 39.6 Mater waves. like light waves. are* probability waves. *From top to bottom the approximate numbers of electrons involved are 7,100 3000, 20,000, and 70,000.*

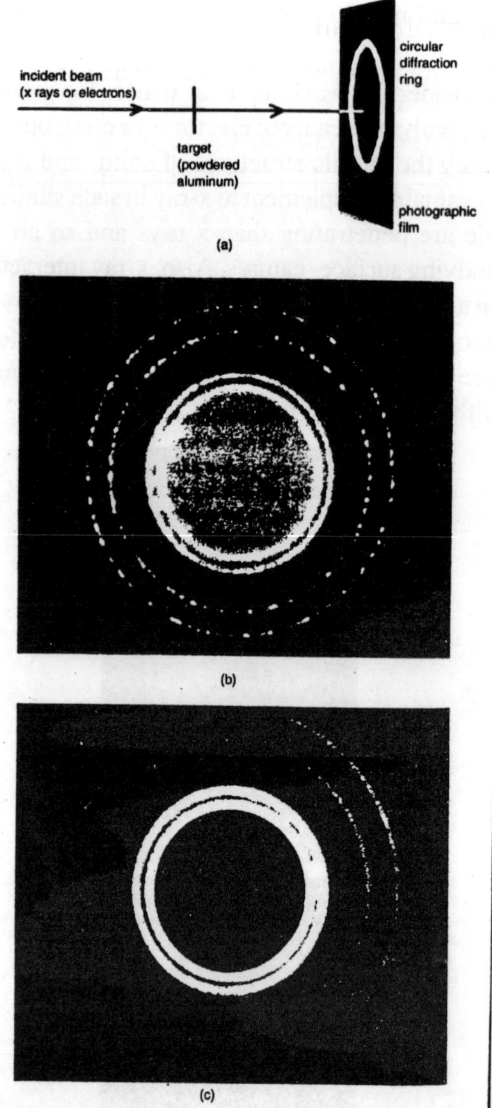

Figure 9.7: ***(a) An experimental arrangement used to demonstrate, by diffraction techniques, the wavelike character of the incident beam. (b) The diffraction pattern when the incident beam is an x-ray beam (light wave). (c) The diffraction pattern when the incident beam is an electron beam is an electron beam (matter wave). Note the basic geometircal identity of the patterns***

Schrodinger's Equation

Any traveling wave, wheather it is a wave on a string, a sound wave or *a* light wave, it is described in terms of some quantity that varies in wavelike fashion. For light waves, for example, this quantity is **E** (*x, y, z, t*) the electric field component of the wave. This value depends on the location and time of observation.

The Schrödinger Equation for a One-electron Atom

The hydrogen atom represents a system in which an electron and proton bonds by electrostatic attraction. Similarly an ionized helium atom and a doubly ionized lithium atom are composed of a positive nucleus and a single electron. By assuming the distance between the particles to be *r*, then the electrostatic potential energy (P) of the system is given by

$$P = -\frac{Ze^2}{4\pi\varepsilon_0 r} \tag{9.20}$$

where Ze = nuclear charge

$-e$ = electronic charge

ε_0 = permittivity of free space

Let us assum the mass of the nucleus located at (*x, y, z,*)is M and mass of electron at (x_2, y_2, z_2,) is *m*. The Schrodinger equation for this two-particle system is

$$-\frac{h^2}{2M}\left(\frac{\partial^2\psi_T}{\partial x_1^2}+\frac{\partial^2\psi_T}{\partial y_1^2}+\frac{\partial^2\psi_T}{\partial z_1^2}\right)$$

$$-\frac{h^2}{2M}\left(\frac{\partial^2\psi_T}{\partial x_2^2}+\frac{\partial^2\psi_T}{\partial y_2^2}+\frac{\partial^2\psi_T}{\partial z_2^2}\right)+P\Psi_T = E\Psi_T$$

where Ψ_T is a function of x_1, y_1, z_1, x_2, y_2, z_2; the subscript *T* of Ψ_T and E_T means total, indicating that they refer to the complete system of nucleus and electron. By replacing the six variables by three coordiantes $n_c y_c$ and z_c of the centre of mass of the system and three co-ordinates of the electron relative to proton by making the substitutions.

$$x_c = \frac{Mx_1 + mx_2}{M+m}$$

$$y_c = \frac{My_1 + my_2}{M+m}$$

$$z_c = \frac{Mz_1 + mz_2}{M+m}$$

$r \sin\theta \cos\varphi = x_2 - x_1$, $r \sin\theta \sin\varphi = Y_2 - y_1$, and $r \cos\theta = z_2 - z_1$. The Schrödinger equation then becomes

$$-\frac{h^2}{2(M+m)}\left(\frac{\partial^2\Psi_T}{\partial x_c^2}+\frac{\partial^2\Psi_T}{\partial y_c^2}+\frac{\partial^2\Psi_T}{\partial z_c^2}\right)-\frac{h^2(M+m)}{2Mm}\left[\frac{1}{r^2}\frac{\partial}{\partial r}\left(r\frac{\partial\Psi_T}{\partial r}\right)\right.$$

$$\left.+\frac{1}{r^2\sin\theta}\frac{\partial}{\partial\theta}\left(\sin\theta\frac{\partial\Psi_T}{\partial\theta}\right)+\frac{1}{r^2\sin^2\theta}\frac{\partial^2\Psi_T}{\partial\varphi^2}\right]-\frac{Ze^2}{4\pi\varepsilon_0 r}\Psi_T = E_T\Psi_T$$

By expressing $\psi_T(x_c, y_c, z_c, r, \theta, \varphi)$ as the product of a function ψ_c of x_c, y_c, z_c alone and a function ψ of r, θ, φ only, this equation can be separated. Upon substituting $\psi_T(x_c, y_c, z_c, r, \theta, \varphi) = \psi_c(x_c\ y_c, z_c)\psi(r, \theta, \varphi)$ in the Schrödinger equation and dividing through by $\psi_c\psi$, it is found that the resulting equation is the sum of two parts, one of which depends only on x_c, y_c, z_c and the other only on *r, r*, θ, φ. Thus each equation should be equal to a constant and the two equations are as follow

$$-\frac{h^2}{2(M+m)}\left(\frac{\partial^2\Psi_c}{\partial x_c^2}+\frac{\partial^2\Psi_c}{\partial y_c^2}+\frac{\partial^2\Psi_c}{\partial z_c^2}\right)=E_{\text{tr}}\Psi_c \qquad (9.21)$$

$$-\frac{h^2(M+m)}{2Mm}\left[\frac{1}{r^2}\frac{\partial}{\partial r}\left(r^2\frac{\partial\psi}{\partial r}\right)+\frac{1}{r^2\sin\theta}\frac{\partial}{\partial\theta}\left(\sin\theta\frac{\partial\psi}{\partial\theta}\right)\right.$$

$$\left.+\frac{1}{r^2\sin^2\theta}\frac{\partial^2\psi}{\partial\varphi^2}\right]-\frac{Ze^2}{4\pi\varepsilon_0 r}=E\psi \qquad (9.22)$$

where $E = Er - E_{\text{tr}}$

The equation 9.21 gives an expression for the motion of the centre of mass, which has *a* mass (M + *m*) and behaves like a free particles. For our purposes the translational energy of the atom is not of interest; E_{tr} can take on any positive value (or zero).

The wave equation of a particle of mass M_m (M + *m*) under the influence of potential function can be written as

$$m_r = \frac{Mm}{M+m} = \frac{m}{1+m/M} \tag{9.23}$$

is just the reduced mass of the electron Indeed, in the limit as *M* goes to infinity, m_r approaches *m* and Equation (9.22) becomes the wave equation of an electron bound to a fixed origin with the potential energy of Equation (9.20).

$$-\frac{h^2}{2m_r}\nabla^2\psi - \frac{Ze^2}{4\pi\varepsilon_0 r}\psi = E\psi \tag{9.24}$$

The Heisenberg Uncertainty Principle

We cannot determined simultaneously both the momentum and position of the particle beyond *a* limit according to the suggestion given by *De broglies* proposal that the motion of a particle with velocity *v* is controlled by the pilot waves with group velocity *v*.

Suppose that we use the two waves to represent the pilot wave for a particle, assuming that the particle is somewhere in the shaded half wave of the envelope. Then the uncertainty in position Δx corresponds to $\lambda_m/2$, from which $\Delta x = \pi/\Delta k$. From $k = 2\pi\lambda = 2\pi p/h$ we have $\Delta k = (2\pi\ \Delta p)/h$. The diffrence between the number of propagation of the two waves in $2\Delta k$ as the uncertainty in the *x* component of the momentum $\Delta pn = (2h\ \Delta k) / 2\pi$ when

$$\Delta x\ \Delta p_x \approx h \tag{9.25}$$

Our assumption of two beating waves of slightly differect frequency is strange because this assumption is far from an optimum way in which to choose the pilot waves for the motion of a particle. It gave us a beat pattern which resembles a row of beads. It is possible to eliminate all but one "bead" by adding other frequencies (or k's) and thus devise a wave packet which can be localized to any Δx we choose. But even when a wave packet of optimum shape is selected, it is impossible to reduce both Δx and Δp_x to zero simultaneously. Indeed, regardless of the wave packet chosen,

$$\Delta x \; \Delta p_x \geq \frac{h}{2} \tag{9.26}$$

This fact was first enunciated in 1927 by *Heisenberg*, who called it the principle of *Unbeslimmlheit*. The ward "Unbestimontheit" is somtimes replaced by the words indeterminacy, indefiniteness or uncertaininty.

This "uncertainity" found by *Heisenberg* is *a* fundamental feature of wave mechanies this uncertainity exists as in valuse of certain mechanical magnitudes such as momentum or position, particle. The indefiniteness in position can be minimized by making the wave packet very small (ψ practically zero except within a very small region); but in that case it can be shown that the packet will spread rapidly because a broad range of propagation numbers is required. Thus, a small packet means a large indefiniteness in momentum and velocity. If we fix the momentum and velocity of particle within narrow limits, there is a large indefiniteness in the position.

In optics, we will find an principle analogus to uncertainity principle. A single sinusoidal wave of light of wavelength λ represents a certain amount of energy that is closely localized in space, but it does not constitute monochromatic light; for, upon passing through a spectroscope, it will spread widely in the spectrum. We must have to use a train of many waves in order to have an approach to mono-chromatic light and this means *a* corresponding dispersal of the energy in space. In general, any wave packet can he expanded in terms of wave trains like just as any patch of light waves can be resolved into monochromatic trains. This amounts to representing ψ by a Fourier integral. When this has been done in a suitable way, the coefficients of the various wave trains in the expansion constitute a probability amplitude for momentum; i.e., the square of the absolute value of any coefficient gives the probability that a suitable observation would reveal the particle as moving in the direction and with the momentum or velocity that is associated with the corresponding wave train. So, a particle does not bear a clearly defined position or *a* clearly defined momentum, at the same time

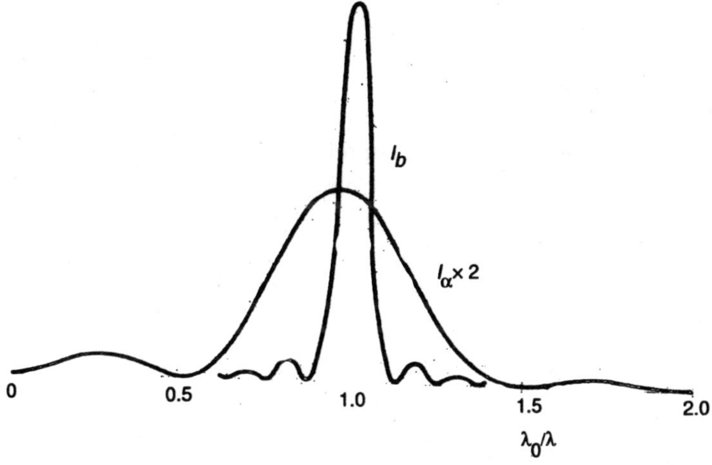

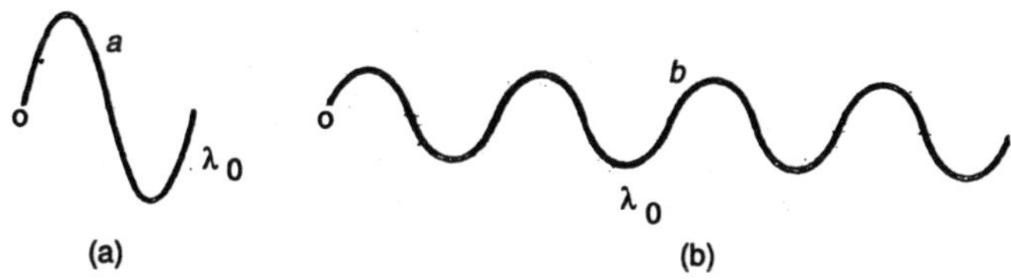

Figure 9.8: ***A single sinusoid (a) of wavelength λ_0 analyzed by a spectrograph has wavelength components over a broad region (curve I_a). The analysis of a four-wave train (b) leads to a much higher peak at λ_0 (curve I_b).***

As the particles cannot have clear momentum and position, so the wave both momentum and position at same time. this conclusion is directly in conflict with the fact that in particle both position and momentum are capable of measurment. *Heisenberg* pointed out, however, that this is possible only because, on the scale of observation used in ordinary physical measurement, the indeterminacy required by Equation (9.26) is so minute as to be lost in the experimental errors.

However the case is different for an electron or *a* molecule. Consider, for example, how an electron might be located with atomic precision. We might use a microscope; but then we should have to use light of extremely short wavelength in order to secure sufficient resolving power. To distinguish positions 10^{-11} m apart, for example, we should have to use γ rays. In such conditions, the effect of light on electron cannot be neglected. If we are to "see" the electron, at least one photon must bounce off it and enter the microscope. In rebounding from the electron this photon will give it a strong Compton kick. Thus at the instant when we locate the electron, its momentum undergoes a discontinuous change. Furthermore, there is an indefiniteness about the magnitude of this change, for it will vary according to the direction in which the scattered photon leaves the scene of action. We cannot limit closely the range of possible directions for the scattered photons that enter the microscope, by stopping down the aperture, without a serious loss of resolving power. When we analyse these conditions it will lead us to uncertanity equation. (9.26).

It again concluded that at the same time an electron or small particles cannot have both definite values of momentum and position. Thus, an assertion that both the position and the momentum of a particle have simultaneously certain precise values is a statement devoid of physical meaning; for, since 1900, it has become increasingly accepted as a principle of physics that only those magnitudes which can be observed, directly or indirectly, have physical significance. Our classical notion of a particle as. something that can move along a sharply defined path, having at each instant a definite position and velocity, is therefore not fully applicable to electrons or protons or atoms or molecules. So, these small bits of matter have properties of particles as well as properties of waves so these are neither absolutetly particles or waves *C.G. Darwin* gave them a name wavicles.

The *Heisenberg* uncertainity principle can be applied to any pair of variable which are canonically conjugate in the hamiltonian formulation of molecules. In addition to Δx and Δp_x we have Δy $\Delta p_y \geq h/2$, Δz $\Delta p_z \geq h/2$ for cartesian coordinates; ΔE $\Delta t \geq h/2$, where ΔE is the uncertainty in energy and Δt that in time; $\Delta\theta$ ΔA_θ $\geq h/2$, where ΔA_θ is the uncertainty associated with the angular momentum A_θ and $\Delta\theta$ the indefiniteness in the corresponding angle; and so forth.

Barrier Tunneling

If someone push a jelly bean along a tabletop on which a book is kept in the way of the jelly bean. You would be very surprised to see the jelly bean appear on the other side of the book instead of bouncing fromm it. Don't expect this to happen for jelly beans. This phenomenon called barrier tunneling happens to electron and other particles with small masses.

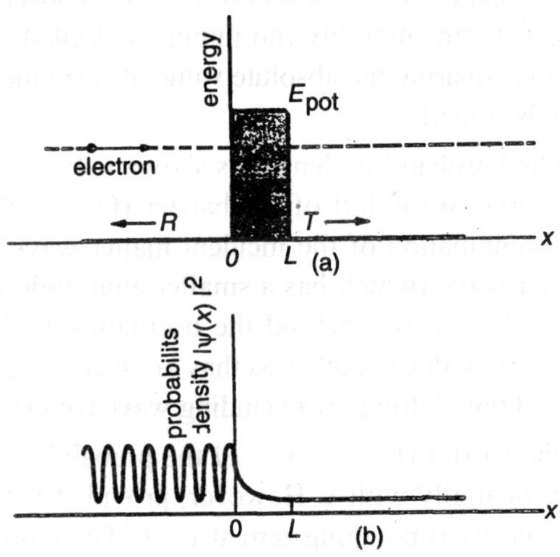

Figure. 9.9: ***(a) An energy diagram showing a potential energy barrier of height E_{pot} and thickness L. An electron with total energy E approaches the barrier from the left. (b) The probability density $|\psi|^2$ of the matter wave representing the electron through the barrier. The pattern to the left of the barrier is a standing matter wave due to the superposition of the incident and reflected matter waves.***

An electron having energy E and moving paralled to x-axis is shown in figurc 9.9(a). Forces act on it such that its pocential energy is zero except when it is in the region $0 < x < L$, where its potential energy has the constant value E_{pot}. This region is defined as *a* potential energy barrier of height E_{pot} and thickness.

As, $E<E_{pot}$, so when an electron moves towards the barrier from left, it wold reflected from the barrier and would move back to the way from which it was coming. Quantum mechanics, however, there is a finite chance that the matter wave associated with the electron will "leak through" the barrier and appear on the other side. Thus, there is a great chance to finding the electron on the far side of the barrier, moving towards right.

By solving Schrodinger's equation separately for the three rigions shown in figure 9.9 (*a*) the wave function ψ (*x*) describing the electron can be found. (1) to the left of the barrier, (2) within the barrier, and (3) to the right of the barrier. The arbitrary constants that appear in the solutions are then chosen so that the values of ψ (*x*) join smoothly (no jumps, no kinks) at $x = 0$ and at $x = L$. After squaring the absolute value of ψ (*x*) the probability density will be found.

The resulted probability density is shown is figure 9.9(*b*). The oscillating curve to the left of the barrier (for $x < 0$) in Figure 9.9(*a*) is a combination of the incident matter wave and the reflected matter wave (which has a smaller amplitude than the incident wave).The reason behind the oscillation is that the two waves, interfere with each other as thay are travelling in opposite directions and thus setting up a standing wave pattern.

Within the barrier (for $0 < x < L$) the probability density decreases exponentially with *x*. However, provided *L* is small, the probability density is not quite zero at $x = L$. The probability density plot of 9.9(*b*) shows *a* transmitted wave with low but constant amplitude to the right of the figure 9.9(*a*), where $0 < x < L$. Thus the electron can be found in this region with an equal but relatively small probability anywhere along the x axis if it happens to tunnel through the barrier.

A transmission co-efficient *T* can be assigned to the incident matter wave and barrier in figure 9.9(*a*). This coefficient gives the probability with which an approaching electron will be transmitted through the barrier, that is, that tunneling will occur. For example—if the value of $T = 0.02$ and we fired 1000 electons at the barrier then 20 of them will tunnel through it and remaining 980 will be reflected.

The **transmission coefficient** T is approximately

$$T \approx e^{-2kL}, \tag{9.27}$$

in which

$$k = \sqrt{\frac{8\pi^2 m\left(E_{pot} - E\right)}{h^2}} \tag{9.28}$$

The value of T is very sensitive to the three variables on which it depends due to the exponential form of equa 9.27. The three variables are particle mass m, barrier thickness L, and energy difference E_{pot}—E.

In technology barrier tunneling has many applications among them the tunnel diode, in which the flow of electrons can be rapidly turned on or off be controlling the barrier height. This can be done very quickly (within 5 ps), so the device is suitable for applications demanding a high-speed response. The 1973 Noble prize was shared be three "tunnelers," *Leo Esaki* (for tunneling in semiconductors), *Ivar Giaever* (for tunneling in supperconductors) and *Brian Josephson* (for the Josephson junction, a rapid quantum switching device based on tunneling). In 1986, *Gerd Binning and Hdinfich Rohrer* were given the Nobel prize for their device scanning tunneling microscope.

The Scanning Tunneling Microscope (STM)

The STM, allows to make detail maps of surfaces. A deviced based on tunneling helps in revealing features on a atomic scale.

A fine metallic tip, mounted at the intersection of three mutually perpendicular quartz rods, is placed close to the surface to be examined. A weak potential is applied between tip and surface. When an electric potential difference is applied across a sample of crystalline quartz, the dimension of the sample change slightlly. This property of crystalline quartz is known as piezoelectricity. This proparty is used to change the length of each of the three rods, smoothly and by tiny amounts, so that the tip can be scanned across the surface (in the *x* and y directions) and also lowered or raised with respect to the surface in the *z* direction.

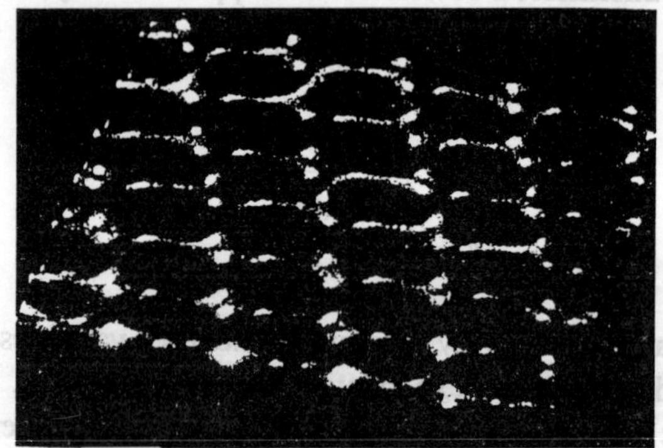

Figure. 9.10: ***The contour of a graphite surface as revealed by a scanning tunneling microscope. The carbon atoms and the hexagonal patterns they form are visible.***

The gap present between tip and surface acts like a potential energy barrier. If the tip is close enough to the surface, electrons from the sample can tunnel through this barrier from the surface to the tip, forming a tunneling current.

When the tip is scanned back and froth an electronic feedback arrangment adjust the vertical position of the tip to keep the tunneling current constant while the operation is going on. This means that the tip-surface separation also remains constant during the scan. The output of the device-for example, these STM are available and used all over the world.

10

Nuclear Physics

Discovering the Nucleus

During the begining of twentieh century the only thing known about the structures of atoms was the fact that they contain electron. In the year 1897, *J. J. Thomson* discovered the electron but in these early days its mass was unknown. Thus it was not possible even to say how many negatively charged electrons a given atom contained. As the electrons were known to be negatively charged and the atoms were neutral, so atoms must also contain some positive charge but their form was not known.

All the *+ve* charge of the atom situated at the centre of the atom forms the nucleus and also it is responsible for the mas of the atom. This was suggested by *Ernest Ruther ford* in 1911. Rutherford's proposal was no mere conjecture but was based firmly on the results of an experiment suggested by him and carried out by his collaborators.

At that time, it was already known that certain radiative elements emit particles spontaneously and transfer themselves into other element. One such element is radon, which emits alpha (α) particles with energies of about 5.5 MeV. Now it is well known that these (α) particles are nuclei of helium atom.

In his experiment, *Rutherford* wanted to bombard the thin piece of foil with α-particles and he wanted to measure the extent of reflection of α-particles by the foil. Those alpha particles which are about 7300 times more massive than electrons, have a charge of +2*e*.

The arrangment of experiment done by *Geiger and Marsden* is shown in Figure 10.1 Their alpha source› was a thin-walled glass tube of radon gas. The experiment involves counting the number of alpha particles that are deflected through various scattering angles ϕ.

The result, shown in Fig 10.2, indicate that the vertical scale is logarithmic. We see that most of the particles are scattered through rather small angles but—

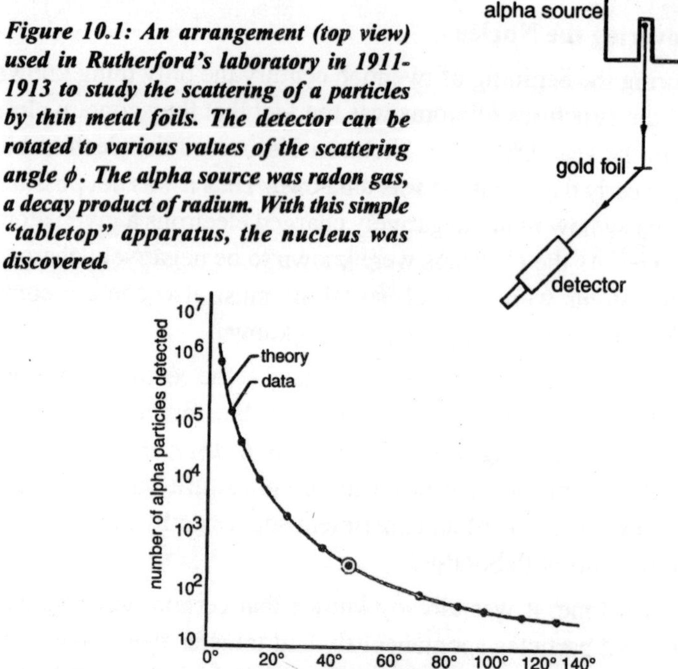

Figure 10.1: An arrangement (top view) used in Rutherford's laboratory in 1911-1913 to study the scattering of a particles by thin metal foils. The detector can be rotated to various values of the scattering angle ϕ. The alpha source was radon gas, a decay product of radium. With this simple "tabletop" apparatus, the nucleus was discovered.

Figure 10.2: The dots are alpha-particle scattering data for a gold foil, obtained by Geiger and Marsden using the apparatus of Figure 10.1. The solid curve is the theoretical prediction., based or the assumption that the atom has a small, massive, positively charged nucleus. Note that the vertical scale is logarithmic, covering six orders of magnitude. The data have been adjusted to fit the theoretical curve at the experimental point that is enclosed in circle.

and this was the big surprise–a very small fraction of them are scattered through very large angles, approaching 180°. In *Rutherford's* words: "***It was quite the most incredible event that ever happened to me in my life.***" It was as impossible as an 18 inch cell is fired at *a* thin tissue paper and the shell is reflected be and hit the person who fired it.

But Rutherford was very much surprised. At the time of these experiments, most physicists believed in the so-called plum pudding model of the atom, which had been advanced by *J. J. Thomson*. In this view the positive charge of the atom was thought to be spread out through the entire volume of the atom. It was thought that the electrom remain vibrating around the fixed points in the sphere of +*ve* charge.

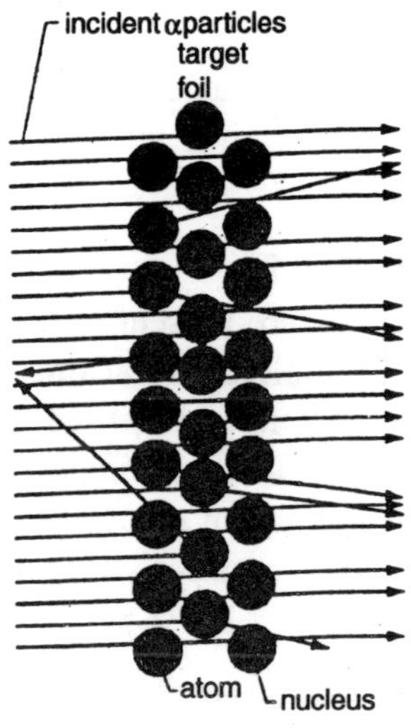

Figure 10.3 : ***The angle through which an alpha particle is scattered depends on how close the particle's incident path lies to an atomic nucleus. Large deflections result only from very close encounters.***

The deflection force acting on α-particles when it travels across the positively charged sphere is too small to deflect the α-particles. The electrons the atom would also have very little effect on the massive energetic alpha particle. So it is presumed that α–particles be themselves strongly deflected.

It was observed by *Rutherford* that there must be a large force to deflect the alpha particle backward. This force would be provided if the positive charge, instead of being spread throughout the atom, were concentrated tightly at its center. So, the attacking α-particles come very close to the +vely charged centre and then there must be a great deflection of +vely charged α-particles from the +vely charged centre.

The path of deflection of α-particles is shown is figure 10.3. As we see, most are either undeflected or only slightly deflected, but a few (those whose incoming paths pass, by chance, very close to a nucleus) are deflected through large angles. From an analysis of the data, *Rutherford* concluded that the radius of the nucleus must be smaller than the radius of an atom by a factor of about 10^4. So the atom has large empty space within it.

Some Nuclear Properties

The properties of atomic nuclei is given in Table 10.1. When the nuclear properties of atomic nuclei is discussed, then these are referred as nuclides.

Some Nuclear Terminology

Protons and Neutrons are present in nucleus. The number of protons in a nucleus (called the **atomic number** or **proton number** of the nucleus) is represented by the symbol Z; the number of neutrons (the neutron number) is represented by the symbol *N*. The mass number of *a* nucleus is represented by total number of proton and neutrons present in it.

$$A = Z + N. \tag{10.1}$$

When we refer to neutrons and protons together then the term nucleons is used.

Nucleides are represented with symbols as given in table 10.1. Consider ^{197}Au for example. The superscript (197) is the mass number *A*. The chemical symbol tells us that this element is gold whose atomic number is 79. By substracting atomic number from the atomic mass of the element, we get the number of neutrons, present in it.

These radionucleides undergo continuous disintegration by

Table 10.1 Some Properties of Selected Nuclides

NUCLIDE	Z	N	A	STABILITY[a]	MASS[b] (u)	SPIN[c]	BINDING ENERGY (MeV/nucleon)
^{1}H	1	0	1	99.985%	1.007 825	½	—
^{7}Li	3	4	7	92.5%	7.016 003	½	5.60
^{31}P	15	16	31	100%	30.973 762	½	8.48
^{84}Kr	36	48	84	57.0%	83.911 507	0	8.72
^{120}Sn	50	70	120	32.4%	119.902 199	0	8.51
^{157}Gd	64	93	157	15.7%	156.923 956	½	8.21
^{197}Au	79	18	197	100%	196.966 543	½	7.91
^{227}Ac	89	138	227	21.8 y	227.027 750	½	7.65
^{239}Tu	94	145	239	24,100 y	239.054 158	½	7.56

[a] for stable nuclides, the isotopic abundance is given; this is the fraction of atoms of this type found in a typical sample of the element. For radioactive nuclides, the half-life is given.

[b] Following standard practice, the reported mass is that of the neutral atoms, not that of the bare nucleus.

[c] Spin angular momentum in units of h.

emitting particles and thus transformed into another element. As it happens, the element gold has 32 isotopes, ranging from^{173}Au to^{204}Au. Only one of them (^{197}Au) is stable, the remaining 31 being radioactive. Thses radionucleides undergo continuous disintegration by ernitting particles and thus transformed into another element.

Organizing the Nuclides

The number of electrons remain same in the neutral atoms of all isotopes of an element, and thus they have same chemical properties. The *nuclear* properties of the various isotopes of a given element, however, are very different. As their periodic properties are also same, so the periodic table is almost useless for the nuclear scientists.

The nuclidic chart can be obtained very easily and it contains the properties of nuclidec. Figure 10.5 shows a section of such a chart, centered on ^{197}Au. Relative abundances are shown for stable nuclides, and half-lives (a measure of decay rate) are shown for radionuclides. The isobars, nuclides of the same mass number, are also shown in these nuclidic charts by stoping lines.

Till 1996, the nuclides had been found with atomic number as high as $z = 112$. Such high-Z nuclides are very unstable and are identified by the products of their radioactive decay. These nuclides are created at very low production rate and these are created in accelerated laboratories.

Nuclear Radii

The unit used for measuring distance on scale of nuclei is femtometer. This unit is often called the fermi; the two names share the same abbreviation. So

$$\text{I femtometer} = 1 \text{ fermi} = 1 \text{ fm} = 10^{-15} \text{ m.} \qquad (10.2)$$

By bombarding the nucleus with electrons and observing the way of deflection of incident electrons the size and shape of the atom can be known. The electrons must be energetic enough (at least 200 MeV) to have de Broglie wavelengths that are smaller than the nuclear structures they are to probe.

The atom has a well defined solid surface but the nucleus hasn't. Furthermore, although most nuclides are spherical, some are notably ellipsoidal. Nevertheless electron-scattering experiments (as well as experiments of other kinds) allow us to assign to each nuclide an effective radius given by

$$R = R_0 A^{1/3} \tag{10.3}$$

in which A is the mass number and $R_0 \approx 1.2$ fm. The volume of nucleus is proportional to the mass number but not proportional to the neutrons and protons forming atomic mass. The volume of nucleus is proportional to R^3.

Nuclear Masses

By using modern mass spectrometer and nuclear reaction techniques atomic masses can be measured with great precision. It is already known that such masses are reported in atomic mass units u, chosen so that the atomic mass (not the nuclear mass) of ^{12}C is exactly 12 u. The relation of this unit to the ***SI*** mass unit is, approximately,

$$1 \text{ u} = 1.661 \times 10^{-27} \text{ kg} \tag{10.4}$$

The mass number A of a nuclide is so named because the number represents the mass of the nuclide, expressed in atomic mass units and rounded off to the nearest integer. The atomic mass of Au 197u is round off figure but its exact or absolute atomic mass is 196.966573u.

The Eeinstein relation *i-e* Q-$\Delta m\ c^2$ is important work *a* day tool. As we saw in this equation is Q the energy released (or absorbed) when the mass of a closed interacting system of articles decreases (or increases) by the magnitude Δm.

It can easily be shown that energy equivalent to 1u is 931.5 *mev*. Thus c^2 can be written as 931.5 MeV/u, and we can use this value to find the energy equivalent (inmillion electron-volts) of any mass or mass or mass difference (in atomic mass units), or conversely.

Nuclear Binding Energies

The energy needed to break the nucleus into its protons and

neutrons is called nuelear binding energy and can be calculated from $Q = \Delta m\, c^2$. If we divide the binding energy of a nucleus by its mass number, we get the binding energy per n\nucleon. Figure 10.6 plots this quantity as a function of mass number. The binding energy per nucleon is low at both high and low mass number. This phenomenon is called drooping.

From this drooping in binding energy curve, we conclude that nucleons are more tightly bound when they are grouped into two middle class nuclides than into a single high-class nuclide. In other words, energy can be released by the **nuclear fission**, or splitting, of a single massive nucleus into two smaller fragments.

Similarly, the drooping of binding energy curve at low mass indicates that when two low class nuclides join together to form a middle class nuclides energy will be released. This process, the reverse of fission, is called **nuclear fusion**. It occurs inside our Sun and other stars and in thermonuclear explosions. Now *a* days much attention is given to controlled nuclear fusion as *a* practical energy source.

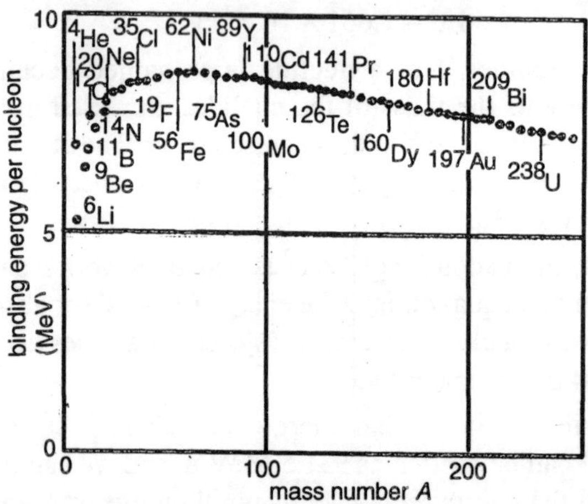

Figure 10.4: ***The binding energy per nucleon for some representative nuclides. The nuclide ^{62}Ni has the highest binding energy per nucleon (8.794 60 ± 0.000 03 MeV/nucleon) of any known stable nuclide. Note that the alpha particle (^4He) has a higher binding energy per nucleon than its neighbors in the periodic table and is thus also particularly stable.***

Nuclear Energy Levels

The energy of nuclei is contained in some quantum sets. That is, nuclei can exist only in discrete quantum states, each with a well-defined energy. Figure 10.6 shows some of these energy levels for ^{28}Al, a typical low-mass nuclide. Note that the energy scale is in millions of electron-volts, rather than the electron-volts used for atoms. The emitted photons typically enters in the gamma ray region of electromagnetic spectrum. When any nucleus emits photons and transit from a higher level to a low level of energy.

Nuclear Spin and Magnetism

An intrinsic nuclear angular momentum and an associated intrinsic nuclear magnetic moment is present in many nuclides. Although nuclear angular momenta are roughly of the same magnitude as the angular momenta of atomic electrons, nuclear magnetic moments are much smaller than typical atomic magnetic moments, by a factor of about 1000.

The Nuclear Force

The electromagnetic force controles the motion of atomic electrons. To bind the nucleus together, however, there must be a

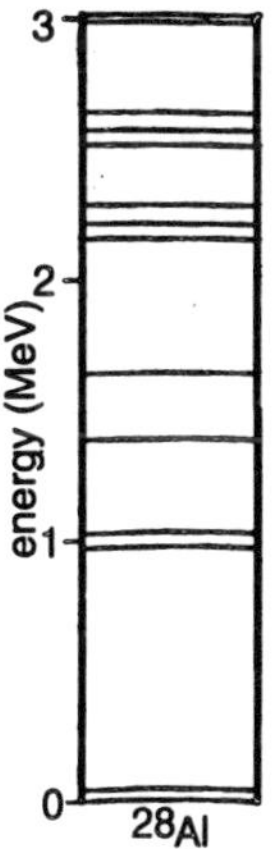

Figure 10.5: ***Energy levels for the nuclide ^{28}Al, deduced from nuclear reaction experiments.***

strong attractive nuclear force of a totally different kind, strong enough to overcome the repulsive force between the (positively charged) nuclear protons and to bind both protons and neutrons into the tiny nuclear volume. As the influence of nuclear force does not exists far beyond the nuclear surface, so the range of nuclear force is very short.

At present, it is believed that the nuclear force binding electrons and protons is a secondary effect of the strong force that binds quarks together to form neutrons and protons. In much the same way, the atractive force between certain neutral molecules is a spillover effect of the Coulomb electric force that acts within each molecule to bind it together.

Radioactive Decay

A large number of the nuclides, found today, are radioactive. These nuclides emit a particle continuosly and transform itself into a new nuclide.

Presence of radioactivity provides support to the view that the laws governing subatomic world are statistical. Consider, for example, a 1 mg sample of uranium metal. It contains 2.5×10^{18} atoms of the very long-lived radionuclide ^{238}U. The nuclei of these particular atoms have existed without decaying since they were created-well before the formation of our solar system.

There is absolutely no way to predict whether any given nucleus in the sample will be among the small number of nuclei that decay during the next second. All have an equal chance. If a sample contains N raqdioactive nuclei than we can express the statistical nautre of the decay process by saying that the rate ($= -dN/dt$) at which nuclei decay is proportional to N:

$$-\frac{dN}{dt} = \lambda N, \tag{10.5}$$

in which λ, the **disintegration constant,** has a characteristic value for every radionuclide. Its *SI* unit is the inverse second (s^{-2}). Equation 10.5 may be integrated to yield

$$N = N_0 e^{-\lambda t} \text{ (radioactive decay)} \tag{10.6}$$

in which N_0 is the number of radioactive nuclei in the sample at $t = 0$ and N is the number remaining at any subsequent time t. Note that lightbulbs (for one example) follow no such exponential decay law. If we life-test 1000 bulbs, we expect that they will all "decay" (that is, burn out) at more or less the same time. The proces of radioactivity follows a different law.

The rate of decay R is more interesting topic than N itself. Differentiating Equation 10.6, we find

$$R = -\frac{dN}{dt} = \lambda N_0 e^{-\lambda t}$$

or $$R = R_0 e^{-\lambda t} \text{ (radioactive decay).} \qquad (10.7)$$

an alternative form of the law of radioactive decay (Equation 10.6). In the equation R_0 ($= \lambda\, N_0$) represent the decay rate at $t = o$ and R represents the decay rate at any subseqent time *i e.* $t \neq 0$.

The total decay rate R of a sample of a radionuclide is called the **activity** of that sample. The unit for activity is named after the scientist Henri Beequerel, who discovered the process of radioactivity.

I becquerel = 1 Bq = 1 decay per second.

An older unit, the curie, is still in common use:

1curie = 1 Ci = 3.7×10^{10} Bq.

An example of the use of these units is the following statement: "The activity of spent reactor fuel rod #5658 on January 15, 1997, was 3.5×10^{15} Bq (= 9.5×10^4 Ci)." Thus, on that day 3.5×10^{15} radioactive nuclei in the rod decayed each second. The disintegration constant λ and the type of radiation they emit has not any effect on the process of measurement.

If a detector is unable to record all the disintegration occuring in the radioactive sample placed near the detector, due to inefficiency of the detector. The reading of the detector under these circumstances is proportional to (and smaller than) the true activity of the sample. This kind of measurement is not expressed in SI unit becaquerel but cont per unit time.

The time after which both N and R are reduced to one half initial values is known as half-life (τ) of the element.

Putting $R = \frac{1}{2}R_0$ in Equation 10.7 and substituting τ for t, we have

$$\frac{1}{2}R_0 = R_0 e^{-\lambda\tau}$$

Solving for τ yields

$$\tau = \frac{\text{In } 2}{\lambda} \tag{10.8}$$

a relation between the half-life τ and the disintegration constant λ.

Alpha Decay

We know that the nucleus of radioactive element ^{238}U decays by emitting an alpha particle which is known to be helium mucleus.

$$^{238}\text{U} \to {}^{234}\text{Th} + {}^{4}\text{He}, \quad Q = 4.25\,\text{MeV}. \tag{10.9}$$

Here Th is the symbol for the element thorium ($Z = 90$). The half-life for this decay is 4.47×10^9 y. Q is the disintegration energy of the process, that is, the amount of energy released during a single decay. We may well ask: If energy is released in every such decay event, why did the ^{238}U nuclei not decay shortly after they were created? Why did they wait so long?. To get the answer of our quesion we must study the mechanism of alpha decay.

Imagine a model in which the alpha particle exist insde the nucleus before its emission. Figure 10.6 shows the approximate potential energy $U(r)$ for the alpha particle and the residual ^{234} Th nucleus as a function of their separation r. The two main causes behind the occurence of energy are- (*i*) a well associated potential with the nuclear force that operates in the interior of nucleus and (*ii*) a Coulomb potential associated with the electric force that acts between the two particle before and after the decay.

The disintegaration energy for the process is shown by the line $Q = 4.25$ *Mev* (Figure 10.6). If we assume that this represents the total energy of the alpha particle during the decay process, then the part of the $U(r)$ curve above this line constitutes a potential energy barrier. This barrier cannot be surmounted· If the alpha particle were able to be at some separation r within the barrier, its potential energy *U* would exceed its total energy *E*. This indicates that the kinetic energy *K* is negative which is an impossible condition.

Now the reason as why the α-particles were not emitted ionidiately by the ^{238}U nucleus can be understood. That nucleus is surrounded by an impressive potential barrier, occupying —if you think of it in three dimensions—the volume lying between two spherical shells (of radii about 8 and 60 fm). This argument will convince everyone and now a new question will arise in our mind. How, since the particle seems permanently trapped inside the nucleus by the barrier, can the ^{238}U nucleus ever emit an alpha particle? The answer is that, as it is learned earlier there is a finite probability that a particle can tunnel through an energy barrier that is classically insurmountable. So the alpha decay occurs due to barrier tunneling.

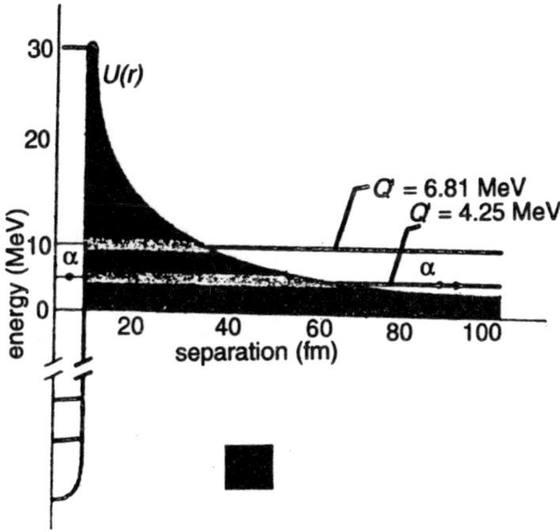

Figure 10.6: ***A potential energy function for the emission of an alpha particle by ^{238}U. The horizontal black line marked Q = 4.25 MeV shows the disintegration energy for the process. The thick gray portion of this line represents separations r that are classically forbidden to the alpha particle. The alpha particle is represented by a dot, both indide the barrier (at the left) and outside it (at the right), after the particle has tunneled through. The horizontal black line marked Q' = 6.81 MeV shows the disintegration energy for the alpha decay of ^{228}U. (Both isotopes have the same potential energy function because they have the same nuclear charge.)***

As the half life period of ^{238}U is time taking, so the barrier should not be leaky. The alpha particle, presumed to be rattling back and forth within the nucleus, must arrive at the inner surface of the barrier about 10^{38} times before it succeeds in tunneling through the barrier. This is about 10^{21} times per second for about 4×10^9 years (the age of Earth)! We are able only to see and count the alpha particles which escapes through the nucleus as we are observing from outside.

By observing other alpha emitters we can support our view regarding alpha decay. For an extreme contrast, consider the alpha decay of another uranium isotope, ^{228}U, which has a disintegration energy Q' of 6.81 MeV, about 60% higher than that of ^{238}U. (The value of Q' is also shown as a horizontal black line in figure 10.6. We have learnt already tht the transmission co-efficient of barrier is very sensitive to small changes in the total energy of the particle waiting to penetrate it. Thus we expect alpha decay to occur more readily for this nuclide than for ^{238}U. Indeed it does. As Table 10.2 shows, its half-life is only 9.1 min! An increase in Q by a factor of only 1.6 produces a decrease in half life (that is, in the effectiveness of the barrier) by a factor of 3×10^{14}. This is sensitivity indeed.

Table 10.2: Two Alpha Emitters Compared

Radionuglide	Q	Half-Life
^{238}U	4.25 MeV	$4.5 \times 10^9 y$
^{228}U	6.81 MeV	9.1 min

Beta Decay

When a nucleus is emitting electrons or a particle having positive charge and mass of electron, then the process of decay is called beta decay. This, like alpha decay. is a spontaneous process, with a definite disintegration energy and half-life. Again like alpha decay, beta decay is a statistical process. governed by Equations 10.6 and 10.7. Here are two examples.

$$^{32}P \rightarrow ^{32}S + e^- + \nu \quad (\tau = 14.3\ d) \tag{10.10}$$

and

$$^{64}\text{Cu} \rightarrow ^{64}\text{Ni} + e^{+} + \nu \;\; (\tau = 12.7\,\text{h}) \qquad (10.11)$$

The symbol ν represents a **neutrino**, a virtually (if not completely) massless, neutral particle that is enitted from the nucleus along with the electron or positron during the decay process. The nutrinos remian unnoticed for a long period because they react very weakly with other matter and their presence cannot be detected.

The two processes given above show the conservation of charge and nucleon number. In the decay of Equation 10.10, for example, we can write for charge conservation

$$(+15e) = (+16e) + (-e) + (0)$$

and for nucleon conservation

$$(32) = (32) + (0) + (0),$$

where we have recognized that neither the electron nor the neutrino is a nucleon and the neutrino has no charge.

Till now we have said that nucleus consist of neutrons and protons, so it seem surprising that it can emit neutrinos, electrons and positron. However, we saw earlier that atoms emit photons, and we certainly do not say that atoms "contain" photons. So we can conclude that photons are made or created during the process of emission.

Now, we can say that photons are with electron, positron and not neutrinos during the process of beta decay. They are created during the emission process. A neutron transforms into a proton within the nucleus according to

$$n \rightarrow p + e^{-} + \nu \qquad (10.12)$$

or a proton transforms into a neutron via

$$p \rightarrow n + e^{+} + \nu \qquad (10.13)$$

Both of these beta–decay processes provide evidence that–as was pointed out–neutrons and protons are not truly fundamental particles. It is very important to note that the mass number of nuclide undergoing decay doesnot changes while one of its nucleons changes its character but the number of that nucleon remain same.

The anount of energy released in alpha decay and beta decay is same.In the alpha decay of a particular radionuclide, every emitted alpha particle has the same sharply defined kinetic energy. (In some cases, a radionuclide may emit more than one group of alpha particles, each group having a sharply defined kinetic energy.) However, in the beta decay of Equation 10.12 for electron emission, the disintegration energy Q is shared-in varying proportions–between the electron and the neutrino. Sometimes the electron gets nearly all the energy, sometimes the neutrino does. In every case, however, the sum of the electron's energy and the neutrino's energy gives a constant value Q. In case of beta decay with electron emission the energy released *i.e.,* Q is shared among the electrons and neutrinos. similarly in case of beta decay with positron emission the energy released is Q and shared among the positron and nutrinos. Thus the energy of the emitted electron is positrons in beta decay may vary from a zero upto a certain maximum K_{max}.

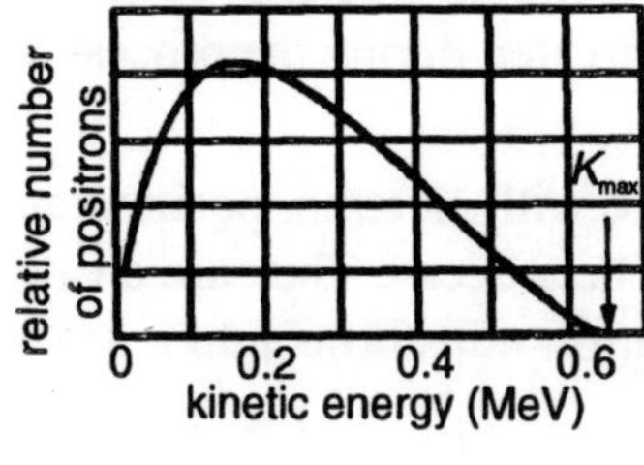

Figure 10.7: ***The distribution of the kinetic energies of protons emitted in the beta decay of ^{64}Cu. The maximum kinetic energy of the distribution (K_{max}) is 0.653 MeV. In all ^{64}Cu decay events, this energy is shared between the positron and the neutrino, in varying proportions. The most probable energy for emitted positron is about 0.15 MeV.***

Figure 10.10 shows the distribution of positron enrgies for the beta decay of ^{64}Cu (see Equation 10.11). The maximum positron energy K_{max} must equal the disintegration energy Q because the neutrino carries away no energy.

when the positron carries away K_{max}. That is.

$$Q = K_{max}. \qquad (10.14)$$

The Nuclear Atom

An atom is largely empty space

In the time of late ninteenth century most scientists accepted that chemical element are formed by atoms but they had have no knowledge about atom. One clue was the discovery that all atoms contain electrons. Since electrons carry negative charges whereas atoms are neutral, positively charged matter of some kind must be present in atoms. But the question was that in what manner these are arranged.

In 1898 *J. J. Thomson* gave his view regarding aoms that atoms are positively charge lumps of matter with electrons embedded in them. Because Thomson had played an important role in discovering the electron, his idea was taken seriously. However, the structure of atom was found to be quite different.

As if someone wants to know the inner matter of *a* fruitcake, he must have to pierce his finger through the cake. *Hans Geiger* and *Ernest Marsden* follow the same rule.At the suggestion of Ernest Rutherford, they used as probes the fast **alpha particles** emitted by certain radioactive elements. The well discussed α-particles are the helium atoms which have lost two electrons and having charge $+2e$.

In their experiment *Geiger and Marsden* used a lead screen with a hole in it. They placed a α-emitting substance in front of the lead screen. A narrow beam of α-particles were produced through the hole of lead screen. This beam was directed at a thin gold foil. A zinc sulfide screen, which gives off a visible flash of light when struck by an alpha particle, was set on the other side of the foil with a microscope to see the flashes.

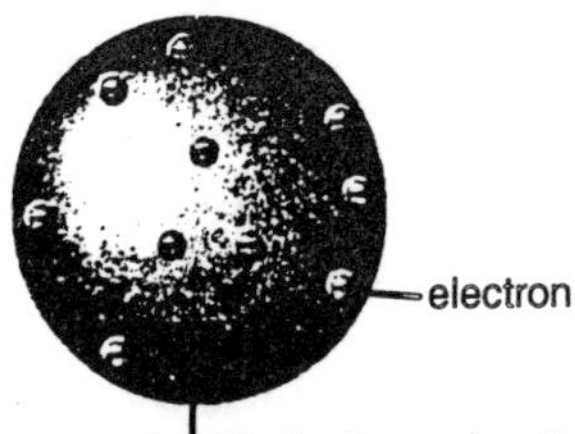

Figure 10.8: ***The Thomson model of the atom. The Rutherford scattering experiment showed it to be incorrect.***

Geiger and Marsden expected that the beam of α-particle would pass the lead screen, through the hole, without any deflection. This follows from the Thomson model, in which the electric charge inside an atom is assumed to be uniformly spread through its volume. Alpha particles which pass through *a* thin foil ought to be deflected very slightly as the force acting on them is very weak.

What Geiger and Marsden actually found was that although most of the alpha particles indeed were not deviated by much, a few were scattered through very large angles. Some were even scattered in the backward direction. *Rutherford* was surprised and said that it is shocking as it seems that he had fired *a* 15 inch shell at a tissue paper and it returns bach and hit at him.

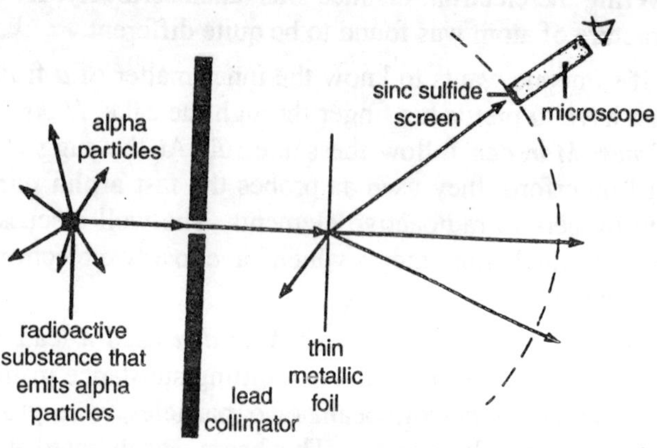

Figure 10.9: ***The Rutherford scattering experiment***

As the mass of alpha particles is heavy and also the speed is high, so the forces causing such great deflection must be very strong. The only way to explain the results, Rutherford found, was to picture an atom as being composed of a tiny nucleus in which its positive charge and nearly all its mass are concentrated, with the electrons some distance away (10.10). With an atom being largely empty space, it is easy to see why most alpha particles go right through a thin foil. However, when an alpha particle happens to come near a nucleus, the intense electric field there scatters it through a large angle. The alpha particles are not affected by the atomic electrons due to their light weight.

All the experiment similar to Geiger and Marsent provide information regarding the nucleus of the atom that composed the targed foils. The deflection of an alpha particle when it passes near a nucleus depends on the magnitude of the nuclear charge.Observation of scattering of alpha particles by various foils and difference in scattering by different foils reeveals the nuclear charges of the atom involved. The nuclear charge of all atoms of one element have same magnitude and this charge differ from element to element in periodic table. The nuclear charges always turned out to be multiples of + *ve*; the number *Z* of unit positive charges in the nuclei of an element is today called the **atomic number** of the element. It is well known that each proton gives a +*ve* charge to nucleus and proton is only +*ve* particle in an atom, so the atomic number of an elements is same as the number of proton present in the nucleus of the atom.

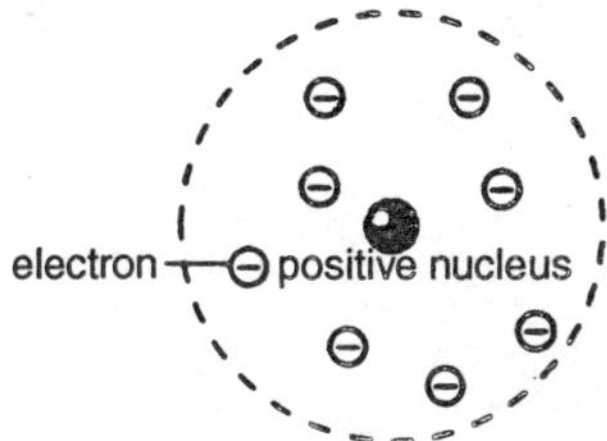

Figure 10.10 : ***The Rutherford model of the atom.***

So the matter is almost empty space. The solid wood of a table, the steel that supports a bridge, the hard rock underfoot, all are simply collections of tiny charged particles comparatively farther away from one another than the sun is from the planets. The particles *i.e*, electrons and the protons constituting the nuclei are only the actual matter in our body. If all these actual matter brought as close to each other as possible then we would become as small that rarely seen by a microscope.

Rutherford Scattering Formula

Rutherford gave a formula for alpha particle scattering by a thin foil based on the nuclear model of atom, as

$$N(\theta)=\frac{N_i ntZ^2e^4}{(8\pi\varepsilon_0)^2 r^2\mathrm{KE}^2\sin^4(\theta/2)} \tag{10.15}$$

The symbols in Equation (10.15) have the following meanings:

$N(\theta)$ = number of alpha particles per unit area that reach the screen at a scattering angle of θ

N_1 = total number of alpha particles that reach the screen

n = number of atoms per unit volume in the foil

Z = atomic number of the foil atoms

r = distance of the screen from the foil

KE = kinetic energy of the alpha particles

t = foil thickness

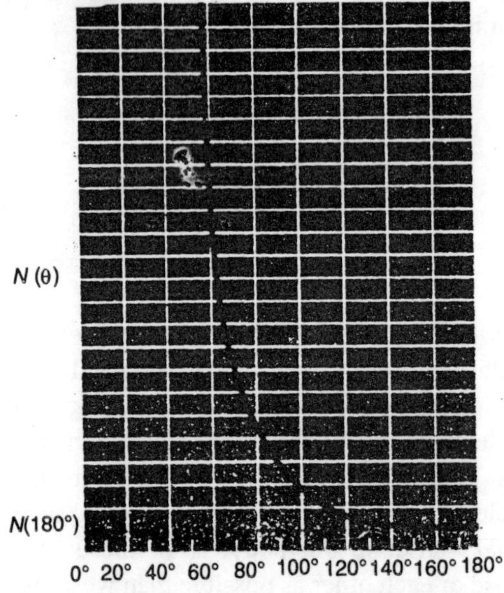

Figure10.11: ***Rutherford scattering.*** N(θ) ***is the number of alpha particles per unit area that reach the screen at a scattering angle of*** θ***; N(180°) is this number for backward scattering. The experimental findings follow this curve, which is based on the nuclear model of the atom.***

Rutherford scattering formula agreed with the measurements of Geiger and Marsden and thus provided support for the nuclear atom hypothesis. This is why Rutherford is credited with the "discovery" of the nucleus. Because $N(\theta)$ *is* inversely proportional to $\sin^4(\theta/2)$ the variation of $N(\theta)$ with θ is very pronounced only 0.14 percent of the incident alpha particles are scattered by more than 1°.

Nuclear Dimensions

While deriving the scattering formula *Rutherford* assumed that the size of target nucleus is very small as compared to the distance R from which the α-particles are attacking the target nucleus. Rutherford scattering therefore gives us a way to find an upper limit to nuclear dimensions.

Now let us find the distance of closest approach R for the most energetic alpha particles used in the experiments. An alpha particle will have its smallest R when it approaches a nucleus head on, which will be followed by a 180° scattering. The initial kinetic energy converted to electric potential energy at the instants of closest approach. Thus

$$\text{KE} = \text{PE} = \frac{1}{4\pi\varepsilon_0}\frac{2Ze^2}{R}$$

since the charge of the alpha particle is $2e$ and that of the nucleus is Ze. Hence

$$\text{Distance of closest approach } R = \frac{2Ze^2}{4\pi\varepsilon_0 \text{KE}} \qquad (10.16)$$

The maximum KE found in alpha particles of natural origin is 7.7 MeV, which is 1.2×10^{-12} J. Since $1/4\Pi\varepsilon_0 = 9.0 \times 10^9$ N. m^2/C^2.

$$R = \frac{(2)(9.0\times10^9 \text{N}\cdot\text{m}^2/\text{C}^2)(1.6\times10^{-19}\text{C})^2 Z}{1.2\times10^{-12}\text{J}}$$

$$= 3.8\times10^{-16} Z \text{ m}$$

The atomic number of gold, a typical foil material, is $Z = 79$, so that

$$R\,(\text{Au}) = 3.0 \times 10^{-14} \text{ m}$$

So the radius of the gold nucleus is 3.0×10^{-14}m which is less than the radius of the atom as a whole.

But in present time particles with higher energies than 7.7 MeV accelerated and it is observed that *Rutherford* scattering formula doesnot stand for these high speed particles. It is found that the radius of the gold atom nucleus is 1/5 of the *R* which is found in above equation.

Neutron Stars

The density of nuclear matter is about 2.4×10^{17} kg/M^3, which is equivalent to 4 billion tons per cubic inch. As mentioned earlier neutron stars are stars whose atoms have been so compressed that most of their. protons and electrons have fused into neutrons, which are the most stable form of matter under enormous pressures. The densities of neutron stars are comparable to those of nuclei: a neutron star packs the mass of one or two suns into a sphere only about 10 km in radius. If the earth were this dense, it would fit into a large apartment house.

Electron Orbits

The planetary model of the atom and why it fails

According to *Rutherford* model of atom, the nucleus is situated in the centre which is very small, massive and +vely charged and it is surrounded by negatively charged electrons at *a* large distance, so the atom appears as a neutral atom. The electrons cannot be stationary in this model, because there is nothing that can keep them in place against the electric force pulling them to the nucleus. The model also suggest that if the electrons are in motion then there would be a stable orbit for them like the planents.

Now let us give *a* look to atomic model of hydrogen which is simplest of them all due to presence of single electron. Let us assume a circular electron orbit for convenience, though it might as reasonably be assumed to be elliptical in shape. The centripetal force

$$F_c = \frac{mv^2}{r}$$

holding the electron in an orbit r from the nucleus is provided by the electric force

$$F_e = \frac{1}{4\pi\varepsilon_0}\frac{e^2}{r^2}$$

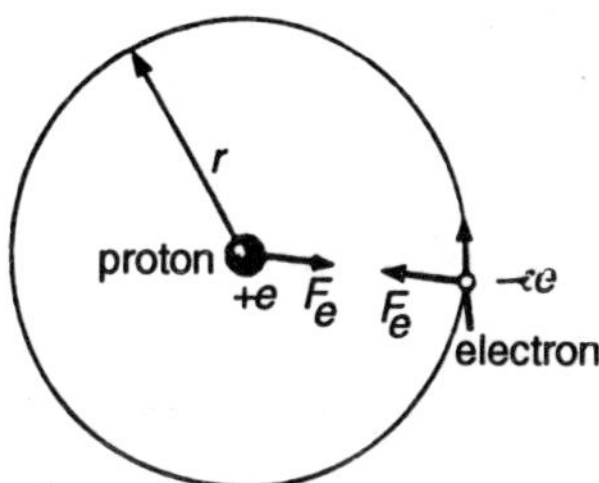

Figure 10.12: ***Force balance in the hydrogen atom.***

between them. The condition for a dynamically stable orbit is

$$F_c = F_e$$

$$\frac{mv^2}{r} = \frac{1}{4\pi\varepsilon_0}\frac{e^2}{r^2} \qquad (10.17)$$

The relation between electron velocity v and its orbit radius r is given by

Electron velocity $$v = \frac{e}{\sqrt{4\pi\varepsilon_0 mr}} \qquad (10.18)$$

Total energy of a hydrogen atom is sum of its kinetic and potential energy that is

$$\text{KE} = \frac{1}{2}mv^2 \quad \text{PE} = -\frac{e^2}{4\pi\varepsilon_0 r}$$

(The minus sign signifies that the force on the electron is in the –r direction.). Hence

$$E = \text{KE} + \text{PE} = \frac{mv^2}{2} - \frac{e^2}{4\pi\varepsilon_0 r}$$

Substituting for v from Equation 10.8 gives

$$E = \frac{e^2}{8\pi\varepsilon_0 r} - \frac{e^2}{4\pi\varepsilon_0 r}$$

Total energy of hydrogen atom

$$E = -\frac{e^2}{8\pi\varepsilon_0 r} \tag{10.19}$$

It means the energy of electron is *–ve*. This holds for every atomic electron and reflects the fact that it is bound to the nucleus. If the energy of an electron is not *–ve* then it would not follow a closed path around the nucleus.

The Failure of Classical Physics

According to classical physics an electron going in a curved path is accelerated and therefore should continuously lose energy and spiraling into the nucleus in a fraction of a second. But in practical, atom does not collapse. This contradiction further illustrates what we saw in the previous chapters: The laws of physics that are valid in the macroworld do not always hold true in the microworld of the atom.

As the particles and waves have many properties in common, so classical physics doesnot give a meaningful analysis of atomic structure. In reality particles and waves have many properties in common, though the smallness of Planck's constant makes the wave-particle duality imperceptible in the macroworld. The usefulness of classical physics decreases as the scale of the phenomena under study decreases, and we must allow for the particle behavior of waves and the wave behavior of particles to understand the atom. In the rest of this chapter we shall see how the Bohr atomic model, which combines classical and modern notions, accomplishes part of the latter task. Our approach towards the atomic structure theory will be meaningful only when we consider the atom from the point of view of quantum mechanics.

Is Rutherford's Analysis Valid?

Now at this point it seems quite surprising that *Rutherford* used the same laws of physics during deriving his scattering formula that fails poorly in the case of atom stability. Might it not be that this formula is not correct and that in reality the atom does not resemble Rutherford's model of a small central nucleus surrounded by distant electrons? This is not a trivial point. It is also interesting that the quatum mechanical analysis of alpha particle scattering by thin formula gives the same formula found by *Rutherford* in his experiments.

To verify that a classical calculation ought to be at least approximately correct, we note that the de Broglie wavelength of an alpha particle whose speed is 2.0 × 10^7m/s is

$$\lambda = \frac{h}{mv} = \frac{6.63 \times 10^{-34}\,\text{J}\cdot\text{s}}{\left(6.6 \times 10^{-27}\,\text{kg}\right)\left(2.0 \times 10^{7}\,\text{m/s}\right)}$$

$$= 5.0 \times 10^{-15}\,\text{m}$$

Atomic Spectra

Each element has a characteristic line spectrum

For a successful atomic theory, the atomic stability is not the only issue. The existence of spectral lines is another important aspect of the atom that finds no explanation in classical physics.

The condensed matter emit *em* radiation at all temprature in which all wavelengths with different intensities are present. The observed features of this radiation were explained by Planck without reference to exactly how it was produced by the radiating material or to the nature of the material. It indicates that we are not observing characteristic behaviour of the atom of a particular element rather a collective behaviour of many interacting atoms.

But to the contrast, the molecules or atoms of a rarified gas are so far from each other that posssibility of collision among them is rare. Under these circumstances we would expect any emitted radiation to be characteristic of the particular atoms or molecules present, which turns out to be the case.

If an atomic gas or vapour is excited by using an electric current & the gas is at a pressure less than atomspheric pressure then the gas would radiate a spectrum which has certain specified wavelengths only. An idealized arrangement for observing such atomic spectra is shown in Figure 10.13; actual spectrometers use diffraction gratings. Figure 10.14 shows the **emission line** spectra of several elements. Every element displays a unique line spectrum when a sample of it in the vapor phase is excited. An ideal spectrometer is a suitable apparatus to identify the composition of an unknown gas.

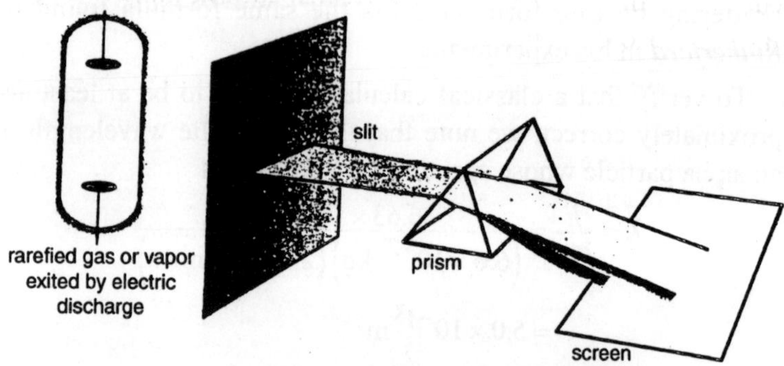

Figure 10.13: ***An idealized spectrometer.***

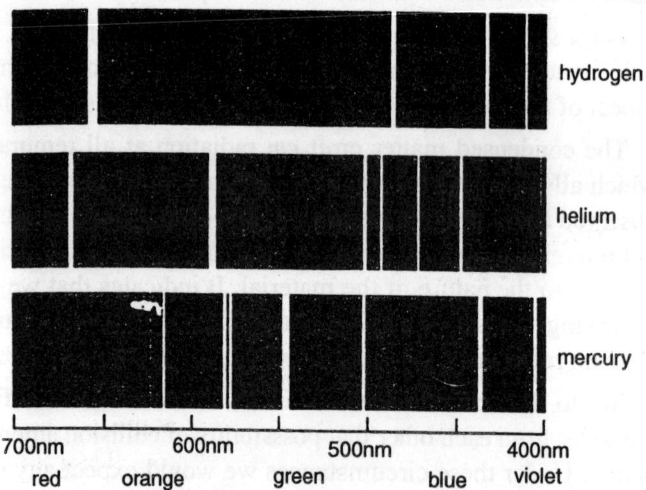

Figure 10.14: ***Some of the principal lines in the emission spectra of hydrogen, helium, and mercury.***

If white light is allowed to pass through a gas, then the gas would certainly absorb light of wavelengths present in its emission spectrum. The resulting **absorption line spectrum** consists of a bright background crossed by dark lines that correspond to the missing wavelengths Figure 10.14; emission spectra consist of bright lines on a dark background. The spectrum of sunlight has dark lines in it because the luminous part of the sun, which radiates very nearly like a blackbody heated to 5800 K, is surrounded by an envelope of cooler gas that absorbs light of certain wavelengths only.Many of the stars give the same spectra.

The nature of the spectrum e.g, number, strength wavelengths etc. depend upon the presence of magnetic and electric fields and also on the temprature and pressure of the source. It is possible to tell by examining its spectrum not only what elements are present in a light source but much about their physical state. By studying the spectrum of a star, the astronomer can predict some facts about the star e.g. which elements its atmosphere contains, if the elements are ionised or not and if the star is going away or towards the earth.

Spectral Series

The discovery of the fact that all wavelengths of a spectrum fall into sets was found almost a century ago. These sets were called spectral series. The first such series was found by J. J. Balmer in 1885, in the course of a study of the visible part of the hydrogen spectrum. Figure 10.16 shows the **Balmer series.** The line with the longest wavelength is 656.3 nm, is designated H_α, the next, whose wavelength is 486.3 nm, is designated H_β, and so on. The lines of spectrum come close to each other with decrease in their wavelength and their intensities also become weaker until the series limit at 364.6 nm is attained beyond which no separate lines can be seen.

Balmer's formula for the wavelengths of this series is

Balmer $$\frac{1}{\lambda} = R\left(\frac{1}{2^2} - \frac{1}{n^2}\right) \qquad n = 3, 4, 5, \ldots \qquad (10.20)$$

The value of R, called Rydberg constant, is given by

Rydberg constant $R = 1.097 \times 10^7 \text{m}^{-1} = 0.01097 \text{ nm}^{-1}$

The H_α line corresponds to $n = 3$, the H_β line to $n = 4$, and so on. To remain in agreement with experiment, the series limit corresponds to $n = \infty$, so that it occurs at a wavelength of $4/R$.

The wavelengths which arc in the visible portion of the hydrogen spectrum form the balmer series. The spectral lines of hydrogen in the ultraviolet and infrared regions fall into several other series. The wavelengths of ultraviolet region will from the Lyman seris. Their formula is given by

$$\text{Lyman} \qquad \frac{1}{\lambda} = R\left(\frac{1}{1^2} - \frac{1}{n^2}\right) \qquad n = 2, 3, 4, \ldots \tag{10.21}$$

Figure 10.15: ***The dark lines in the absorption spectrum of an element correspond to bright lines in its emission spectrum.***

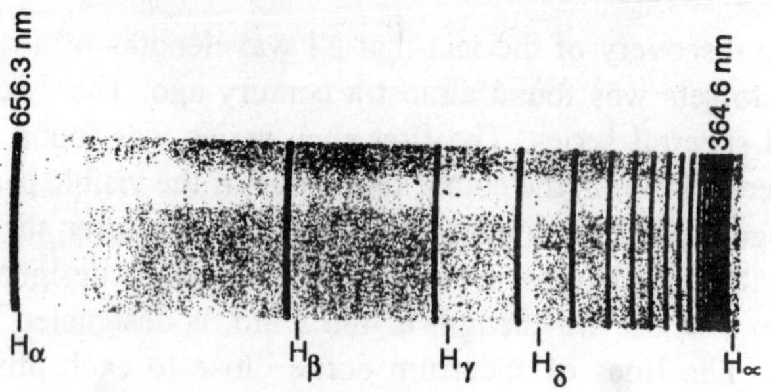

Figure 10.16: ***The Balmer series of hydrogen. The H_α line is red, the H_β line is blue, the H_γ and H_δ lines are violet, and the other lines are in the near ultraviolet.***

In the infrared, three spectral series have been found whose lines have the wavelengths specified by the formulas

$$\text{Paschen} \qquad \frac{1}{\lambda} = R\left(\frac{1}{3^2} - \frac{1}{n^2}\right) \qquad n = 4, 5, 6, \tag{10.22}$$

$$\text{Brackett} \qquad \frac{1}{\lambda} = R\left(\frac{1}{4^2} - \frac{1}{n^2}\right) \qquad n = 5, 6, 7, \ldots \tag{10.23}$$

$$\text{Pfund} \qquad \frac{1}{\lambda} = R\left(\frac{1}{5^2} - \frac{1}{n^2}\right) \qquad n = 6, 7, 8, \ldots \tag{10.24}$$

These spectral series of hydrogen are plotted in terms of wavelength in Figure 10.17; the Brackett series evidently overlaps the Paschen and Pfund series.Starting from equation 10.20 upto equation 10.24 the value of R remain same in all equation.

The occurence of such regularities in hydrogen spectrum as well as spectrum of some other complex elements present a test for any theory of atomic structure.

The Bohr Atom

Electron waves in the atom

Neil Bohr was the first scientist who put forward a successful theory of atom in 1913. The concept of matter waves leads in a

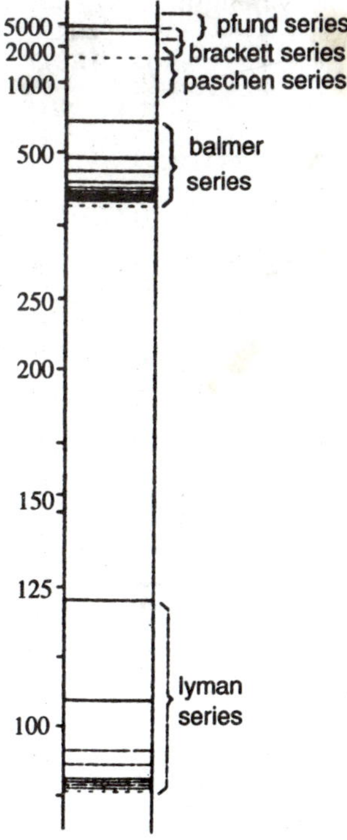

Figure 10.17: ***The spectral series of hydrogen. The wavelengths in each series are related by simple formulas.***

natural way to this theory, as de Broglie found, and this is the route that will be followed here. Bohr himself used a different approach, since de Broglie's work came a decade later, which makes his achievement all the more remarkable.But the results of both experiments were same.

At first, we take a look at wave behaviour of electron in an orbit around hydrogen nucleus. The de Broglie wavelength of this electron is

$$\lambda = \frac{h}{mv}$$

where the electron velocity v is that given by Equation (10.18):

$$v = \frac{e}{\sqrt{4\pi\varepsilon_0 mr}}$$

Hence

Orbital electron wavelength

$$\lambda = \frac{h}{e}\sqrt{\frac{4\pi\varepsilon_0 r}{m}} \qquad (10.25)$$

Energy Levels and Spectra

A photon is emitted when an electron jumps from one energy level to a lower level

The different orbit of an atom has electron with different energies. The electron energy E_n is given in terms of the orbit radius r_n by Equation

$$E_n = \frac{e^2}{8\pi\varepsilon_0 r_n}$$

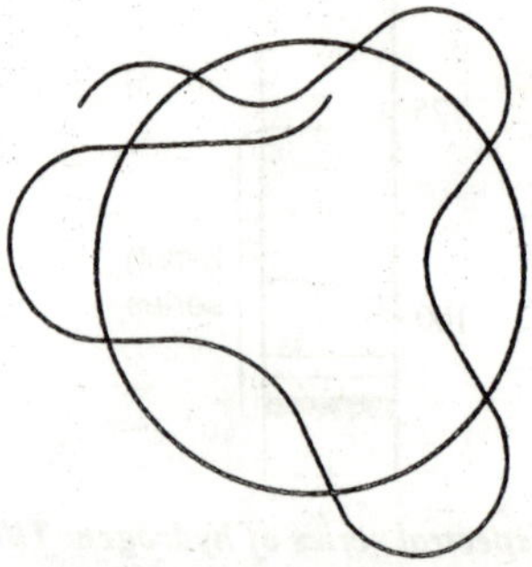

Figure 10.18: ***A fractional number of wabelengths cannot persist because destructive interference will occur.***

Substituting for r_n from Equation (10.17), we see that

$$\textit{Energy levels } E_n = -\frac{me^4}{8\varepsilon_0^2 h^2}\left(\frac{1}{n^2}\right) = \frac{E_1}{n^2} \quad n = 1, 2, 3, \ldots \qquad (10.26)$$

$$E_1 = -2.18 \times 10^{-18}\,\text{J} = -13.6\,\text{eV}$$

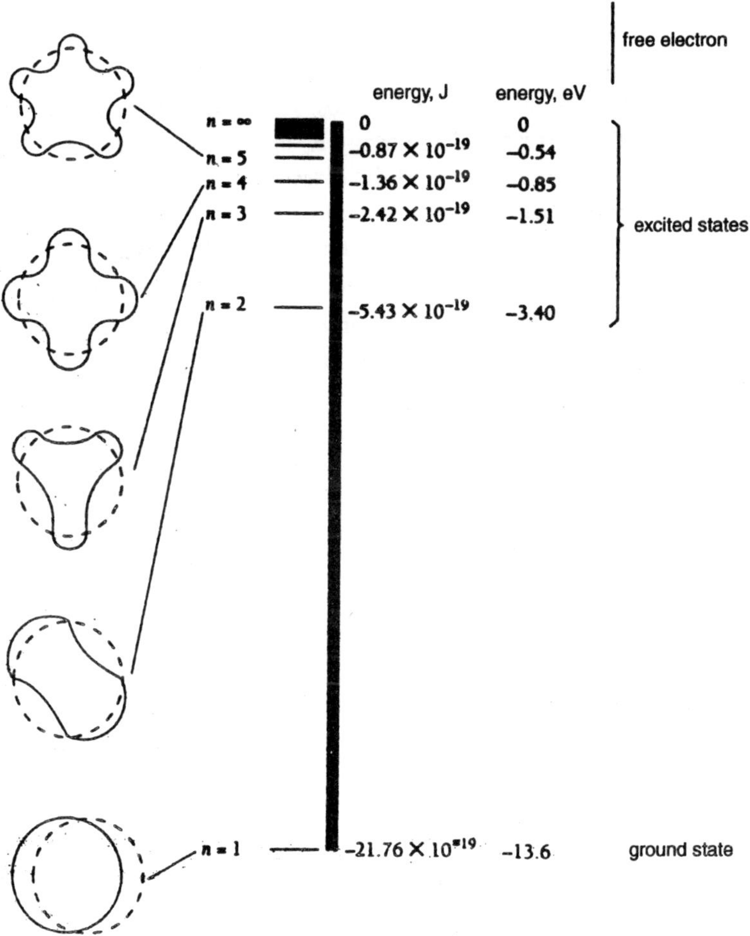

Figure 10.19: ***Energy levels of the hydrogen atom.***

The energies specified by Equation (10.26) are called the **energy levels** of the hydrogen atom and are plotted in Figure 10.19 These levels are all negative, which signifies that the electron does not have enough energy to escape from the nucleus. An atomic electron can have only these energies and no others. It can be compared with a person who is standing on a ladder and can't stand in between.

The level whic has lowest energy electron is called ground state of the atom and the level E_2, E_3 and E_4 and others which has higher energy is called excited states. As the quantum number n increases, the corresponding energy E_n approaches closer to 0. In the limit of $n = \infty$, $E_\infty = 0$ and the electron is no longer bound to the nucleus to form an atom. As said earlier, if the electron has positive energy then it means that it is free and is no longer bound to the nucleus. Obviously such situation does not exist.

The energy, required to break an electron from the nucleus in its ground state, is called its ionisation energy. The ionization energy is accordingly equal to $-E_1$,the energy that must be provided to raise an electron from its ground state to an energy of E = 0, when it is free. In the case of hydrogen, the ionization energy is 13.6 eV since the groundstate energy of the hydrogen atom is – 13.6 eV.

11

Quantum Theory of the Hydrogen Atom

Quantum Numbers

Three dimension, three quantum numbers

We are familiar with, one of the conditions that a wave function–and hence Φ, which is a component of the complete wave function ψ—must obey is that it have a single value at a given point in space. It is clear that ϕ and $\phi + 2\pi$ both identify the same meridian plane from figure 11.1. Hence it must be true that $\Phi(\phi) = \Phi(\phi + 2\pi)$ or

$$Ae^{im_l\phi} = Ae^{im_l(\phi+2\pi)}$$

which can happen only when m_1 is 0 or a positive or negative integer (±1, ±2, ±3, . . .) Here m_1 represents the magnetic quantum number of hydrogen atoms.

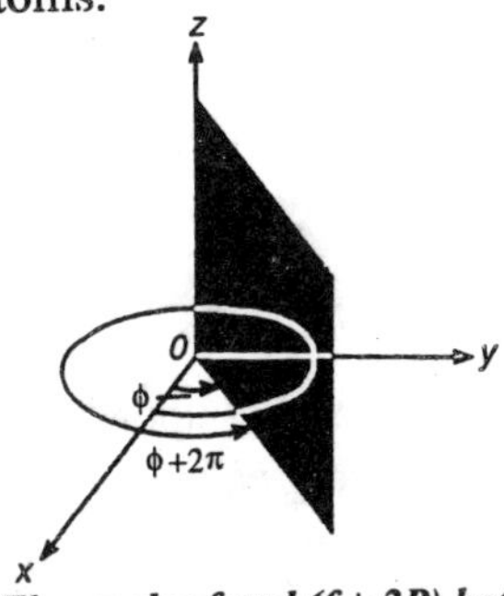

Figure 11.1 : ***The angles f and (f + 2P) both identify the same maridian plan.***

If the constant 'l' is an iteger equal to or greater than $| m_l |$, then the differential equation for θ (θ) has a solution, the absolute value of m_l. This requirement can be expressed as a condition on m_l in the form

$$m_l = 0, \pm 1, \pm 2, \ldots, \pm l$$

Here 'l' represents the orbital quantum number. While solving the equation for radial part $R(r)$ of the hydrogen atom wave function ψ, it is necessary to fulfill certain conditions. This condition is that E be positive or have one of the negative values E_n (signifying that the electron is bound to the atom) specified by

$$E_n = -\frac{me^4}{32\pi^2\varepsilon_0^2 h^2}\left(\frac{1}{n^2}\right) = \frac{E_1}{n^2} \qquad n = 1, 2, 3, \ldots \tag{11.1}$$

The above formula for enetgy level of hydrogen atom is same as given by *Bohr*.

During solving the equation for radial part $R(r)$ of the hydrogen atom wave function ψ, one more condition must fulfilled that is the principal quantum number 'n' must be $\geq l + 1$. This requirement may be expressed as a condition on l in the form

$$l = 0, 1, 2, \ldots, (n-1)$$

Hence we may tabulate the three quantum numbers n, l, and m together with them permissible values as follows:

Principal quantum number $n = 1, 2, 3, \ldots$

Orbital quantum number $1 = 0, 1, 2, \ldots, (n - l)$ (11.2)

Magnetic quantum number $m_l = 0, \pm 1, \pm 2, \ldots, \pm l$

It is worth noting again the natural way in which quantum numbers appear in quantum-mechanical theories of particles trapped in a particular region of space

The dependence of R_1 θ and ψ upon the quantum numbers n, l, m can be shown by following equantion

$$\psi = R_{nl}\theta_{lm_1}\Phi_{m_l} \tag{11.3}$$

The wave functions R, θ, and Φ together with ψ are given in Table 11.1 for $n = 1, 2,$and 3.

Table 11.1 Normalized Wave Functions of the Hydrogen Atom for $n = 1, 2,$ and 3*

n	l	m_l	$\Phi(\phi)$	$\Theta(\theta)$	$R(r)$	$\psi(r, \theta, \phi)$
1	0	0	$\frac{1}{\sqrt{2\pi}}$	$\frac{1}{\sqrt{2}}$	$\frac{2}{a_0^{1/2}} e^{-r/a_0}$	$\frac{1}{\sqrt{\pi} a_0^{3/2}} e^{-r/a_0}$
2	0	0	$\frac{1}{\sqrt{2\pi}}$	$\frac{1}{\sqrt{2}}$	$\frac{1}{2\sqrt{2} a_0^{3/2}}\left(2 - \frac{r}{a_0}\right) e^{-r/2a_0}$	$\frac{1}{4\sqrt{2\pi} a_0^{3/2}}\left(2 - \frac{r}{a_0}\right) e^{-r/2a_0}$
2	1	0	$\frac{1}{\sqrt{2\pi}}$	$\frac{\sqrt{6}}{2}\cos\theta$	$\frac{1}{2\sqrt{6} a_0^{3/2}} \frac{r}{a_0} e^{-r/2a_0}$	$\frac{1}{4\sqrt{2\pi} a_0^{3/2}} \frac{r}{a_0} e^{-r/2a_0} \cos\theta$
2	1	± 1	$\frac{1}{\sqrt{2\pi}} e^{\pm i\phi}$	$\frac{\sqrt{3}}{2}\sin\theta$	$\frac{1}{2\sqrt{6} a_0^{3/2}} \frac{r}{a_0} e^{-r/2a_0}$	$\frac{1}{8\sqrt{\pi} a_0^{3/2}} \frac{r}{a_0} e^{-r/2a_0} \sin\theta\, e^{\pm i\phi}$
3	0	0	$\frac{1}{\sqrt{2\pi}}$	$\frac{1}{\sqrt{2}}$	$\frac{2}{81\sqrt{3} a_0^{3/2}}\left(27 - 18\frac{r}{a_0} + 2\frac{r}{a_0^2}\right) e^{-r/3a_0}$	$\frac{1}{81\sqrt{3\pi} a_0^{3/2}}\left(27 - 18\frac{r}{a_0} + 2\frac{r^2}{a_0^2}\right) e^{-r/3a_0}$
3	1	0	$\frac{1}{\sqrt{2\pi}}$	$\frac{\sqrt{6}}{2}\cos\theta$	$\frac{4}{81\sqrt{6} a_0^{3/2}}\left(6 - \frac{r}{a_0}\right)\frac{r}{a_0} e^{-r/3a_0}$	$\frac{\sqrt{2}}{81\sqrt{\pi} a_0^{3/2}}\left(6 - \frac{r}{a_0}\right)\frac{r}{a_0} e^{-r/3a_0} \cos\theta$
3	1	± 1	$\frac{1}{\sqrt{2\pi}} e^{\pm i\phi}$	$\frac{\sqrt{3}}{2}\sin\theta$	$\frac{4}{81\sqrt{6} a_0^{3/2}}\left(6 - \frac{r}{a_0}\right)\frac{r}{a_0} e^{-r/3a_0}$	$\frac{1}{81\sqrt{\pi} a_0^{3/2}}\left(6 - \frac{r}{a_0}\right)\frac{r}{a_0} e^{-r/3a_0} \sin\theta\, e^{\pm i\phi}$
3	2	0	$\frac{1}{\sqrt{2\pi}}$	$\frac{\sqrt{10}}{4}(3\cos^2\theta - 1)$	$\frac{4}{81\sqrt{30} a_0^{3/2}} \frac{r^2}{a_0^2} e^{-r/3a_0}$	$\frac{1}{81\sqrt{6\pi}\, a_0^{3/2}} \frac{r^2}{a_0^2} e^{-r/3a_0}(3\cos^2\theta - 1)$
3	2	± 1	$\frac{1}{\sqrt{2\pi}} e^{\pm i\phi}$	$\frac{\sqrt{15}}{2}\sin\theta\cos\theta$	$\frac{4}{81\sqrt{30} a_0^{3/2}} \frac{r^2}{a_0^2} e^{-r/3a_0}$	$\frac{1}{81\sqrt{\pi} a_0^{3/2}} \frac{r^2}{a_0^2} e^{-r/3a_0} \sin\theta\cos\theta\, e^{\pm i\phi}$
3	2	± 2	$\frac{1}{\sqrt{2\pi}} e^{\pm 2i\phi}$	$\frac{\sqrt{15}}{4}\sin^2\theta$	$\frac{4}{81\sqrt{30} a_0^{3/2}} \frac{r^2}{a_0^2} e^{-r/3a_0}$	$\frac{1}{162\sqrt{\pi}\, a_0^{3/2}} \frac{r^2}{a_0^2} e^{-r/3a_0} \sin^2\theta\, e^{\pm 2i\phi}$

* The quantity $a_0 = 4\pi\varepsilon_0\hbar^2 / me^2 = 5.292 \times 10^{-11}$ m is equal to the radius of the innermost Bohr orbit.

Principal Quantum Number

Quantization of energy

It is very important to note the signification given by hydrogen atom quantum numbers in terms of the classical model of the atom. This model, corresponds exactly to planetary motion in the solar system except that the inverse-square force holding the electron to the nucleus is electrical rather than gravitational. There are two quantities in planetary motion which are conserved that are scalar totalenergy and vector angular momentum of each planet.

The total energy of planent must be negative in order to assure that it remain permanently trapped in the solar system. In the quantum theory of the hydrogen atom the electron energy is also a constant, but while it may have any positive value (corresponding to an ionized atom), the only negative values the electron can have are specified by the formula $E_n = E_1/n^2$. The principal quantum number n describes the quantization of electron energy in hydrogen atom.

Schrodinger's equation is also able to give the theory of planetary motion and it gives a similar energy restrictions However, the total quantum number n for any of the planets turns out to be so immense that the separation of permitted levels is far too small to be observable. Due to these quantum number complexicity, classical physics fail to describe the theory for atomic structure.

Orbital Quantum Number

Quantization of angular momentum magnitude

To interpret the orbital quantum number 'l' is less obvious. Let us look at the differential equation for the radial part $R(r)$ of the wave function ψ.

$$\frac{1}{r^2}\frac{d}{dr}\left(r^2\frac{dR}{dr}\right)+\left[\frac{2m}{h^2}\left(\frac{e^2}{r\pi\varepsilon_0 r}+E\right)-\frac{l(l+1)}{r^2}\right]R=0 \quad (11.4)$$

So, the above equation only depends upon the motion of electron, toward or away from the nucleus. However, we notice the presence of E, the total electron energy, in the equation. The kinetic energy of electron is the total energy E of the electron this

KE is concerned only to orbital motion and not with the radial motion. Here is an controversy which can be removed. The kinetic energy KE of the electron has two parts, KE_{radial} due to its motion toward or away from the nucleus, and $KE_{orbital}$ due to its motion around the nucleus. So the electric energy represents the potential energy of electron.

$$U = -\frac{e^2}{4\pi\varepsilon_0 r} \tag{11.5}$$

Hence the total energy of the electron is

$$E = KE_{radial} + KE_{orbital} + U = KE_{radial} + KE_{orbital} - \frac{e^2}{4\pi\varepsilon_0 r}$$

Inserting this expression for E we obtain, after a slight rearrangement,

$$\frac{1}{r^2}\frac{d}{dr}\left(r^2\frac{dR}{dr}\right) + \frac{2m}{h^2}\left[KE_{radial} + KE_{orbital} - \frac{\hbar^2(l+1)}{2mr^2}\right]R = 0 \tag{11.6}$$

We want a differential equation for $R(r)$, which have function of the radius vector r, and what we want can be achieved only when last two terms in bracket cancel each other.

We therefore require that

$$KE_{orbital} = \frac{\hbar^2(l+1)}{2mr^2} \tag{11.7}$$

Since the orbital kinetic energy of the electron and its angular momentum are respectively

$$KE_{orbital} = \frac{1}{2}mv^2{}_{orbital} \qquad L = mv_{orbtial}$$

The equation for the orbital kinetic energy can be written as—

$$KE_{orbital} = \frac{L^2}{2mr^2}$$

Hence, from Equation (6.20),

$$\frac{L^2}{2mr^2} = \frac{\hbar^2(l+1)}{2mr^2}$$

Electron angular momentum

$$L=\sqrt{l(l+1)}\hbar \tag{11.8}$$

The values for 'l' orbital quantum number is restricted as

$$l = 0, 1, 2,... ,(n-1)$$

the electron can have only those particular angular momenta L specified by Equation 11.8. Like total energy E, angular momentum is both conserved and quantized.

$$\hbar = \frac{h}{2\pi} = 1.054 \times 10^{-34}\ \text{J}\cdot\text{s}$$

The above equation has natural unit of angular momentum.

The quantum number describing angular momentum, in macroscopic planetary motion, is so large that the separtation into discrete angular momentum states cannot be experimentaly obserred. For example, an electron (or, for that matter, any other body) whose orbital quantum number is 2 has the angular momentum

$$L = \sqrt{2(2+1)}\hbar = \sqrt{6}\hbar$$

$$= 2.6 \times 10^{-34}\ \text{J.s}$$

But the actual angular momentum of earth is given by 2.7×10^{40} J.S

Designation of Angular Momentum States

. The angular momentum states for electron is specify by the letters *S, P, etc*, and *S* corresponds to $l = o$, *P* corresponds to $l = 1$ and so on.

Angularmomentum states	$l =$	0	1	2	3	4	5	6 ...
		s	*p*	*d*	*f*	*g*	*h*	*i* ...

This peculiar code originated in the empirical classification of spectra into series called sharp, principal, diffuse, and fundamental which occurred before the theory of the atom was developed. So an electron with angular momentum state *S* has *no* angular momentum and the electron with *P* state has an angular momentum of $\sqrt{2}\,\hbar$ and so on.

Table. 11.2: Atomic Electron States

	$l = 0$	$l = 1$	$l = 2$	$l = 3$	$l = 4$	$l = 5$
$n = 1$	1*s*					
$n = 2$	2*s*	2*p*				
$n = 3$	3*s*	3*p*	3*d*			
$n = 4$	4*s*	4*p*	4*d*	4*f*		
$n = 5$	5*s*	5*p*	5*d*	5*f*	5*g*	
$n = 6$	6*s*	6*p*	6*d*	6*f*	6*g*	6*h*

The total quantum number, of an electron, together with letter representing the angular momentum state inform the widely used notation for the atomic electron. In this notation a state in which $n = 2, l = 0$ is a 2s state, for example. and one in which $n = 4, l = 2$ is a 4*d* state. Table 11.2 gives the designations of electron states in an atom through $n = 6, l = 5$.

Magnetic quantum Number

Quantization of angular momentum direction

The magnitude of electron's angular momentum 'L' is determined by the orbital quantum number '*l*'. However, angular momentum, like linear momentum, is a vector quantity, and to describe it completely means that its direction be specified as well as its magnitude. (The vector **L**, we recall, is perpendicular to the plane in which the rotational motion takes place, and its sense is given by the right-hand rule: When the fingers of the right hand point in the direction of the motion, the thumb is in the direction of **L**.

Now, let us think that what possible significance can a direction in space have for a hydrogen atom. The answer becomes clear when we reflect that an electron revolving about a nucleus is a minute current loop and has a magnetic field like that of a magnetic dipole. Hence an atomic electron that possesses angular momentum interacts with an external magnetic field **B**. The magnetic quantum number m_l specifies the direction of **L** by determining the component of **L** in the field direction. We call this phenomenon as space quantization.

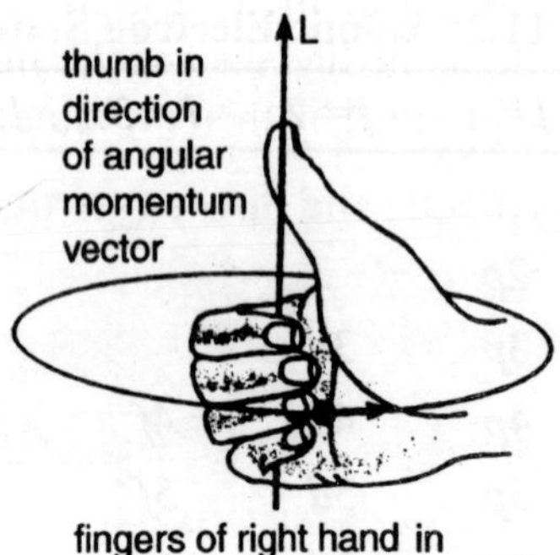

Figure 11.2: ***The right-hand rule for angular momentum.***

Let us consider that the magnetic field direction os parallel to the *Z*-axis, then the component of '*l*' in this direction will be Space quantization

$$L_z = m_1 \hbar \qquad m_1 = 0, \pm 1, \pm 2, \ldots, \pm l \tag{11.9}$$

The possible values of m_l for a given value of *l* range from $+l$ through 0 to $-l$, so that the number of possible orientations of the angular-momentum vector **L** in a magnetic field is $2l + 1$. When $l = 0$, L_Z can have only the single value of 0; when $l = l$, L_z may be $\hbar$, 0, or $-\hbar$; when $l = 2$, L_Z may be $2\hbar$, $\hbar$, 0, $-\hbar$, or $-2\hbar$; and so on.

The phenomenon called space quantization of hydrogen atoms angular momentum is explained in figure 11.3 An atom with a certain value of m_l will assume the corresponding orientation of its angular momentum **L** relative to an external magnetic field if it finds itself in such a field. We see that '*L*' can never be exactly parallel or antiparallel to *B* as the L_z remains always smaller than the quantity $\sqrt{l+(l+1)K}$.

The direction of the *Z* axis will remain arbitary inabsence of an external magnetic field. What must be true is that the component of **L** in any direction we choose is $m_l h$. What an external magnetic field does is to provide an experimentally meaningful reference direction. A magnetic field is not the only such reference direction possible. As in hyerogen molecule, the line present between the two hydrogen atoms is also as meaningful as the magnetic field, and along the line the angular momentas component of H atom are determined by their m_l values.

The Uncertainty Principle and Space Quantization

Let us think that why only one component of **L** is quntized. The answer is related to the fact that **L** can never point in any specific direction but instead is somewhere on a cone in space such that its projection L_Z *is* $m_l h$. Were this not so, the uncertainty principle would be violated. If **L** were fixed in space, so that L_x and L_y as well as L_z had definite values, the electron would be confined to a definite plane. For instance, if L were in the z direction, the electron would have to be in the xy plane at all time. It is possible only when the momentum component P_2 of electon is infinitely uncertain, whichis not possible.

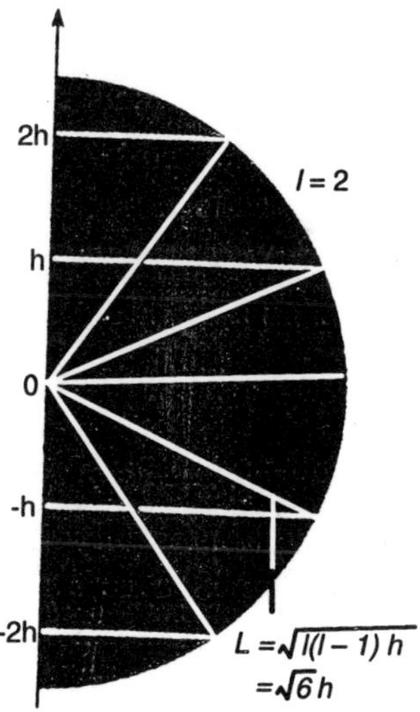

Figure 11.3: ***Space quantization of orbital angular momentum. Here the orbital quantum number is l = 2 and there are accordingly 2l + 1 = 5 possible values of the magnetic quantum number with each value corresponding to a different orientation relating to the z axis.***

However, since in reality only one component Lz of **L** together with its magnitude L have definite values and $|L| > |L_z|$ the electron is not limited to a single plane. Thus there is a built in uncertainty in the electron's z coordinate. According to figure 11.5, the direction of L remains changing, so the average values of L_x and $L_y = 0$, while the value of L_z remains fixed as $m_l \hbar$.

Electron Probability Density

No definite orbits

According to *Bohr's* atomic model the electrons remain moving around the nucleus in a circular path. This model is pictured in a spherical polar coordinate system in figure 11.6. It implies that if a suitable experiment were performed, the electron would always be found a distance of $r = n^2 a_0$ (where n is the quantum number of the orbit and a_0 is the radius of the innermost orbit) from the nucleus and in the equatorial plane $\theta = 90^\circ$, while its azimuth angle ϕ changes with time.

Two modifications were suggested by the atomic theory of hydrogen in *Bohrs* model.

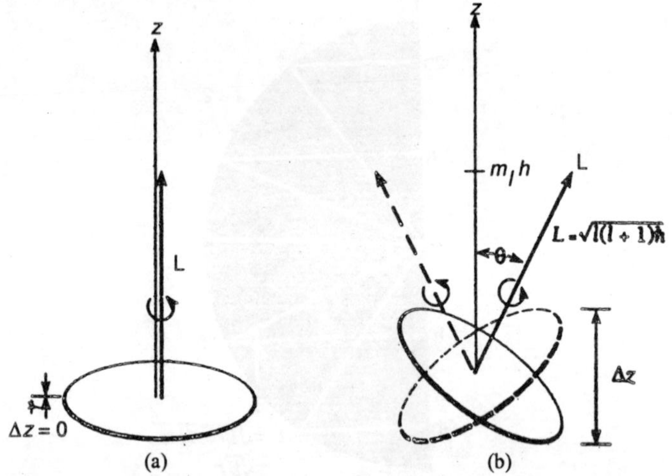

Figure. 11.4: ***The uncertainty principle prohibits the angular momentum vectro L from having a definite direction in space.***

The absolute or exact value for r, θ and ϕ cannot be assigned rather only probabilities to finding the electron at various positions. This imprecision is, of course, a consequence of the wave nature of the electron.

As the probability density $|\psi|^2$ is independent of time and varies from place to place, so it seems imposible to think that electron revolve around the nucleus.

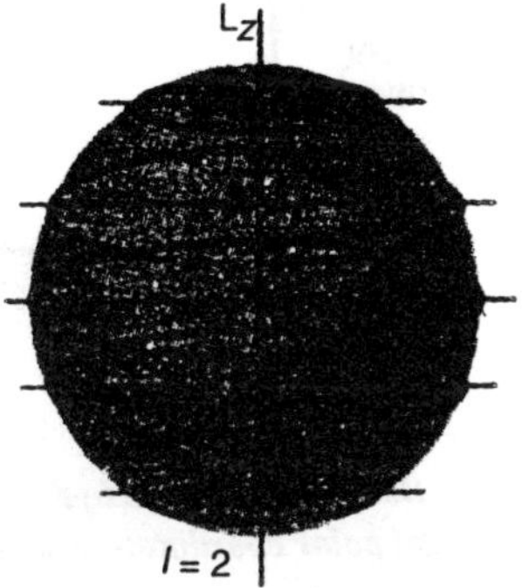

Figure. 11.5: ***The angular-momentum vectro L precesses constantly about the z axis.***

The probability density $|\psi|^2$ that corresponds to the electron wave function $\psi = R\,\theta\,\Phi$ in the hydrogen atom is

$$|\psi|^2 = |R|^2 |\theta|^2 |\Phi|^2 \tag{11.10}$$

The product of the function and its complex conjugate is usually replaced by the square of nay function which is complex.

From Equation (11.1) we see that the azimuthal wave function is given by

$$\Phi(\phi) = Ae^{im_l\phi}$$

The azimuthal probability density $|\Phi|^2$ is therefore

$$|\Phi|^2 = \Phi^*\Phi = A^2 e^{im_l\phi} e^{im_l\phi} = A^2 e^0 = A^2$$

The likelihood of finding the electron at a particular azimuth angle ϕ is a constant that does not depend upon ϕ at all. About the z-axis regardless of the quantum state it is in, the electron probability density is symmetrical and the electron has the same chance of being found at one angle ϕ as at another.

The radial part R of the wave function, in other way to Φ not only varies with r but does so in a different way for each combination of quantum nubers n and l. Figure 6.8 contains graphs of R versus r for $1s$, $2s$, $2p$, $3s$, $3p$, and $3d$ states of the hydrogen atom.

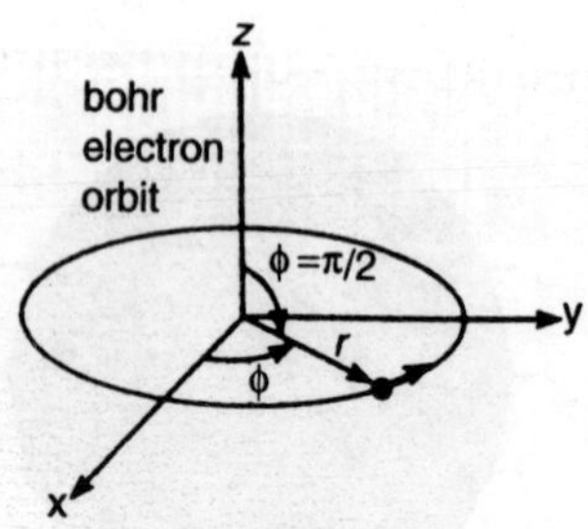

Figure. 11.6: ***The Bohr model of the hydrogen atom in a spherical polar coordinate system.***

Evidently R is a maximum at $r = 0$—that is, at the nucleus itself-for all s states, which correspond to L = 0 since $l = 0$ for such states. The value of R is zero at $r = 0$ for states that possess angular momentum.

Probability of Finding the Electron

At the point r, θ, ϕ the probability density of electron is proportional to $|\psi|^2$. But the actual probability of finding in it the infinitensimal volume element dv ther is $|\psi|^2 dv$.

In spherical polar coordinates (Figure 11.8),

$$dV = (dr)(r \ d\theta)(r \sin\theta \ d\phi)$$

Volume element $\qquad = r^2 \sin\theta \ dr \ d\theta \ d\phi \qquad (11.11)$

The actual probability $P(r)\ dr$ of finding the electron in a hydrogen atom somewhere in the spherical shell between r and $r + dr$ from the nucleus because θ and Φ are normalized function.

$$P(r)dr = r^2 |R|^2 dr \int_0^{\pi} |\theta|^2 \sin\theta \, d\theta \int_0^{2\pi} |\Phi|^2 d\phi$$

$$= r^2 |R|^2 dr \qquad (11.12)$$

Equation 11.12 is plotted in figure 11.10 for the same states whose radial function R were shown in figure 11.7. The curves are quite different as a rule. We note immediately that P is not a maximum at the nucleus for s states, as *R* itself is, but has its maximum a definite distance from it.

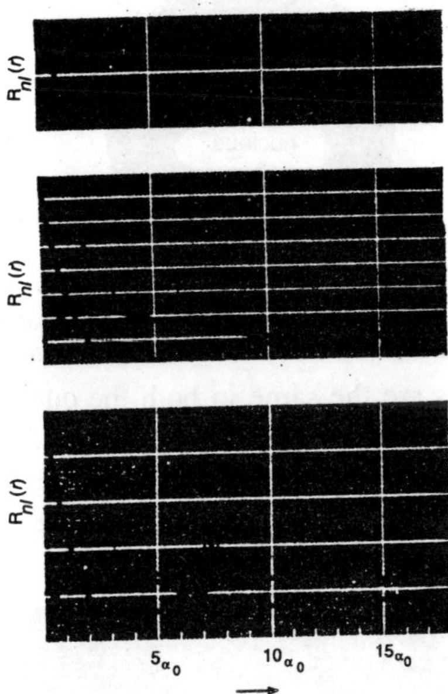

Figure 11.7 : ***The variation with distance from the nucleus of the radial part of the electron wave function in hydrogen for various quantum states. The quantity*** $a_0 = 4\pi\varepsilon_0 h^2/me^t = 0.053$ ***nm is the radius of the first Bohr orbit.***

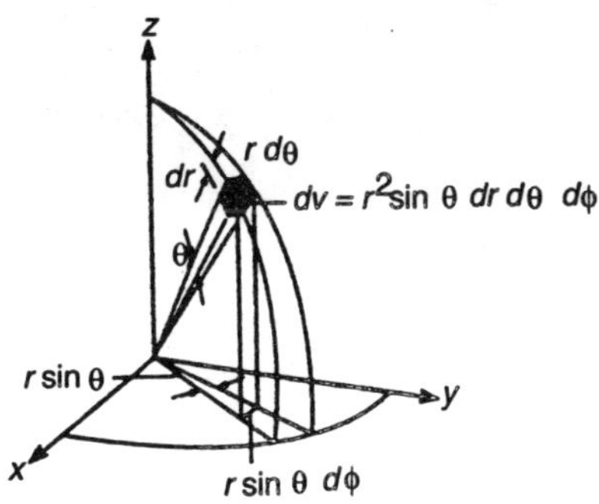

Figure. 11.8: ***Volume element*** **dv** ***in spherical polar coordinates.***

For an electron situated in is, the most probable value of *r* find out to be exactly a_o, which represent the orbital radius of *a* ground state electron in the *Bohr model*. However, the average value of

r for a is electron is $1.5a_0$, which is puzzling at first sight because

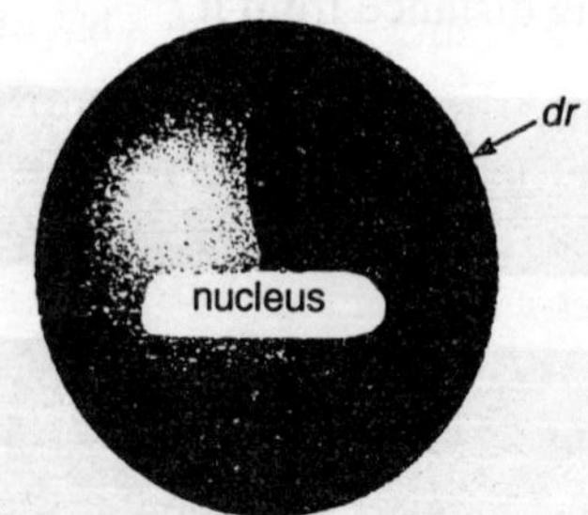

Figure. 11.9: ***The probability of finding the electron in a hydrogen atom in the spherical shell between* r *and* r + dr *from the nucleus is P*(r) dr.**

the energy levels are the same in both the quantum-mechanical and Bohr atomic models. This confusion can be cleared by recalling that the electron energy depends upon $1/r$ and not on r itself, so the average value of $1/r$, for a is electron is exactly $1/a_0$.

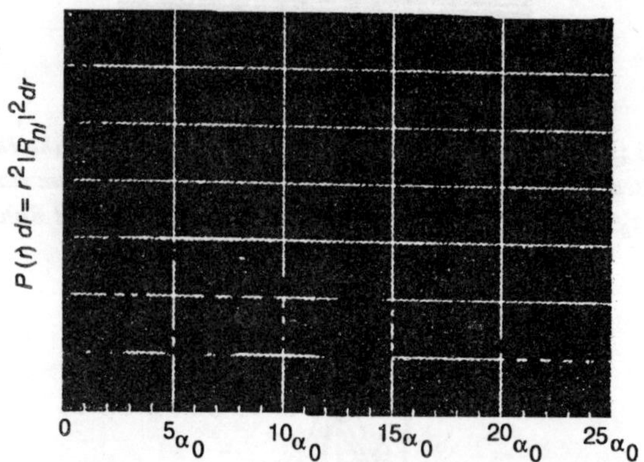

Figure. 11.10: ***The probability of finding the electron in a hydrogen atom at a distance between r and r + dr from the nucleus for the quantum states of Figure 6.8.***

Thus $$\left\langle \frac{1}{r} \right\rangle = \left(\frac{1}{\pi a_0^3}\right)\left(\frac{a_0^2}{4}\right)(2)(2\pi) = \frac{1}{a_0}$$

Angular Variation of Probability Density

The only case, in which function-doesnot vary with zenith angle θ is that when $l = m_l = 0$ which are *s* states. The value of $|\Theta|^2$ for an *s* state is a constant; $\frac{1}{2}$, in fact. This means that since $|\Phi|^2$ is also a constant, the electron probability density $|\psi|^2$ is spherically symmetric: it has the same value at a given *r* in all directions. Electrons in other states, however, do have angular preferences, sometimes quite complicated ones. Some of such preferences are given in figure 11.11, in which electron probability densities as function of *r* and θ are shown for several atomic states. Since $|\psi|^2$ is independent of ϕ, we can obtain a three dimensional picture of $|\psi|^2$ by rotating a particular representation about a vertical axis. When this is done, we see that the probability densities for *s* states are spherically symmetric whereas those for other states are not. In chemistry, the pronounced labe pattern characteristic of many of the states are prooved significant as they help in determining the way in which adjacent atoms in an molecule interact.

The figure 11.11 also indicate a similarity of the quantum mechanical states with the Bohr molel. The electron probability-density distribution for a 2*p* state with $m_1 = \pm 1$, for instance, is like a doughnut in the equatorial plane centered at the nucleus. Calculation shows the most probable distance of such an electron from the nucleus to be $4a_0$—precisely the radius of the Bohr orbit for the same principal quantum number $n = 2$. Similar correspondences exist for 3*d* states with states with $m_l = \pm 2$, 4*f* states with $m_l = \pm 3$ and so on. In each of these cases the angular momentum is the highest possible for that energy level, and the angular-momentum vector is as near the *z* axis as possible so that the probability density is close to the equatorial plane. So Bohr model help to determine the most probable location of the electron in one of the several possible staes in each energy level.

Radiative Transitions

What happens when an electron goes from one state to another

While deriving the hydrogen atom theory, Bohr was obliged to postulate that the frequency ν of the radiation emitted by an atom, which is dropped from an energy level E_m to a lower energy level E_n is given by

$$\nu = \frac{E_m - E_n}{h}$$

It is not hard to show that this relationship arises naturally in quantum mechanics. For our convinience, we would imagine a case in which atom can move only is *n* direction.

As discussed in previous chapter, we know that time dependent wave function ψ_n of an electron in an state of quantum number *n* and energy E_n is the product of time dependect wave function ψ_n and a time varying function whose frequency is

$$\nu_n = \frac{E_n}{h}$$

Hence $\quad \Psi_n = \psi_n e^{-(iE_n/h)t} \qquad \Psi_n^* = \psi_n^* e^{+(iE_n/h)t}$ (11.13)

For such an electron, the expectation value (x) of its position is given by

$$\langle x \rangle = \int_{-\infty}^{\infty} x\Psi_n^* \Psi_n dx = \int_{-\infty}^{\infty} x\psi_n^* \psi_n e^{[(iE_n/\hbar)-(iE_n/\hbar)]t} dx$$

$$= \int_{-\infty}^{\infty} x\psi_n^* \psi_n dx \tag{11.14}$$

The expectation value (x) is constant in time since ψ_n and ψ_n^* are, by definition, functions of position only. The electron does not oscillate, and no radiation occurs. So, according to quantum mechanics, a system in a specific quantum state does not radiate as observed.

Now we will concentrate on an electron shifting from one energy level to another. A system might be in its ground state *n* when an excitation process of some kind (a beam of radiation, say, or collisions with other particles) begins to act upon it. Subsequently we find that the system emits radiation corresponding to a transition from an excited state of energy E_m to the ground state. We conclude that at some time during the intervening period the system existed in the state m. There is a question that what is the frequency of the radiation. For a electron that can exist in both states m and n, the wave function will be as

$$\Psi = a\Psi_n + b\Psi_m \tag{11.15}$$

where a^*a is the probability that the electron is in state n and b^*b the probability that it is in state m. Of course, it must always be true that $a^*a + b^*b = 1$. Initially $a = 1$ and b 0; when the electron is in the excited state, $a = 0$ and $b = 1$; and ultimately $a = 1$ and $b = 0$ once more. If the electron is either in m *or* n state, in that case, no radiation will occur rather radiation will occur when the electron is in the midst of its transfer from one state to another.z electromagnetic wave will radiate.

The expression value (n) that corresponds to the composite wave function of equation 11.15 is

$$\begin{aligned}\langle x\rangle &= \int_{-\infty}^{\infty} x\left(a^*\Psi_n^* + b^*\Psi_m^*\right)\left(a\Psi_n + b\Psi_m\right) dx \\ &= \int_{-\infty}^{\infty} x\left(a^2\Psi_n^*\Psi_n^* + b^* a\Psi_n^*\Psi_m + \right. \\ &\qquad\left. a^* b\Psi_n{}^*\Psi_m + b^2\Psi_m^*\Psi_m\right)\end{aligned} \tag{11.16}$$

Here, as before, we let $a^*a = a^2$ and $b^*b = b^2$. As the first and 2nd variable donot vary w_1 with time, so only third and fourth variable will give their contribution in the time variation of x.

With the help of Equation (11.13) we expand Equation (11.16) to give

$$\begin{aligned}\langle x\rangle = a^2\int_{-\infty}^{\infty} x\psi_n^*\psi_n dx + b^* a\int_{-\infty}^{\infty} x\psi_m^* e^{+(iE_m/\hbar)t} dx\,\Psi_n e^{-(iE_n/\hbar)tdx} \\ + a^* b\int_{-\infty}^{\infty} x\psi_n^* e^{+(iE_n/\hbar)t}\psi_m e^{-(iE_m/\hbar)t} dx + b^2\int_{-\infty}^{\infty} x\psi_m^*\psi_m dx\end{aligned} \tag{11.17}$$

Because

$$e^{i\theta} = \cos\theta + i\sin\theta \text{ and } e^{-i\theta} = \cos\theta - i\sin\theta$$

the two middle terms of Equation (11.17), which are functions of time, become

$$\cos\left(\frac{E_m - E_n}{\hbar}\right)t\int_{-\infty}^{\infty} x\left[b^* a\psi^*_m\psi_n + a^* b\psi^*_n\psi_m\right] dx$$

$$+i\sin\left(\frac{E_m - E_n}{\hbar}\right)t\int_{-\infty}^{\infty} x\left[b^* a\psi^*_m\psi_n - a^* b\psi^*_n\psi_m\right) dx \qquad (11.18)$$

The real part of this result varies with time as

$$\cos\left(\frac{E_m - E_n}{\hbar}\right)t = \cos 2\pi\left(\frac{E_m - E_n}{h}\right)t = \cos 2\pi\nu t \quad (11.19)$$

The electron's position therefore oscillates sinusoidally at the frequency

$$\nu = \frac{E_m - E_n}{h} \qquad (11.20)$$

The expectation value of the electron's position will remain constant when the electron is in slate *n* or state *m*. When the electron is undergoing a transition between these states, its position oscillates with the frequency ν Such an electron, of course, is like an electric dipole and. radiates electromagnetic waves of the same frequency ν. This result is the same as that postulated by Bohr and verified by experiment. Thus quantum mechanics doesnot need to make any assumption to give the equation 11.20.

Selection Rules

Some transitions are more likely to occur than others

We can find the frequency ν_1 without knowing the values of probabilities $a * b$ as function of time or the electron wave function ψ_m and ψ_n. We need these quantities, however, to calculate the chance a given transition will occur. The general condition necessary for an atom in an excited state to radiate is that the integral

$$\int_{-\infty}^{\infty} x\psi_n\psi^*_m dx \qquad (11.21)$$

not be zero, since the intensity of the radiation is proportional to it. When the value of this integral is finite, then those transitions are known as allowed transition, but if the value of this integral is zero, then those transitions are called forbidden transition.

Three quantum numbers are needed to specify the initial and final states involve in a radiative transition, for a hydrogen atom. If the principal, orbital and magnetic quantum numbers of the initial state are n', l', m_l', respectively, and those of the final state are n, l, m_l and the coordinate u represents either the x, y, or z coordinate the condition for an allowed transition is

$$\text{Allowed transitions} \int_{-\infty}^{\infty} u\Psi_{n,l,m_l}\Psi^*_{n',l',m_l'}dv \neq 0 \qquad (11.22)$$

where the integral is now over all space. E.g. if u is used to replace x, the radiation would be that produced by dipole antenna lying on the x-axis.

As we know wave functions $\psi_{n,\,l,\,m'}$ for a hydrogen atom, we can evaluate equanion 11.22 for $u = x$, $u = y$ and $u =$ z for all pairs of states differing in one or more quantum numbers. When this is done, it is found that the only transitions between states of different n that can occur are those in which the orbital quantum.number l changes by +1 or –1 and the magnetic quantum number m_l does not change or changes by + 1 or – 1. That is, the condition for an allowed transition is that

Selection rules
$$\Delta l = \pm 1 \qquad (11.23)$$
$$\Delta m_l = 0, \pm 1 \qquad (11.24)$$

The change in total quantum number n is not restricted. Both the above equations *i.e*, 11.23 and 11.24 are also known as selection rules for allowed transition.

The selection rule requiring that l change by ± 1 if an atom is to radiate means that an emitted photon carries off the angular momentum $\pm\hbar$ equal to the difference between the angular momenta of the atom's initial and final states. A left or right polarized electromagnetic wave is the classical analog of a photon with angular momentum $\pm\hbar$, so this notion is not unique with quantum theory.

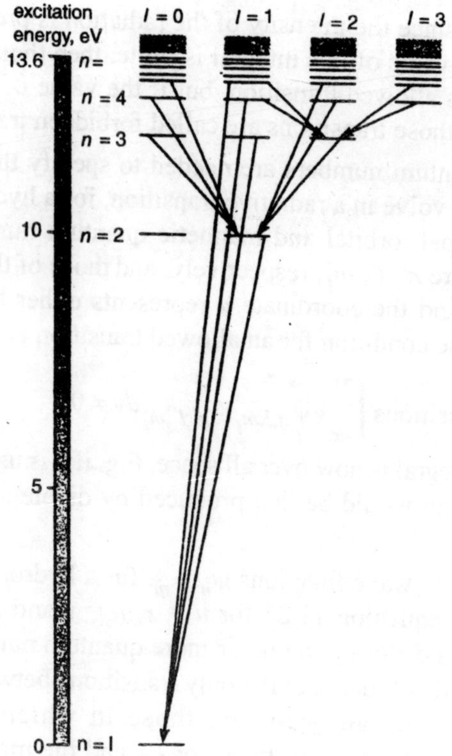

Figure. 11.11: ***Energy-level diagram for hydrogen showing transitions allowed by the selection rule $\Delta l = \pm 1$. In this diagram the vertical axis represents excitation energy above the ground state.***

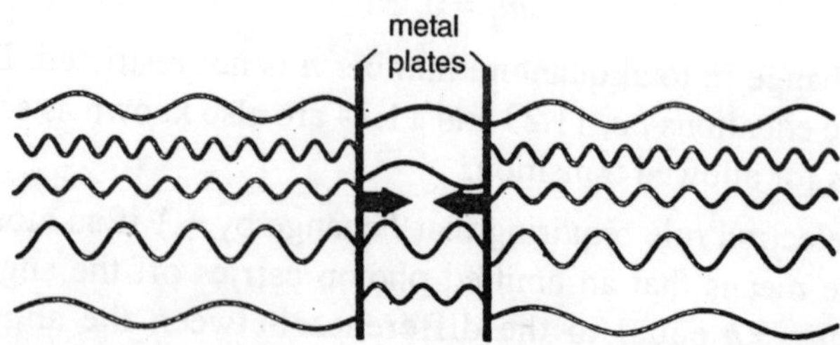

Figure. 11.12: ***Two parallel metal plates exhibit the Casimir effect even in empty space. Virtual photons of any wavelength can strike the plates from the outside, but photons trapped between the plates can have only certain wavelengths. The resulting imbalance produces inward forces on the plates.***

Quantum Electrodynamics

Now we will analyse the radiative transitions in an atom which is based on a mixture of classical and quantum physics. As we have seen, the expectation value of the position of an atomic electron oscillates at the frequency v of Equation 11.20 while passing from an initial eigenstate to another one of lower energy. Classically such an oscillating charge gives rise to electromagnetic waves of the same frequency v, and indeed the observed radiation has this frequency. However, classical concepts are not always reliable guides to atomic processes, and a deeper treatment is required. A kind of these deeper treatments is called quantum electro-dynamics, according to which the radiation emitted durring a transition from state m to state n is in the form of single photon.

Quantum electrodynamics also explains the mechanism that causes the spontaneous transition of an atom from one energy state to a lower one. All electric and magnetic fields turn out to fluctuate constantly about the **E** and **B** that would be expected on purely classical grounds. Such fluctuations occur even when electromagnetic waves are absent and when, classically, $\mathbf{E} = \mathbf{B} = 0$. Due to these fluctuations, an atom spontaneously emitts photons in its excited state.

These vaccum fluctuations are also consider as a set of virtual photons which are as much short lived that they donot violet the energy conservation because of the uncertainity principle in the form $\Delta E\ \Delta t \geq \hbar / 2$. These photons, among other things, give rise to the **Casimir effect**. Figure 11.12. Only virtual photons with certain specific wavelengths can be reflected back-and-forth between two parallel metal plates, whereas outside the plates virtual photons of all wavelengths can be reflected back-and-forth between two parallel metal plates, whereas outside the plates virtual photon of all wave lengths can be reflected by them. The resulting force is very small but is detectable and aslo tends to push the plates together.

Zeeman Effect

What happens to an atom in a magnetic field

A magnetic dipole has an anount of potential energy U*m*, in an external magnetic field, that depends upon both its magnitude of magnetic moment *N* and orientation of the moment with respect to the field.

The torque τ on a magnetic dipole in a magnetic field of flux density **B** is

$$\tau = \mu B \sin\theta$$

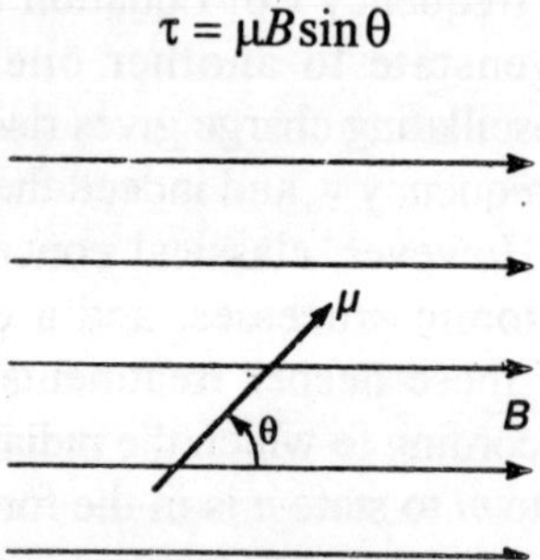

Figure. 11.13: ***A magnetic dipole of moment μ at the angle θ relative to a mangetic field B.***

where θ is the angle between $\boldsymbol{\mu}$ and **B**. The torque is a maximum when the dipole is perpendicular to the field, and zero when it is parallel or antiparallel to it. To calculate the potential energy U_m we must first establish a reference configuration in which U_m is zero by definition. (Since only *changes* in potential energy are ever experimentally observed, the choice of a reference configuration is arbitrary.) It is convenient to set $U_m = 0$ when $\theta = \pi/2 = 90^\circ$, that is, when $\boldsymbol{\mu}$ is perpendicular to **B**. So, the potential energy at any other orientation of *N* is equal to the external work which is done in order to rotate the dipole from $\theta_0 = \pi/2$ to the angle θ that corresponds to that orientation. Thus.

$$U_m = \int_{\pi/2}^{\theta} \tau \, d\theta = \mu B \int_{\pi/2}^{\theta} \sin\theta \, d\theta$$

$$= -\mu B \cos\theta \qquad (11.25)$$

When $\boldsymbol{\mu}$ points in the same direction as **B**, then $U_m = -\mu B$, its minimum value. It is supported by the fact that a magnetic dipole tries to align itself with an external magnetic field.

In an hydrogen atom, the magnetic moment of an orbital electron depend upon its angualar momentum L. Hence both the magnitude of **L** and its orientation with respect to the field determine the extent of the magnetic contribution to the total energy of the atom when it is in a magnetic field. The magnetic moment of a current loop is

$$\mu = IA$$

where I is the current and A the area it encloses. The electron that performs f rev/* in a circular orbit with radius r is equivalent to a current of $-ef$, thus its magnetic moment will be

$$\mu = -ef\,\pi r^2$$

As we know, that the linear speed y of the electron is $2\pi fr$, So its angular momantum will be

$$L = mvr = 2\pi mfr^2$$

Comparing the formulas for magnetic moment $\boldsymbol{\mu}$ and angular momentum L shows that

Electron magnetic moment

$$\mu = -\left(\frac{e}{2m}\right)\mathbf{L} \tag{11.26}$$

for an orbital electron (Figure 11.14). Gyromagnetic ratio is that quantity which involves only the charge and mass of the electron *i.e.* $-l/2m$. The minus sign means that $\boldsymbol{\mu}$ is in the opposite direction to **L** and is a consequence of the negative charge of the electron. While the above expression for the magnetic moment of an orbital electron has been obtained by a classical calculation, quantum mechanics yields the same result. In a magnetic field, the magnetic potential of an atom is given by

$$U_m = \left(\frac{e}{2m}\right)LB\cos\theta \tag{11.27}$$

According to figure 11.3 the angle θ between L and Z direction can have only values specified by

$$\cos\theta = \frac{m_l}{\sqrt{l(l+1)}}$$

with the allowed values of L specified by.

$$L = \sqrt{l(1+1)}\hbar$$

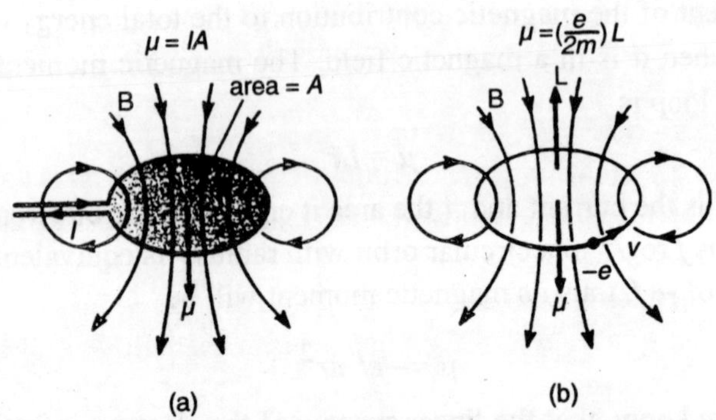

Figure. 11.14: ***Magnetic moment of a current loop enclosing area A. (b) Magnetic moment of an orbiting electron of angular momentum L.***

When an atom of magnetic quantum number m_1 is in a magnetic fields *B,* its magnetic energy will be given as

Magnetic energy $$U_m = m_l\left(\frac{e\hbar}{2m}\right)B \qquad (11.28)$$

The quantity $e\hbar / 2m$ is called the **Bohr magneton:**

Bohr magneton

$$\mu_B = \frac{e\hbar}{2m} = 9.274 \times 10^{-24}\,\text{J / T} = 5.788 \times 10^{-5}\,\text{eV / T} \qquad (11.29)$$

The energy of a particular atomic state depends on the value of m_1 and n when the atom is in magnetic field. A state of total quantum number n breaks up into several substates when the atom is in a magnetic field, and their energies are slightly more or slightly less than the energy of the state in the absence of the field. This phenomenon leads to a "splitting" of individual spectral lines into separate lines when atoms radiate in a magnetic field. The magnitude of the field is mainly responsible for the spacing of the lines.

In 1896, the Dutch Physicist *Pieter Zeeman* first observed the phenomenon of splitting of the spectral lines by a magnetic field and this phenomenon is named after him as Zeeman effect. The

Zeeman effect is a vivid confirmation of space quantization.

As m_1 can have the $2l + 1$ values of $+l$ through 0 to $-l$, a state

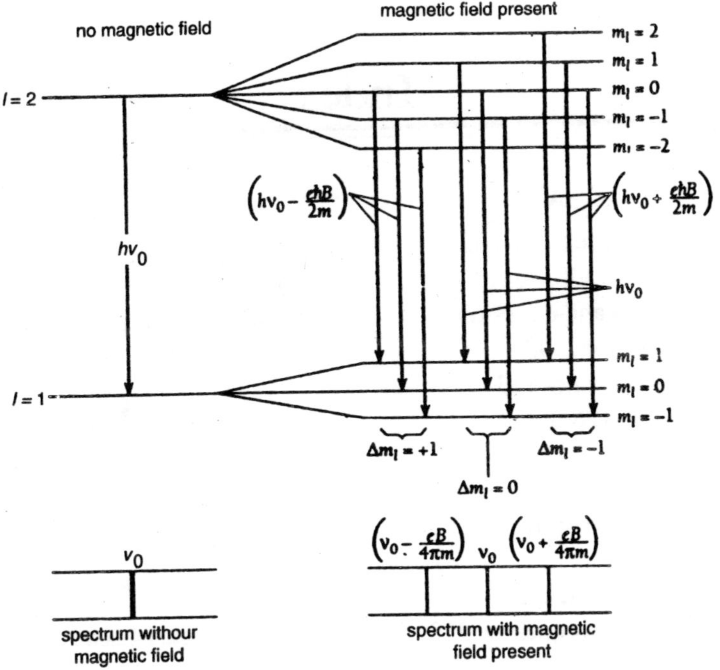

Figure.11.15: ***The normal Zeeman effect.***

of given orbital quantum number l is split into $2l + l$ substances that differ in energy by $N_b B$ when the atom in a magnetic field. However, because changes in m_l are restricted to $\Delta m_l = 0, \pm 1$, we expect a spectral line from a transition between two states of different l to be split into only three components, The phenomenon of Zeeman effect normally have the splitting of the spectral line of frequency ν_o into three component whose frequencies are given by

$$\nu_1 = \nu_0 - \mu_B \frac{B}{h} = \nu_0 - \frac{e}{4\pi m} B$$

Normal Zeeman effect $\qquad \nu_2 = \nu_0 \qquad$ (11.30)

$$\nu_3 = \nu_0 + \mu_B \frac{B}{h} = \nu_0 + \frac{e}{4\pi m} B$$

Index